Online GIS and Spatial Metadata

Second Edition

Online GIS and Spatial Metadata

Second Edition

Terry Bossomaier
Brian A. Hope

with the contribution of Christoph Karon

CRC Press
Taylor & Francis Group
Boca Raton London New York

CRC Press is an imprint of the
Taylor & Francis Group, an **informa** business

CRC Press
Taylor & Francis Group
6000 Broken Sound Parkway NW, Suite 300
Boca Raton, FL 33487-2742

First issued in paperback 2019

© 2016 by Taylor & Francis Group, LLC
CRC Press is an imprint of Taylor & Francis Group, an Informa business

No claim to original U.S. Government works

ISBN-13: 978-1-4822-2015-5 (hbk)
ISBN-13: 978-0-367-87433-9 (pbk)

Visit the Taylor & Francis Web site at
http://www.taylorandfrancis.com

and the CRC Press Web site at
http://www.crcpress.com

Contents

Preface

This book is a second edition, in a field which has grown rapidly since 2000. The technologies in use, combined with quality datasets now available online, are providing quality solutions to an increasing number of business and everyday problems. The software in use with Online GIS, however, is often hidden from users so that solutions are simple and easy to use.

This book provides details of how these technologies work, going *under the hood* to explain the details of their implementation.

This is not a book about GIS, *per se*. There are many other excellent books that already cover the basic theory and practice more than adequately. However, for programmers and others who are new to geographic information, we do provide a brief primer on the basics.

Conversely, for GIS people new to Web technology, we have provided an introduction to many of the protocols and standards now in use with Online GIS. The technical details are mainly to show readers examples of real scripts, markup and other elements that make the technology work. We have tried to avoid turning the book into a technical manual. We hope that our accounts of the various topics are clear and lucid enough to help non-technical readers understand all the issues. We encourage anyone who finds source code daunting to simply skip over the scripts.

Another goal of the book was to give the reader a global perspective of where Online GIS is currently. We have included many references and links to online resources that we hope readers find useful. We have also included some future directions and applications of Online GIS.

If you are interested in learning how Online GIS works, and how you can implement your own apps using Online GIS, this book is a great place to start.

List of Figures

List of Tables

Acknowledgments

The authors want to thank the many people who have contributed to the quality of this book. In particular, we would like to thank Christoph Karon who contributed Chapter 12.

This book owes a huge debt to David Green, first author of the first edition. David was a pioneer of online GIS and his ideas and insights still permeate the book.

In the intervening years, many staff at the Land and Property Information (formally Department of Lands) in NSW Australia, notably Don Grant, Des Mooney and Warrick Beacroft, have been immensely supportive of spatial research.

Don Grant AM, NSW Surveyor General and General Manager of the Department of Lands for many years, and one-time director of the Federation Internationale de Geometres was a tireless supporter of spatial information education and research at Charles Sturt University. Along with Ian Williamson AM, then professor at the University of Melbourne, he led the United Nations workshop which subsequently culminated in the Bathurst Declaration. The conference that followed was a key driver for the first edition of the book.

We are also greatful to Carolyn Leeder who provided essential research assistance, and checked all the minute details which go into producing a book of this kind. We thank several other colleagues for their assistance with reviews of examples, in particular, Simon Reynolds for his code reviews and Daniel Miller for his review of the mobile security content.

Finally, we are grateful to the following organisations for granting permission to reproduce some Web pages from their online services:

- Xerox, PARC Alto Research Center

- US Census Bureau

- GeoPlatform.gov, US

- City of Vancouver, Canada

- Toronto City, Canada

- Data.Gov.uk

- Virtual.Tourist.com

- Scottish Government

- Google

- OpenStreetMap

- Nations Online

- Pierce County, US

- Walmart, US

- Trade and Investment, NSW

- New York State, US

- Canadian Spatial Data Infrastructure

- Data.Gov.uk

- Data.gov.au

- Data.Govt.nz

- Microsoft's Bing Maps API

- London City

- NSW Land and Property Information

- Geoscience Australia

- National Oceanic and Atmospheric Administration

1

Introduction

CONTENTS

Since the first edition of this book, the advances in Online GIS have been significant. There are the obvious changes to technology including internet speed and availability. However, there is also a greater willingness for organisations to share their data online. Global initiatives such as Web 2.0, Gov 2.0 and the Semantic Web have fueled this push for better access to information, as have changes in society where users demand easy and immediate access to a wide range of spatial information through Web services. They want this data anywhere, anytime. There is the power of the Internet and global search engines, and the value added when a spatial context is given to data. Online GIS is no longer the domain of the spatial specialist, but now supports many aspects of life and work, and the diversity of online resources makes the need for metadata even greater than it was a decade ago.

1.1   Geographic Information Systems

Geographic Information Systems (GIS) are computer programs for acquiring, storing, interpreting, and displaying spatially organised information. GIS had their origins in many different disciplines, including electronic cartography, geological surveys, environmental management, and urban planning. They have now become an essential tool in all of these professions, as well as many others.

During this second decade of the new millennium, a great transition is taking place in the way geographic information systems function. Instead of being isolated in standalone machines, a new environment is gradually being created in which geographic information is stored and accessed over the Internet. In this book we argue that this transition has implications that go far beyond a different format. One of the effects is the greater accessibility that the Internet offers. However, even more significant is the potential to combine information from many different sources in ways that were never previously possible. To work in the new environment, managers, developers and users all need to learn the basic technology involved.

Many commercial developments in online mapping are now available, and one book could hardly describe or catalogue them in any detail. Thus the goal herein is more to examine the issues underlying their development. Before getting to the main content of the book, there are a couple of underlying topics we need to briefly consider. On the software side, there is object-oriented programming (§1.2) and then there is the nature of GIS itself (§1.3).

1.2   GIS and Object-Oriented Technology

The simplest way to deliver a map online would be to store it as a file, say a JPEG image, and allow users to download it via ftp or some more modern facility such as a dropbox. But we often want something more sophisticated, perhaps a fraction of the map, or part of the map with special annotations, such as local fish restaurants. In this case we don't want direct access to the map data: we want a process which owns the data and will process it to meet our needs.

This idea, that a process, or method, intervenes between us and the data, is the essence of object-oriented programming (OOP). Objects own data and allowed access to it only through their own methods. This is intrinsically a secure approach: only specific functions in a program can access particular data elements. From this central idea grew a massive programming paradigm, an organisation, the Object Management Group (OMG) (OMG 1997) and new software practices such as design patterns (Gamma et al. 1995).

A very great deal of Online GIS software is object oriented, but a full discussion would take us way off course. But there are some key ideas from OOP which appear throughout XML schemas:

Inheritance is the reusing of characteristics and methods of some sort of parent class. For example, an apricot is a stone fruit, like a plum or damson. All three examples have a skin, a fleshy interior and a big stone in the middle. Apricot and damson inherit the basic characteristics of stone fruit. Similarly, stone fruits are examples of fruit in general, and so on.

Class and Object define the properties of something, such as a fruit. An object is an instantiation of the class. Thus lemon is an object of the class of citrus fruits. Alternatively lemon could be a class itself of which individual lemons (say as entries in a harvest show) are instances. There is a loose parallel here between XML documents (objects), which are the substance of Chapter 5, and XML schemas (§5.5), which describe classes of XML documents.

Abstraction is used frequently in the XML schemas used for spatial data. We shall see some examples in Chapter 9 and especially §9.3. The idea is that we have some class from which other classes inherit, but of which there are no objects. Thus we might make the class fruit abstract: the concrete instances of fruit are specific fruits such as citrus, orange, and so on.

1.2.1   The Document Object Model

As we shall see, markup documents, HTML, XML or SGML, have a tree (or forest) structure, with a root node containing a hierarchy of sub-nodes. This so-called **Document Object Model**, usually referred to by its abbreviation, DOM, has a long history dating back to the early days of the web, but formal technical specifications are available from the W3C.[1] The most recent final recommendation is DOM Level 3 from 2004,[2] but DOM Level 4 is in the state of the last call working draft in April 2015.

The DOM is an API (Application Programming Interface (§4.7)). It is language independent, although it grew out of JavaScript (§4.3) and java requirements. Consider the following HTML fragment:

```
<html>
<head>
<title>My Vacation 2015</title>
</head>
<body>
<h1>Vacation 2015</h1>
<p>
Our holiday this year was a Mediterranean cruise, starting at
Marseille in France, followed by a winery tour of the Bordeaux
    region.
</p>
<table>
<tr><th>Date<th>Depart From <th>Arrive At <th>Transport</tr>
<tr><td>10 May <td>Sydney<td>Paris<td>Air France</tr>
<tr><td>14 May <td>Paris<td>Marseilles<td>TGV</tr>
<tr><td>15 May <td>Marseilles<td>MarseillesMSC Fantasia cruise
    ship<td></tr>
<tr><td>30May <td>Marseilles<td>Bordeaux<td>TGV</tr>
<tr><td>6 June <td>Bordeay<td>Paris<td>TGV</tr>
```

[1] http://www.w3.org/DOM/
[2] http://www.w3.org/TR/2004/REC-DOM-Level-3-Core-20040407/

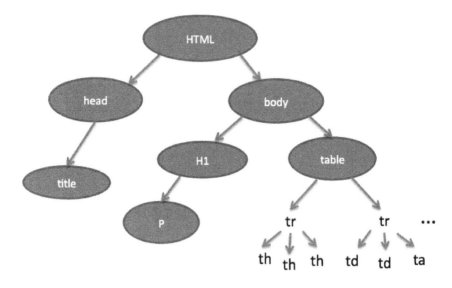

FIGURE 1.1: DOM tree for the vacation Web page

```
<tr><td>8 June <td>Paris<td>Sydney<td>Air France</tr>
</body>
</html>
```

This can be represented as a DOM tree, as shown in Figure 1.1.

Expressing the document in a tree this way enables software to find and modify selected parts of the document, which is one of the primary roles of JavaScript, ubiquitous among today's Web pages. Since XML is now used for other things besides display documents, the DOM is important for much of the sophisticated processing now applied to XML documents. The DOM specification is very complicated, and, as in many cases in the book, we have to skip over the many details.

1.3 A Brief Primer on the Nature of GIS

1.3.1 GIS Data

Geographic data consists of layers. A layer is a set of geographically indexed data with a common theme or type. Examples might include coastlines, roads, topography, towns, public lands. To form a map, the GIS user selects a base map (usually a set of crucial layers, such as coastlines or roads) and overlays selected layers. Layers are of three types:

- Vector layers consist of objects that are made up of points (e.g. towns), lines (e.g. roads) or polygons (e.g. national or state boundaries).

- Data of this kind is usually stored in database tables. Each record in the table contains attributes about individual objects in space, including their location.

- Raster layers consist of data about sites within a region. Examples include satellite images, digitised aerial photographs and quadrat samples. The region is divided into a grid of cells (or pixels), each representing an area of the land surface. The layer contains attributes for each cell. For instance, in a satellite image the attributes might be a set of intensity measurements at different light frequencies or a classification of the land features within the cell (e.g. forest, farmland, water).

- Digital models are functions that compute values for land attributes at any location. For instance, a digital elevation model would interpolate a value for the elevation, based on values obtained by surveying. In practice, digital models are usually converted to vector or raster layers when they are to be displayed or otherwise used.

As mentioned above, the data in vector layers is often stored in database tables. This enables a wide range of searching and indexing by attributes. The data may also be indexed spatially. For instance, a quadtree is a construct that divides a region successively into four quadrants, down to any desired scale. Any item is indexed by the quadrants in which it lies. This procedure allows for rapid searching based on *spatial location.*

Overlays form a classic analytic tool in GIS. Some online services allow data layers to be overlaid on various backgrounds. For instance, the Victorian Butterfly Database [3] allows the user to plot the point distributions of species against a variety of environmental classifications, each of which provides clues about factors controlling the species distributions.

1.3.2 GIS Functionality

Although most commonly associated with map production, GIS encompass a wide range of forms and functions. The most common (but far from only!) outputs from a GIS are maps. These maps are either the result of a simple plot of layers for a given region, or they display the results of a query. Queries involving GIS can be classified as first or second order in nature.

First-order questions involve features within a single layer. Examples include the following:

- Where is object X?, e.g. Where is Sydney?

- How big (or long) is object X? What is the area of Europe?

- Where are all objects having a certain property? Where are oil wells in Alberta?

---

[3] http://museumvictoria.com.au/bioinformatics/butter/

Second-order questions involve overlays of two or more layers, such as:

- What objects are found within areas of a given type?, e.g. What mines lie within a state or district?

- What is the area of overlap between two kinds of objects? For example, what areas of rainforests lie within national parks? What farms does a proposed highway run through?

- What environmental factors influence distribution of a rare species?

- The process of overlay often involves the construction of a new data layer from existing ones. Suppose that two layers consist of polygons (e.g. electoral districts and irrigation regions). Then the overlay would consist of polygons formed by the intersection of polygons in the two layers. For instance, in the example given, these polygons would show the areas within each electorate that were subject to irrigation.

Two important kinds of functions within a GIS are data analysis and models. Data analysis tools include a host of techniques for comparing and analysing data layers. Some examples include:

- Kriging (e.g. estimating size of ore bodies) and other methods of interpolation (e.g. elevation);

- Spatial correlation of variables (e.g. incidence of medical conditions versus pollution levels by district);

- Nearest neighbour analysis (e.g. the numbers of seabird species that nest close together);

- Fractal dimension of shapes (e.g. forest boundaries).

Models in GIS are usually layers obtained by manipulation of other layers. For instance, operations on maps of topography, streams and rainfall, and soil type might be inputs to a model whose output is a map of soil moisture. Traditionally these layers have all been handled by a single integrated package. Yet at the same time, it is clear that these datasets might come from different places. The online paradigm would allow each of these components to be dynamically accessed from different sites.

1.3.3 The Object Model of GIS

GIS introduced a whole new way of looking at spatial data. Although cartography had been operating digitally for years before GIS appeared on the scene, the emphasis was solely on producing maps. A road map, for instance, would consist of lines to be drawn on the page. Names for roads and towns formed a separate set of textual data. There was no direct link between (say) the point representing a town and the name of that town. These associations became apparent only when the map was printed.

As scientists started to use spatial data for research purposes, the drawbacks of the map-oriented approach rapidly became obvious, as the following examples illustrate:

- The data for the rivers in a region were stored as dozens of separate line segments. The data for individual rivers consisted of many isolated line segments, with no indication that the segments formed a single feature. The rivers were broken into segments because the map required gaps wherever other features (e.g. towns, roads) were to be drawn over the top.

- In a gazetteer of towns, each location was found to have a small error. These errors turned out to be offsets that were added to allow each place name to be plotted in a convenient space on the map, not at the actual location.

The biggest change that GIS introduced was to store geographical features as distinct objects. So a road would be stored as an object with certain attributes. These attributes would include its name, its type (i.e. a road), its quality (e.g. a highway) and a sequence of coordinates that represent the path that the road follows across the landscape.

In this object-oriented approach (§1.2), each geographic object belongs to some class of objects. Thus the objects Sydney and Bathurst would both belong to the class city, and each of them is termed an instance of the city class. The notion of classes is a very natural one in GIS. It also has implications for the way geographic data is treated.

Each object has attributes associated with it. For the city class, these would at least include a name and a location (i.e. latitude and longitude). However, it might also include many other attributes that we could record about a city, such as its population, the state in which it lies, the address and phone number of the city offices, and so on. The importance of defining a class is that for any other city that we might want to add to the GIS, we have a precise list of data that we need to supply.

The object approach to GIS has many advantages. The first is that it *encapsulates* everything we need about each geographic entity. This not only includes data, but also extends to methods of handling that data. So, for instance, the city object might include a method for drawing a city on a map. In this case it would plot a city as a circle on a map, with the size of the circle being determined by the population size of the city.

Another advantage of objects is that we can define relationships between different classes of objects. For instance, the class of cities that we described above really belongs to a more general class of objects that we might term inhabited places, perhaps called `inhabitedPlaces`, using the camel case which is common in object-oriented programming. This general class would not only include cities but also towns, military bases, research stations, mining camps, farms and any other kinds of places that humans inhabit. The class inhabitedPlaces would form a typical abstract class, with cities, towns, and so on being concrete classes inheriting it.

The city class inherits some of its attributes from the inhabitedPlaces. These basic properties would include the location, as well as (say) the current population. Other properties of the city class, such as contact details for the city offices, would be unique to it, and would not be shared by all inhabited places. The inhabited places actually form a hierarchy, with capital cities inheriting properties from cities, cities inheriting attributes and methods from towns, and towns inheriting attributes from inhabited places. Thus a capital city has six attributes, all but one of which it inherits from the more general classes to which it belongs.

Object models are usually described in terms of classes of objects and the relationships between them. The Uniform Modelling Language (UML) provides a standard approach for specifying object models of data and processes (Larman 1998).

Another important type of link between classes of objects is a whole-part relationship. A map, for instance, consists of several elements. There are the border, the scale and possibly several layers, each containing many elements.

A further interesting area of activity is the object-oriented database. Traditional relational databases store data as tables and relationships between them. Until the surge of interest in multimedia and the web over the last two decades, they had been more or less limited to textual information. In attempting to store other material such as video or images, considerable changes were required in storage and query mechanisms.

Around the same time, the idea of storing data as complete objects with an associated object-query language arose. In the relational model a complex object might be broken up and split over many tables. In the object model the object (data and methods) is stored as an entity. An example might be a map and associated methods for extracting distances between places and many other things. Although relational databases are still dominant, the *xquery* model for XML is essentially an object-oriented approach (§5.6).

The Object-Oriented database model is certainly very attractive for GIS-style data. However, the commercial reality has been different. Enterprise databases store vast amounts of information, require exceptional robustness and stability and embody much experience in tuning and optimisation. Object-Oriented started a long way behind, and the relational databases have taken on board some of their advantages such as stored procedures, while hybrid object-relational databases have evolved. Thus Object-Oriented databases are still a relatively small part of the market.

1.4   The Rise of the Internet

The Internet Krol (1992) is a vast communications network that links together more than computers all over the world. CISCO estimates that there will be 25 billion connected devices in 2015, which will double by 2020 (Evans 2011).

The rapid growth of Internet activity over the decade has produced a literal explosion of information. From the user's point of view this process has emphasised several crucial needs:

- *Organisation*. Ensuring that users can obtain information easily and quickly. Various projects have developed general indexes of pointers to network information services (especially Internet search engines). However, these struggle to cope with the sheer volume of available information. Self-organization, based on user interests and priorities, is one practical solution to help in the search and discovery of information.

- *Stability*. Ensuring that data sources remain available and that links do not go stale. Rather than gathering information at a single centre, an important principle is that the site that maintains a piece of information should be the principal source. Copies of (say) a dataset can become out of date very quickly, so it is more efficient for other sites to make links to the site that maintains a dataset, rather than take copies of it. This idea is fundamental to the discussion of Information Networks in Chapter 7.

- *Quality*. Ensuring that information is valid, that data are up to date and accurate, and that software works correctly.

- *Standardisation*. Ensuring that the form and content of information make it easy to use.

The rise of the Internet, and especially the World Wide Web during the 1990s, revolutionised the dissemination of information. Once an organisation had published information online, anyone, anywhere could access it. This capacity meant that people could access relevant information more simply than in the past. They could also access it faster and in greater quantities. It also raised the potential for organisations to provide wider access to information resources that normally require specialised software or hardware. For example, by filling in an online form, remote users can query databases. Geographic Information Systems (GIS) that previously required specialised, and often expensive, equipment, can now be accessed remotely via a standard Web browser.

The volume of information available on the World Wide Web has grown exponentially since 1992, when the National Centre for Supercomputer Applications (NCSA) first released a multimedia browser (Mosaic). This explosion was driven first by data providers who recognised the potential audience that published information could attract. As the volume of information grew, users began to drive the process by demanding that information be available online.

The explosion of online information creates a problem: finding one item amongst billions is worse than finding a needle in a haystack. Potential solutions for searching include the use of intelligent agents that continually sift and record relevant items and the promotion of metadata standards to make documents self-indexing. Information Networks (Chapter 7) provide a way of organising sources of information. Almost akin to the invention of fire, Sergei Brin and Larry Page came up with a novel search algorithm, which rapidly pushed their search engine Google into near total dominance, creating one of the largest and most successful tech companies.

Internet policy is driven by the Internet Society (ISOC)[4], formed by Internet pioneers Vint Serf and Bob Kahn. It includes technical committees to vet and oversee the

---

[4] http://www.internetsociety.org

development and implementation of new standards and protocols. The World Wide Web is now governed by the World Wide Web Consortium, W3C, and has been actively developing many of the standards that we touch on in this book.

1.4.1 Advantages of Distributed Information Systems

Perhaps the greatest impact of the Internet is the ability to merge information from many different sources in seamless fashion (Green 1994). This ability opened the prospect of data sharing and cooperation on scales that were formerly impossible. It also brings the need for coordination sharply into focus.

As a geographic information system, the Web has some important advantages. One of the greatest practical problems in the development of GIS is the sheer volume of data that needs to be gathered. Simply gathering datasets from suppliers can be a long drawn out process. Most systems require specialised data that the developers have to gather themselves. Inevitably, the lack of communication between developers leads to much duplication of effort.

The Internet has the potential to eliminate these problems. In principle, the suppliers of individual datasets or data layers could distribute their products online. This approach would not only speed up development of any GIS, but it would also help to eliminate duplication. In effect, online distribution of data would have the effect of increasing the potential volume of information available to developers and greatly increase innovation. It would distribute the workload amongst many organisations and provide location-based apps that add value to consumers. This would also simplify the updating of information, as it could be made available online as soon as it is updated. This underpins Information Networks (Chapter 7) and the idea of *mashups* (§13.4.2).

Another advantage of online GIS is that it expands the potential pool of GIS developers and users. As we shall see in Chapter 2, there are many options for placing GIS online. The result is that the overall costs can be much lower for developers. It is not only possible, but also cost effective to implement small GIS that are designed for specific applications which only require limited functions, such as providing a geographic interface to online documents or databases.

For some purposes, it is not even necessary for online publishers to implement a GIS at all. Instead, GIS services can be provided by leasing information from a specialist GIS site. For instance, suppose that a travel agency that provides information about (say) tourist attractions, accommodation and so on, wants to provide street maps to show the location of each site in its database. This is a very useful service for travellers who are trying to find their hotel in a strange city. Rather than developing a series of maps itself, the agency could arrange many new commercial prospects for GIS, chiefly through on-selling of geographic services. We return to some of these new commercial models in Chapters 11 and 12.

For GIS users the prospects are equally exciting. The Internet brings GIS within reach of millions of users who previously could not afford the necessary equipment and specialised software. Apart from the availability of numerous free services, there is also the potential for access to a fully fledged GIS on demand. For instance, instead of buying an entire GIS themselves, users could buy GIS services from providers as they require them. For regular users, these services could take the form of subscription accounts to an online commercial system. On the other hand, irregular, one-time users could buy particular services in much the same way as they might previously have bought a paper map.

The Internet is an ideal medium for collaboration. It makes possible communication and information sharing on scales that were hitherto unheard of. However, the explosive growth of the Internet at first led to enormous confusion. Organisations duplicated services and facilities in inconsistent ways. This pattern was exacerbated by commercial interests, which saw the Internet as an extension of their traditional competitive marketplace. The essential obstacle to information sharing is the tension between self-interest and cooperation. To resolve these issues, organisations and nations needed to agree on protocols and standards for data description (metadata), data recording, quality assurance, custodianship, copyright, legal liability and for indexing (see Chapter 7).

1.4.2   Examples of Distributed Information

The ability of the World Wide Web to link information from many different sources creates a synergy effect in which the overall information resource may be greater than the sum of its parts. One of the earliest demonstrations of this power was a service called The Virtual Tourist, discussed in §2.3.4.1.

There are many examples, in many different fields of activity, that highlight the advantages of distributed, online information systems. Many of these have been in fields that relied on the Internet from the beginning.

One such field is biotechnology. The vast online resources are compilations of genomic and protein sequences, enzymes, type cultures and taxonomic records. There are also many large repositories of useful software and online services for processing and interpreting data. Most of the prominent sites provide a wide range of other information as well, including bibliographies, electronic newsgroups and educational material. All of the resources are accessible online. Some of the larger facilities, such as the European Molecular Laboratory,[5] are actually networks of collaborating organisations. The field has even reached the stage where many funding agencies and scientific journals actually require researchers to contribute their data to online repositories as a condition of acceptance of a grant or paper.

Many kinds of environmental information are already online. As scientists, we have been involved ourselves in some of the efforts already in progress to compile online information about many kinds of environmental resources, such as forestry and global biodiversity. In each case, the new possibilities are leading people and organisations to rethink the way they do things, to look at the broader geographic context, and to initiate schemes to enhance international cooperation.

---

[5] http://www.embl.de/

1.5   Overview of the Book

This book has three goals: the first is to give readers an overview of the basic technology involved in online geographic information systems; the second is to outline models for how the development of online geographic information might be coordinated; and finally the use of metadata to glue everything together.

Firstly, we have to understand the mechanisms of publishing GIS data over the Web. We then have to move on to consider how data is organised, accessed, searched, maintained, purchased and processed online. This involves us tackling some fairly complex standards, which are currently redefining the way in which online GIS will operate in the future. Key to these standards is XML (Chapter 5), which underlies just about all of the current Web standards.

With all of the above details in place, we will then look at the content of the metadata standards for GIS information and proceed to look at how organisations are delivering geospatial data online in a serious way.

fixedOne area we do avoid is detailed discussion of existing packages for Web GIS. These tools change rapidly, but come with their own documentation and quite often with third party tutorials. But what we want to do here is to look under the hood, at the common building blocks and glue which subserve all these tools.

The early chapters (Chapters 2, 3 and 4) introduce the technical methods involved in developing and implementing geographic information systems on the Internet. They are intended to provide potential developers with technical details and examples to help them understand the issues. They also provide the technical background that readers need to appreciate many of the issues discussed in later chapters. Chapter 2 outlines the main technical issues surrounding online GIS and gives an overview of how far Online GIS has evolved since the first edition. Chapter 3 describes the kinds of facilities and operations that can be implemented at the server side, on a Web server. Chapter 4 describes GIS tools that can be implemented at the client to operate on a standard Web browser.

In Chapter 5 we get to the scaffolding for Online GIS and spatial metadta, XML, which began life as a flavour of SGML. For many years, the SGML international standard languished, little used outside a few big organisations such as the US military. Partly this was due to its complexity and the high cost of processing tools. Partly the superficial attractiveness of WYSIWYG publishing pushed it into the shadows. But as the Web grew, the need for better organisational and searching methods came to the fore. HTML, the language of the Web, is in fact an SGML DTD, and it became apparent that a simpler version of the full SGML standard would be advantageous. So, XML came about and has vacuumed up most of the other Web standards.

SGML also became the choice for writing spatial metadata standards, but has now been largely replaced by XML. Chapter 6 discusses the detail of four XML based standards now in common use for Online GIS and spatial metadata.

The remaining chapters examine technical issues involved in coordinating the development and deployment of geographic information in the Internet's distributed environment. These issues include many technical matters, but they also concern the ways in which human collaboration impinge on technology.

Chapter 7 begins the discussion by outlining the nature of Information Networks, systems in which information is distributed across many different sites. Chapter 8 outlines some of the interoperable standards involved in distributed information systems. One of the biggest such initiatives in the spatial arena has been the Open GIS Consortium (OGC), which is essentially designing vendor independent standards for many spatial operations.

Chapter 8 looks at the conceptual framework of metadata, by studying the RDF and similar standards for the Web. Chapter 9 follows this by describing several metadata standards in use around the world for spatial metadata and by giving current examples of SDI Metadata Portals, Catalogues and Clearinghouses. This chapter builds on the XML work in Chapter 5, the XML examples in Chapter 6 and the metadata fundamentals in Chapter 8. Chapter 10 looks at ways in which distributed information can be built into data warehouses, and introduces basic ideas in data mining within such systems.

Finally, two fast growing areas come in for special discussion: Mobile GIS (Chapter 11) and Location Based Services (Chapter 12). Chapter 13 then discusses some of the emerging new technologies, future directions and trends associated with this fast growing industry of Online GIS.

2

GIS and the Internet

CONTENTS

As discussed in the opening chapter, GIS is no longer the domain of the scientific user. In all its many forms, Online GIS is now literally in the hands of the user and is available on demand to solve an unimaginable number of location based problems. In this chapter, we discuss some of the challenges that need to be considered by organisations and individuals seeking to make geographic information available over the Internet.

2.1   The Advantages of Online GIS

The World Wide Web has become a standard platform for the delivery and integration of geographic information. As a result, access to geographic information online has exploded over the last decade. Many government organisations are sharing spatial information online to reduce cost and create efficiencies. Scientific users of the traditional geographic information systems can now integrate with other geographic information online using dynamic access to external datasources to mashup (§13.4.2) with their existing mapping systems. Individuals can also search on and query information using their standard browsers and mobile devices and these services are available in many locations on demand. This federated model for accessing geographic information online provides significant advantages to organisations and individuals. These advantages include:

1. *Worldwide access.* An information system on the Web is accessible from anywhere in the world.

2. *Standard interface.* Every Web user has a browser, meaning that any system that uses the Web is accessible by everyone, without the need for costly, specialised equipment or software.

3. *Worldwide content.* The global nature of these information systems now provide users with access to geograhpical data from all over the world.

4. *Reduced data replication.* Information can be accessed live from the data custodian, leaving other businesses to focus on their own organisational information needs.

5. *Reduced capital investment.* Organisations can reduce their internal infrastructure and storage costs, choosing to access data dynamically as required.

6. *Faster, more cost-effective maintenance.* Information can be accessed from its source and can be mashed up (§13.4.2), reducing the need to maintain spatial information where it is not an organisation's core business.

This chapter sets out to do the following:

- Examine some of the technical issues involved in developing an online GIS and some of the options for dealing with them (§2.2).

- Briefly look at examples of published geographic information systems and services that are online (§2.3).

- Demonstrate some simple methods for implementing basic types of geographic information online for integration with your existing information systems (§2.7).

- Discuss some of the challenges still facing the future development of Online GIS (§2.6).

2.2 Considerations When Deploying over the Internet

The Internet, and in particular the World Wide Web, is a powerful environment for all kinds of computing. As such it has provided many opportunities for web based service provision of geographic information, however, not without its challenges that do not exist in a stand-alone GIS environment. In the following discussion of online GIS, we shall for the most part gloss over methods of implementing standard GIS operations, some of which we cover in Chapters 3 and 4. Instead we focus on some of the technical issues involved in placing GIS online. The main issues that have to be addressed arise from:

1. The nature of the Web environment.
2. The separation of the user interface from geographic data and processing.
3. GIS processing optimisation.

2.2.1 The Web Environment

The Web uses a number of protocols for its communication. The main protocol is still the HyperText Transfer Protocol (HTTP (§3.1)), which it uses to communicate queries and responses across the Internet. HTTP is a client-server protocol. This means that a *client* – the user's browser program – transmits a query to a Web server, which then sends back its response. In HTTP 1.0, these transactions are carried out on a connectionless, single query basis. Even if a user makes a series of queries to the same server, the server normally retains no history of past queries, and no client session is established. This is in contrast to several other Internet protocols, such as FTP and *ssh*, where the server establishes a dialogue with clients who *log in*.

Historically, interactive GIS software that runs on a dedicated machine makes the implicit assumption that the program's current state is a direct result of a user's interactions with it. Remote GIS cannot do this under HTTP 1.0, which necessitated work arounds such as using hidden fields and caching to maintain a known state, discussed further in §2.7.3.1. HTTP and associated client/server software have the advantage of being a generic technology. Any service provided via HTTP is immediately available to anyone, on any type of machine or device, who is running a suitable client program.

Another important aspect of the hypertext propocols is the role of the Unified Resource Locator, or URL. This *link*, which is a fundamental part of using the Web, is a powerful way to pass contents between servers and clients. The variables and values that URLs can pass can be accessed programmatically and may include tokens that are used to validate HTTP requests. The content of the HTTP headers are also powerful ways to provide additonal information relating to each request. They can be used to track the source of HTTP requests and tokens can be hidden inside these headers as well. Client and server programs alike can access these headers and can take advantage of the cross site security.

Modern Internet browsers have now replaced the traditional applets and plugins, removing the need for users to install any applets or plugins at their clients. The most common Internet browsers include Internet Explorer, FireFox, Safari and Chrome. Important features of these Internet browsers and clients include:

- They permit browsing of all of the main network protocols including HTTP, HTTPS, FTP.

- They permit both text formatting and images that are embedded directly within text, thus providing the capability of a true electronic book.

- They integrate freely available third party display tools and plugins for image data, sound, Postscript, animation and other new forms of media.

- The third party tools and plugins include powerful debuggers such as Firebug that can show the executed requests and the results.

- They permit seamless integration of a user's own local data, without the need of a server, with information from servers anywhere on the Web.

- The *forms interface* (§4.3.1) allows users to interact with documents that appear as forms, including buttons, menus, and dialog boxes which can pass complex queries back to the server.

- The *imagemap interface* (§4.2) allows users to search for and to query map features interactively, allowing users to find out information about map features being displayed in GIS-like fashion.

- The authorisation feature provides various security options, such as restricting access to particular information.

- The SQL gateway allows servers to pass queries to databases where such gateways are already implemented for many databases.

- The ability to run scripts or programs on the server and to deliver the results to the Web.

- The ability to include files dynamically and thus build up and deliver documents on the fly.

These generic capabilities of browsers and client applications provide enough functionality to do spatial queries and searches, render vector and raster information, as discussed in Chapter 4. There is also a number of other technologies now that provide much improved user experience and GIS capabilities than the earlier online GIS systems. Some of these technolgies include:

- Application Programming Interfaces (§2.3.4)

- Java and JavaScript classes and libraries (§4.3)

- HTML5 and Canvas (§5.3)

- Image Pyramid Caches (§2.2.2) and (§10.8.3)

- Mobile Standards (Chapter 11)

- Image Streaming Protocols (§2.2.2)

As more geographic information is published onto the Web, there are consider-able benefits to organisations publishing the information to know who is accessing the data and for what purpose. In addition to sourcing data usage patterns to feed into their business analytics, there are instances where an organisation may wish to limit access to some sensitive information or to provide mechanisms to limit access to freely available geographical information where users are found to be breaching any terms and conditions. To provide these access controls, many organisations are now using the HyperText Transfer Protocol Security (HTTPS) to provide service management for Online GIS services.

2.2.2   Separation of the User Interface

The Web environment separates a user interface from the site of most data processing. This separation poses problems for any operation that relies on a rapid response from the system. Responses required may be to simply *pan* sideways or *zoom* out efficiently, requiring fast data retrieval and rendering, or to run a spatial query to snap two lines together to answer a complex spatial query quickly. GIS systems had became efficent with these operations, with the data and user interface on the same machine or intranet, so it was reasonable to expect these operations to perform just as well over the Internet for any Web-based system to gain user acceptance. Consider the following examples:

1. A common GIS operation is to define a polygon using the point and click operation of the computer mouse. Standard Web browsers treated a single mouse click as a prompt to transmit a query to the server. The delay be-tween responses from the server was far too long to maintain a coherent procedure. So Web-clients needed a number of different ways of handling mouse events.

2. GIS operations are often computationally intensive due to the ways in which users wanted to query geospatial data. Spatial operators existed in traditional GIS, such as *juxtaposition, next to, inside, inside and crossing*, and users wanted to perform the same operations in their Online GIS sys-tems. Ways had to be found to make these operations efficent enough to be functional in Online GIS systems.

3. Geographic queries are often context sensitive. For instance, a user may wish to have pop-up menus on a map that quickly display data about the properties of objects at the location of data in the user interface. Add to

this, the spatial and temporal aspects of geospatial information, and traditional GIS systems were returning valuable context-based location information to inform business decisions. Users wanted this same capability in an Online GIS.

4. High resolution imagery was required at the user interface for analytical work. But it was not practical to download all of the image file to the client, as this could take days and consume all the bandwidth available to the users. So ways had to be developed to make this download simple and reduce the dependency on downloading all data to render a small area.

There are a number of considerations for Online GIS, many of these around the optimisation of online processing, and there have been a number of technical developments since 2000 that provide solutions to these issues. Early Online GIS focused on the delivery of vector data to the client, to support client side querying and rendering. This made it necessary to increase client functionality, with client side languages such as JavaScript improving client processing significantly. Clients now process functions initiated by mouse clicks and mouse events at the client, without making a round trip to the servers and back again to perform an operation.

Spatial operators are a big part of GIS. Some of these were written to run on a client, such as *point in polygon*; however, it was not possible to download all of the spatial data to the client to enable client side processing of the complex spatial queries. So, many of these operations had to remain as server side processes, usually within the spatial database. This meant that spatial databases had to support a wide range of spatial operations as either *spatial data blades* or natively within the database itself, and several of them now do.

In the case of high resolution imagery, it was not practical to download large images over the Internet to render at the user interface, so new file formats and streaming protocols were needed to minimise the amount of data being passed over the Internet to the client. There have been two significant advances in imagery for Online GIS. The first has been the development of image streaming protocols, such as ECWP[1] and JPIP.[2] These protocols stream the ECW format developed by Earth Resource Mapping and JP2 (JPEG2000) files from the OpenJPEG Project, delivering highly compressed imagery over the Internet and optimising the delivery based on scale and extent of the user map viewing location. The second, and just as significant, has been the creation of a Pyramid File Format. This file format consisted of a number of different image resolutions, or layers, as part of its image file, that could be serviced out as a combination of adjoining image tiles to cover the user map viewing location and at an image resolution that matched the screen pixel resolution of the user interface. Google Maps data is a great example of a Pyramid File Format. Finally, and most relevant to this book, has been the development of all those types of open standards that enable this user interface separation to work effectively over the Web.

---

[1] \http://www.hexagongeospatial.com/products/data-management-compression/ecw/

[2] \http://www.openjpeg.org/jpip/doc/html/

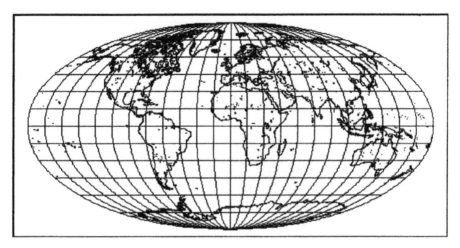

FIGURE 2.1: The Xerox PARC Map Viewer was the first online mapping system. Courtesy of Xerox, Palo Alto Research Center.

2.3 The Evolution of Online GIS

The first edition of this book was published in 2002 during the early evolution of Online GIS. At that time, these first implementations of Online GIS fell into one of two categories: spatial query systems, and map building and rendering programs. Many of those online systems incorporated both. In more recent times, the functionality and performance have dramatically improved, as has the infrastructure. The following section provides some historical context from these early systems through to some more recent and functional Online GIS examples.

2.3.1 Spatial Map Viewers

The first online map-building program was the Xerox PARC Map Viewer, built by Steve Putz (1994). The original system drew simple vector maps of any part of the world, at any resolution (Figure 2.1). Originally the service was entirely menu driven. That is, users selected options to build new maps (e.g., zooming in by a factor of 2) one at a time from the options provided. More recent versions have added new features and enhanced the interface.

Another early inovation in spatial map viewers that brought standard GIS tools and functionality online was the US Census Bureau TIGER mapping system, shown in Figure 2.2. TIGER had been launched by the Census US Census Bureau (1994), generating 45,000 to 50,000 maps per day and providing geocoded census related data, without requiring specialist GIS software. This system introduced maps to a much broader audience through exposing US census data to anyone who had access to the Internet and was a leading example of Online GIS around the time of the first

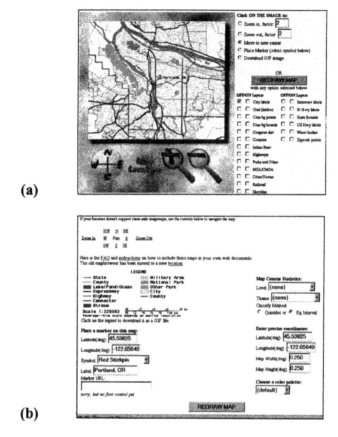

(a)

(b)

FIGURE 2.2: Early interactive map display for the TIGER mapping system provided by the US Census Bureau. (a) The interactive map display. (b) Further form fields for customising outputs.

edition of this book. This TIGER (Topologically Integrated Geographic Encoding and Referencing database) Mapping System for the USA included some zoom and pan tools and was one of the most prominent early online mapping facilities.

The TIGER Map Service was retired by the US Census Bureau in 2010 and replaced by a beta version of TIGERweb. This online portal[3] is now in production (Figure 2.3) and contains TIGERweb map viewing applications, links to Web services and TIGERweb Decennial applications for viewing census data without the need to view spatial information.

---

[3] http://tigerweb.geo.census.gov/tigerwebmain/TIGERweb_main.html

FIGURE 2.3: TIGERweb contains adminstrative boundaries and their associated census information, covering the US, where this example shows Central Park in New York with highlighted boundary areas.

2.3.2 Spatial Data Infrastruture Metadata Portals

There are a significant number of Online GIS systems and Web services that are now available. These systems and services include downloadable applications that link directly to online map and imagery services, many of which are free to use. Moreover, these disparate services can be mashed up with spatial data from other services that has been collected or maintained at different scales and to different accuracies and currencies. For this reason, it becomes important to users of this data to have a clear understanding of where the data has come from and the quality that can be expected from the data.

Spatial Data Infrastructure (SDI) Metadata Portals (Chapter 9) are one way for users to search and discover Online GIS services. These online portals include text based searches and map viewers to visualise the geospatial datasets available. Searches can also be location based, where users can zoom to a location of choice and query what services may be available at that location. Service types, such as Web Map Services (WMS)[4] (§10.8.1) and Web Feature Services (WFS)[5] (§10.8.2),

---

[4] An OGC standard that returns a raster image of a given map extent.
[5] An OGC standard to return vector data for a given map extent.

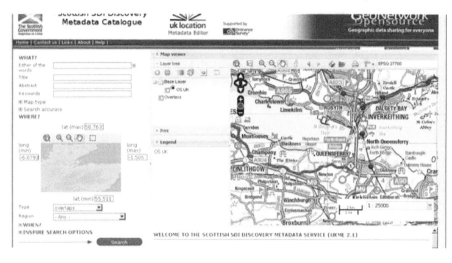

FIGURE 2.4: Scottish Spatial Data Infrastructure Metadata Portal.

are often supported at these portals and standards based metadata records from OGC (§9.3) are commonplace.

There are a number of well-known SDI Metadata Portals now on the Web. Figure 2.4 shows an example of an online SDI Metadata Portal, as published by the Scottish government. We will have a close look at some of the leading online portals in Chapter 9, highlighting the shift towards Data.gov portals, catalogues and clearinghouses since the Open Data Movement by leading government organisation around the world.

2.3.2.1 The Need for Metadata

The desire to coordinate data across many different sites led to the need for ways to label and identify data resources. Metadata is data about data and it plays a crucial role in recording, indexing and coordinating information resources (Cathro 1997). Look at any book. On its cover you will nearly always find the name of the book, the name of its author and the name of the publisher. These pieces of information are metadata. Inside the front cover you will find more metadata: the copyright date, the ISBN, the publisher's address, and more.

When you are reading the book, or if it is sitting on a bookshelf by your desk, then this metadata is rarely important. But if you want to find it in a large library, which might hold a million books, then you depend entirely on that metadata to find it. One of the worst things that can happen in a library is for a book to be put back in the wrong place. If that happens, then no one can find it without a long and painful search through the shelves.

As it is for books, so it is for online data. Metadata is crucially important for storing, indexing and retrieving items of information from online sources. Later in this book (especially Chapters 8 and 9), we will look in detail at the ways in which meta-data is structured and used. In other chapters we will look at the ways in which meta-data helps to create large-scale information resources online. Metadata has tended to be about content, creation and ownership, as we shall see. But online geospa-tial datasets and services may require additional metadata, in terms of who may be granted access and how the data may be used.

This book highlights the importance of this metadata to users. Trusting online content from data custodians and authoritative sources requires accurate metadata about the information being accessed to ensure that it is fit for the purpose for which it is being used.

2.3.3 Spatial Web Services

Web services play a significant role in any online system, from bank transactions to managing email servers. Spatial Web services are Web services that provide infor-mation specific to spatial datasets (or the datasets themselves) and are now available in many different service types and data formats to enable the online integration of rich spatial datasets at many different clients. These spatial Web services can be used to publish geospatial data in a standard way that can be understood by computer sys-tems and automatically mashed up with other datasets and services. The spatial Web services enable the creation of online marketplaces that are discoverable through Web service calls and they also provide online capabilities to search, query, geo-locate and validate data accuracy across disparate datasets.

2.3.4 Spatial Application Programmer Interfaces

The enormous growth of the IT industy has seen a similar growth within it of Ap-plication Programmer Interfaces, usually just referred to by the acronym API. In the early days of computing, reusable code came in the form of libraries, usually bina-ries specific to particular machines, architectures or chipsets. But the modern trend is to provide not the library, but details of the interface. Different developers or ven-dors may then create different implementations, provided that they meet the interface specification.

One of the features of using APIs in Online GIS is the ability for organisations to embed mapping components from other organisations into their own websites. This enables the developers of websites to focus on their own content. There is a lot of information on the Internet now describing spatial APIs which are available for use. The SIX Maps API[6] from Land and Property Information in NSW, Australia is a map and image viewer built on the JavaScript API from Esri Inc. This viewer also uses JavaScript tools from the DoJo Toolkit[7] and connects to map services, image services and Web services to access the data that it mashes up. This client application (Figure 2.5) has a number of search and query capabilities that use the

---

[6] http://maps.six.nsw.gov.au
[7] http://www.dojotoolkit.org

map and Web services. It also has a number of client side GIS tools that can be used to measure and select features and was an early implementer of drag and drop information where users could simply drag images and text files from the file system onto the map canvas and the API would geo-locate these image and data within the text files.

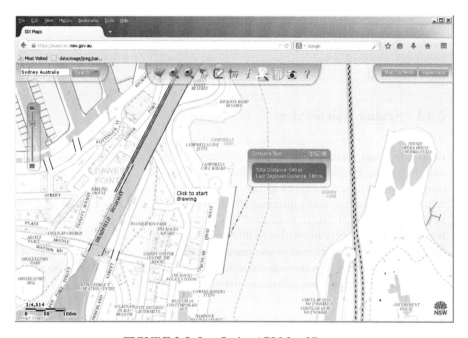

FIGURE 2.5: JavaScript API Map Viewer

Companies such as Google also offer a number of APIs, the most notable being the Google Maps API[8] that can be used by organisations to create their own websites. Reuse of the libraries of existing Online GIS capabilities is an essential way to reduce development times. APIs often link in to robust code bases that provide functionality that is often maintained by communities of developers.

In early 2015, however, there is a dark cloud hovering over the widespread use of APIs. Oracle and Google are engaged in a tense legal battle as to whether APIs can be copyrighted and require licensing (and possibly) fees to use. Until now their use has been free, but subtle legal reasons may mean this is about to change (Sprigman 2015). APIs will be further covered in Chapter 4 when we consider client-side programming.

2.3.4.1 Embedded Viewer in Virtual Tourist

The Virtual Tourist (Plewe 1997) was a service that aimed to provide tourists and the tourist industry with detailed information about every country in the world. Starting

---
[8] https://developers.google.com/maps/

from a map of the world, users could click on successive maps until they had selected a country or region of their choice. The Virtual Tourist website was one of the earliest demonstrations of the power of online hypermedia to create systems that are more than the sum of their parts.

FIGURE 2.6: The Virtual Tourist website with an embedded Google Maps component showing three-dimensional data over Banff National Park in Canada as activated from metadata records on the Virtual Tourist website.

The original service (which has since been commercialised), placed a central indexing site on top of literally thousands of contributing sites that provided sources of information about individual countries. A second, but equally important lesson was that it is easier to keep information up to date if detailed information is published online by the organisations that maintain it, rather than by a central agency that would be hard pressed to keep the information current.

The current version of this early Online GIS website is shown in Figure 2.6,[9] where the maps are not the focus of the website, but an embedded component that is linked to the other content of the Web pages.[10]

2.3.4.2 Embedded Viewer in Nations Online

Another example of a website with embedded Online GIS is the Nations Online website.[11] This website contains a wide range of geographic data from around the globe and is an ideal resource for geography and history students looking for maps. The website uses an embedded Google Maps API map canvas and search engines, as shown in Figure 2.7, *Google and the Google logo are registered trademarks of Google Inc. Used with permission.* to support its online maps and geography information.

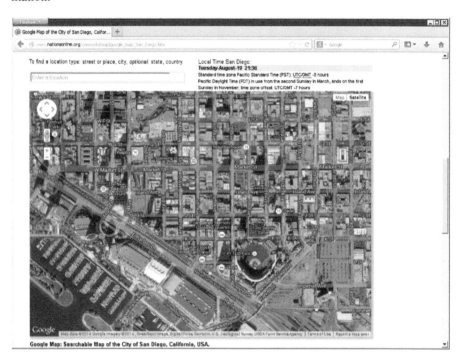

FIGURE 2.7: An example Online GIS website which has embedded the Google Maps API as a map base for its other geospatial information.

The use of these APIs means that the project has focused on developing its own map resource overlays and textual information on a wide range of topics, leaving some of the Online GIS rendering to code developed elsewhere.

---

[9] Google and the Google logo are registered trademarks of Google Inc. Used with permission.
[10] http://www.virtualtourist.com
[11] https://www.google.com.au/q=nations+online+project

2.3.5 Three-Dimensional Globes

There have been four well-known 3D globe viewers available on the Internet over the last decade. The most popular of these is the Google Earth application (Figure 2.8).
[12] This form of an Online GIS may have less GIS functionality than other 2D Online GIS systems; however, users get a real sense of looking at the world when they use it. In many ways, 3D global viewers simply make sense to non-GIS specialists.

FIGURE 2.8: The simple interface of Google Earth that shows Online GIS data for the entire globe.

There are a number of globe implementations now online that provide collaboration points for several data types, including photography from the map viewer extent and links to other data themes, such as natural disasters. Improvements in data content of these viewers, such as better digital terrain models (surface models) and improvements in aerial photography quality will no doubt lead to an increase in use of these global based Online GIS systems.

---

[12] Google and the Google logo are registered trademarks of Google Inc. Used with permission.

2.4   Interactive Online GIS

User interaction with Online GIS has also improved greatly over the last decade since its beginning, owing to many factors, including faster Internet, improvements in computer hardware and software, a significant increase in the amount of data available online and a willingness of users to participate in Online GIS. Here we will give an overview of the changes in interactive maps over the last two decades.

2.4.1   An Early Interactive Map

One issue for map users is to be able to plot their own data on a map. The Map Maker service (Steinke et al. 1996) developed at Charles Sturt University, Australia, addressed the need of field researchers (and others) to be able to produce publication quality plots of regions with their own sites or other locations added. The first trial was in 1993 and this service was made publicly available in 1995. The base layers which generated by the GMT system, which were a freeware package of Generic Mapping Tools (Wessel and Smith 1991).

While the website no longer exists and functionality was rudimentary based on today's standards, users were able to plot their own data on custom designed maps (Figure 2.9) by entering basic data content into a form field including the set of coordinates and labels for points that they wished to include. The Map Maker also allowed a limited degree of searching for major cities.

2.4.2   Environment Australia's Species Mapper

Environment Australia's service Species Mapper (ERIN 1995) provided a system for querying its database of species distributions and plotting them on maps. However, the service went further than that. The service also generated and plotted models predicting the potential geographic distribution of individual species. To achieve this (using the BIOCLIM algorithm of Busby et al. (2014)), the server carried out the following sequences of steps (here slightly simplified) in real time:

1. Query database to retrieve records of species locations.

2. For each species location, interpolate values of essential climatic variables.

3. Calculate the climatic envelope bounding all the species records.

4. At the resolution specified, identify all other sites in the landscape that fall within the climatic envelope.

5. Plot the sites identified on a base map.

6. Deliver the map to the user.

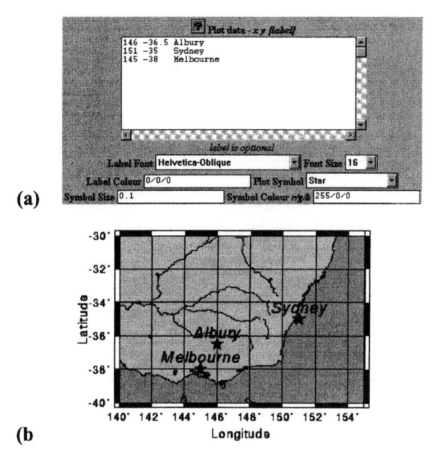

FIGURE 2.9: Adding user data to a map built online using CSU's Map Maker service. (a) Form for inserting user data. (b) The resulting map with the towns added.

2.4.3 Pierce County Interactive Database Access

In the state of Washington, USA, Pierce County provided an early public GIS online. Called MAP-Your-Way™, the system provided a flexible and widely accessible interface for many of the county's public databases. This interactive system allowed users to create their own maps online on demand.

Placing the system online means that anyone could generate customised maps for any combination of features for any part of the county. The service was implemented using vendor developed mapping tools and the site included a disclaimer that covered the following crucial matters: data limitations, interpretations, spatial accuracy, liability, and warranty against commercial use.

Since those early days, Pierce County has upgraded its online interactive mapping

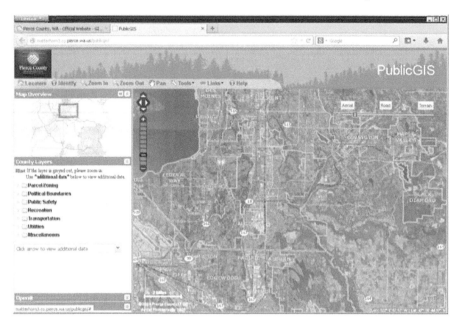

FIGURE 2.10: Pierce County's public GIS information system in 2014.

system (Figure 2.10) and it now provides a feature rich public GIS system for online users.[13]

2.4.4   Crowd Sourcing and Geoprocessing with Online GIS

There are two trends in Online GIS that are getting a lot of attention from spatial organisations and online user communities. The first is a process known as crowd sourcing,[14] which involves the publication of online tools for users outside an organisation to review existing data held by the custodians, providing feedback on the quality of the data, highlighting errors and suggesting changes. Crowd sourcing also provides a mechanism for external users to capture new data to be added to custodial datasets. Including metadata as a key component of any changes or new data, such as who captured it, when the data was captured and how was it validated, will be an important part when assessing the success of crowd sourcing geospatial data. It was unclear at the time of writing if standards now exist that can be used to define these metadata records; however, Dublin Core (§8.2) may be a suitable standard to use.

The second trend, which is being driven by software vendors, is the development of online geo-processing tools that can be used in Online GIS. These tools include spatial modelling and spatial analysis tools. One popular example of online geo-processing has been transport routing to provide directions to a location, but there

---

[13] https://www.co.pierce.wa.us/index.aspx?NID=491

[14] One crowd sourcing organisation can be found at http://www.crowdsourcing.org

are other tools that create profiles for the rise and fall of these routes, adding a terrain perspective to the data. More complex geo-processes tools are also available that provide image processing in Online GIS systems. Some products provide a number of functions for processing of remote sensing data, which can produce data for crop analysis, soil moisture and salinity, and there are a few products on the market that are now providing client side object detection and feature extraction.

While these software solutions are still maturing and performance of the processing tools is still a major consideration, they do give the reader some insight into the potential for interactive Online GIS and confirm that the future of Online GIS remains bright.

2.5   The Rise of Online Data Warehouses

We saw above how the Virtual Tourist linked together information from many sources. However, that was a case of creating a unifying umbrella for online resources that had already appeared. In many other cases, the organisation and development of online data sources required a more systematic and coordinated approach.

One example was cooperation amongst sites dealing with related topics. This process led to the formation of Information Networks (Chapter 7) dealing with a wide range of topics. This process still continues and has led to technical innovations to try to promote cross-site coordination. The first priority was to develop methods of indexing information. This goal led to numerous indexing and production standards, such as XML (Chapter 5 and Chapter 6) and others.

There have also been a variety of software developments to enhance cross-site queries and information sharing. The first of these was the notion of a Web crawler, an automatic software agent that trawls the Web recording and indexing everything it finds. Many Internet search engines have used this kind of software. Later developments have included collaboration whereby a single query can spawn searches on many different sites, after which the results are pooled and transmitted back to the user.

However, the ultimate need is to develop systems that return not indices of links to data, but the data itself. That is, they draw on data from many different sites and combine it into a single set of information for the user. This need gives rise to the idea of a distributed Data Warehouse. Data warehouses are simply assemblies of many different databases that are combined into a single information system. With the spread of electronic commerce and large-scale data collection systems, Data warehouses are now common. A distributed Data Warehouse is just a Data Warehouse that is spread across several sites on the Internet. In some contexts (such as environmental information, biotechnology, etc.) there is a trend toward global information systems that integrate similar kinds of information worldwide, creating what is now termed Big Data. We will look at Data Warehouses more closely in Chapter 10.

It is this kind of information that is needed for Online GIS. No single agency can store and maintain all information, in complete detail, for every part of the world. Perhaps the ultimate Online GIS would be a global information system that integrated all kinds of geographical data from all sites into a single universal atlas of the world. Unlike a traditional paper atlas, there is in principle no reason why such a system needs to be restricted to a particular scale, or to particular data layers.

2.6   Differences between Stand-Alone and Online GIS

The traditional model for GIS assumes that the system consists of a single software package, plus data, on a single machine. This model no longer meets the realities of many GIS projects, which today are often multi-agency, multi-disciplinary, multi-platform, and multi-software. Large numbers of contributors may be involved, and there is often a large pool of potential users. These users may require not only maps, but also many forms of multi-media output (e.g., documents). Moreover, they are likely to require access to the most current data available, and not to copies that may be months, or even years, old. A central practical issue, therefore, is how to provide widespread, device-independent, access to GIS for large numbers of contributors and users.

The chief difference between an Online GIS and traditional systems is the separation of user interface, data storage and processing. In stand-alone GIS all of these elements are normally present on a single machine. In Online GIS the elements are normally spread across several machines. This separation of GIS elements creates its own special issues and problems. One is the need to transfer data from one site to another. This leads to a need to cut down the number of transfers where possible. Another problem is that keeping track of what the user is currently doing becomes a non-trivial matter. Normally these details are immediately available because they are stored in computer memory while the GIS software is running. However, across a network they need to be stored so that the software can be initiated with the correct state. In the remainder of this chapter, we look at some of the options for dealing with this separation of elements. In the chapters that follow, we look at some of the issues in the course of implementing GIS in an online environment.

2.7   Options for Implementing Online GIS

With traditional GIS, there were several ways in which GIS could be placed online and perhaps the most telling is the way in which the user interface was handled. There were essentially two approaches. One was to take an existing stand-alone GIS and enable it to run online. The other was to provide GIS functionality via standard Web browsers and browser plugins. More recently, as a result of mobile computing

(Chapter 11) and Location Based Services (Chapter 12), lightweight GIS is being made available in downloadable applications as well.

2.7.1 Internet Enable an Existing Stand-Alone GIS

In this approach the user interface was a stand-alone GIS, which handled all (or most) of the processing involved. However, the system is provided with the capability to access files across the Internet. This was akin to the downloadable application, with local data (i.e. data residing on the user's machine) being combined with data from other sites. The above approach had several advantages:

- The Online GIS retains all the power, speed and functionality of a traditional system.

- Users can continue to use a system that they are familiar with.

  However, there were also some disadvantages:

- Users needed to acquire and install specialist software (and often hardware), which may be expensive.

- The system was not universally available. Its audience was still limited to those who have the necessary software.

- The GIS being redeveloped was not optimised for Web processing and connectivity.

2.7.2 GIS Functionality in Standard Web Browsers

In this approach, a GIS was built to use existing standard browsers as its user interface. Normally this means that most of the data, and most of the processing, is delivered via a server, which then sends the results to the client for display. The main advantages of this approach were:

- The system was potentially available to any Web user, although browser plugins were sometimes required to get suitable levels of performance or to do complex calculations at the browser client.

- Simple GIS can be implemented more quickly and easily than using the power of a full system. For instance, if all you want is for users to be able to make geographic queries of a database (e.g., identifying sites that fall within the selected region), then just a few geographic operations are required. The manager therefore need only install software to handle those operations. The rest of the processing can be handled within the standard Web browser by using JavaScript and GIS functions available from a number of freely available APIs.

  Some disadvantages, however, were:

- There remains a strong dependency for any Online GIS system on the availablity and performance of the Internet connection. While many sites work well where there are strong Internet connections, some regions of the world still do not have suitable Web networks.

- Many interactive processes, which are taken for granted on a stand-alone system, become difficult to sustain in a distributed environment.

Besides the above two extreme approaches there are a number of hybrid implementation models that can be used. For example, an Online GIS could use a standard Web browser for the starting interface, and for routine search and selection functions. However, for specialised geographic operations and display, it might pull up a GIS display package as a helper application. Some GIS vendors have developed technology for enabling these hybrid systems.

2.7.3   Technical Aspects of Implementing Online Systems

Where existing GIS software is to be Internet enabled, then the Internet queries can be seen as akin to reading files in a traditional system, except that the files are stored somewhere else on the Internet, rather than on the user's hard disk. The rest of the system operates just as any stand-alone GIS would. Thus the main issue is how to build and send queries across the Internet, and how to handle the replies. We shall take up this issue in Chapter 5, where we address the issue of querying XML documents (§5.6). Since this type of GIS applies essentially to existing software, it is mainly an issue for commercial developers.

Developing GIS that use the Web as their interface is a much more general concern. At the time of writing, there are already many online GIS that use the Web as their interface. For the most part, these still require an amount of custom coding to bring the service online. There are a number of open source projets and vendor products that will assist when building Online GIS. It is therefore important for potential developers to have a clear understanding of the issues involved.

The following two chapters (Chapter 3 and Chapter 4) examine issues and methods for implementing GIS that use the Web as the main platform. In Chapter 3 we shall look at GIS processing on a Web server. Chapter 4 looks at processes that need to be handled by the Web client.

A fundamental practical issue for the development of Online GIS is how it should relate to other services. Systems that are built in ad hoc and idiosyncratic fashion will not be consistent with other Online GIS services. There are many advantages in developing GIS services that are consistent with those at other sites. As we shall see later, standards now exist that promote consistency and provide a basis for Web mapping tools that simplify the development of GIS services online. Future projects that are implementing Online GIS should always consider the use of the standards as a way of future proofing their solutions and improving interoperability of their solutions.

2.7.3.1 Connectionless Interaction

Perhaps the most basic issues associated with HTTP 1.0 are those arising from connectionless interaction. Web browsers and the HTML specification have been developed as a stateless interface into WWW resources. Therefore, under HTTP 1.0, no context is preserved at either the client or server end, with each data transaction treated independently from any other. Later releases do allow preservation of context, but the stateless nature of the Web has dictated a number of technological compromises. This preservation of state is important to Online GIS.

One method is to embed hidden state variables within Web documents. The HyperText Markup Language (HTML) provides several methods by which values for state variables can be transferred between server and client.

First, the HTML forms interface includes provision for hidden fields (§3.4). We can use these to record a user's current state and essential portions of their interaction with the program. In effect, the server builds a script to reconstruct the current position as part of each interaction and embeds it in the document that it returns to the user. This information is not a simple log; previous panning and zooming can be ignored, for example.

A second method is the use of cookies (see §4.5). Amongst other things, cookies provide a way to identify users when they reconnect to a service. So, for instance, a user's preferences can be stored from session to session, allowing some management of the browser session.

A third procedure is caching. Creating a particular map, for instance, may require a series of operations that would soon grow impossibly tedious and time consuming if repeated on each step. When interacting with a dedicated system, we avoid this problem by saving the result of a series of steps. This is done either implicitly, as binary data linked with the user's session, or explicitly, as a file under the user's name. In caching, the server not only delivers information back to the user, but also saves it to disk for a finite time. The file is then available as a starting point for the next query, if there is one, or can be sent to other users who can replicate the session in their own system (sharing of information on current Online GIS systems, such as Google Maps, does use a cache that can be shared. This cache is in the form of a URL).

2.8 Summary and Outlook

- This chapter started out with some information on why Online GIS offers additional benefits to traditional Online GIS in §2.1 and then introduced some issues which should be considered with deploying spatial data over the Internet (§2.2).

- §2.3 then provides some historical context to the way Online GIS has evolved since the first edition of this book, and gives a snapshot of some of the changes that have taken place as Online GIS transitioned towards a home in mainstream computing.

- The interactive nature of Online GIS is then discussed (§2.4), with some examples of how people are starting to use these technologies to solve business problems, linking back to the first edition.

- Several differences between how traditional GIS can be implemented as Online GIS are then discussed in §2.6. The move toward Data Warehouses for geospatial data (§2.5) is also introduced; for discussion in more detail (see Chapter 10).

- Finally, §2.7 introduces some of the functionality and technologies that factor in online implementation of traditional geospatial data stores.

This chapter provides more of a philosophical introduction to Online GIS, setting the roadmap of subsequent chapters, and should give the reader a sense of how much Online GIS has grown over the last decade, whetting their appetite to find out more.

3

Server-Side GIS Operations

CONTENTS

Server-Side GIS operations play a critical role in the publication of feature-rich GIS datasets for online access. These operations have been evolving, albeit at a seemingly slower rate than client-side changes. The running of GIS operations on client machines is seen as a way of distributing processing as a form of grid computing. However, much of what happens at the server side is still critical to support the speed and viability of Online GIS systems.

3.1 Web Servers

In this chapter and the one that follows, we examine issues arising in the development and use of standard Web systems as a medium for GIS. Since we can run GIS packages, such as ArcInfo, over widespread client server systems such as X-Windows, we might ask what the Web protocol has to offer. And, as we shall see in this chapter, Web-based GIS is not without a few problems. Since the Web server's primary role is to deliver Web pages, server-side GIS operations have to be carried out by secondary programs (usually via the Common Gateway Interface (CGI), which we describe below).

A Web server is a program, or daemon, which runs the HTTP protocol, hence the server is called an HTTP Daemon (HTTPD). The original Web protocol, HTTP, was designed for delivery of static text and simple images, and was not optimised for much else. The gains come in low cost, package independence and in cross-platform Web availability. But any program can package information up and transmit it according to HTTP. This is one strategy available for GIS packages. The other advantage is close integration with a website which may have information and resources going way beyond the spatial. An essential feature of geographic information systems is that they allow the user to interpret geographic data. That is, they incorporate features for data processing of various kinds.

Some of these operations include spatial database queries, map building, geo-statistical analysis, spatial modelling and interpolation. In an online GIS, the question immediately arises as to where the above processing is carried out: on the Web server, which is remote from the user and supplies the information, or at the Web client, which receives the information and displays it for the user. In this chapter we look at issues and methods involved in carrying out GIS processing by a Web server. Chapter 4 covers client-side operations.

What GIS operations can and should be run on a server? Only a few operations cannot be run effectively on a server. These consist chiefly of interactive operations, such as drawing, that require rapid response to a user's inputs. Other operations, such as querying large or sensitive databases, must be carried out on the server. However, for most other operations, the question of whether to carry it out on the server or on the client machine is not so clear cut. The two most crucial questions when deciding whether an operation should be performed by the server or the client are

- Is the processing going to place too large a load on the server? Any busy Web server will be accessing and transmitting several files per minute. The processing needed to (say) draw a map may take only a few seconds, but if requests for this operation are being received constantly, then they could quickly add up to an unmanageable load.

- Does the volume of data to be sent to the client place too great a load on the network? For instance, delivering large images constantly might slow response for the user.

Over the last decade, the huge increase in processing speed has made both client-side and server-side processing a lot faster. Broadband speed has also increased for many end users, but there is a caveat – WiFi. Much online GIS now takes place on mobile phones (Chapter 11), where the communication is WiFi, which is often of lower bandwidth. This has led to new metadata specifications (mobileOK, §8.4.2,

Chapter 11) through which the server can know if it is talking to a mobile device and modify the data transmission accordingly.

3.1.1 Popular Web Service Protocols

The **Hypertext Transfer Protocol** (HTTP) is one of the main communications protocols used on the World Wide Web. It passes hypertext links from a browser to a server and allows the requested documents and images to be passed back to the browser. The server, or HTTP daemon (HTTPD), manages all communication between a client and the programs and data on the website.

Another popular protocol is the **Simple Object Access Protocol** (SOAP). This protocol is a popular communications protocol for exchanging well-structured data between applications over the Internet. Based on XML, SOAP communicates over HTTP and provides a container that can hold a number of different formats of data, making it both simple to use and extensible. One of the advantages of using SOAP is that it can communicate with Web services on different operating systems and programming languages.

Representational State Transfer (REST) is a Web services architecture style that works like the Web. One of its key strengths is that it makes websites usable by machines. It leverages the power of the HTTP application protocol, the URI naming standard and XML. The design of RESTful Web services is simple to use, is versatile and more scaleable than systems designed using Remote Procedure Calls (RPC) (Richardson and Ruby 2007). REST is now used for a significant number of websites where there is a need for machine communication between servers and clients.

Irrespective of the form of Web services implemented, Web services and mashups have now turned the programmable Web into a powerful distributed platform for GIS.

3.1.2 Hypermedia

The World Wide Web has turned the Internet into a medium for hypermedia. The term *hypermedia* is a contraction of hypertext and multimedia. Hypertext refers to text that provides links to other material. *Multimedia* refers to information that combines elements of several media, such as text, images, sound and animation. *Hypertext* is text that is arranged in non-linear fashion. Traditional, printed text is linear: you start at the beginning and read through all the passages in a set order. In contrast, hypertext can provide many different pathways through the material. In general, we can say that hypertext consists of a set of linked objects (text or images). The links define pathways from one text object to another. This is of course the way everyday Web pages are constructed.

Electronic information systems have led to a convergence of what were formerrly very different media into a single form, known as multimedia. Film and sound recordings, for instance, have now become video and audio elements of multimedia publications. In multimedia publications, one or another kind of element tends to dominate. In traditional publications, text tends to be the dominant element, with images provided to illustrate the written account. One exception is the comic book, in which a series of cartoon pictures provides the storyline, with text provided as support.

Vision in humans is the dominant sense, so in multimedia, visual elements (particularly video or animation) often dominate. However, in online publications, the speed of the network transmission is still a major consideration for narrow band delivery, such as WiFi. [1] At present full-screen video is not practical, except across the very fastest network connections.

It was the introduction of browsers with multimedia capability that made the World Wide Web a success. However, audio and video elements were not formerly supported by any Web browsers. These require supporting applications that the browser launches when required. Early examples included the programs `mpegplay` for MPEG video files and `showaudio` for sound. Each kind of information requires its own software for generating and editing the material. Now, with the huge popularity of sites such as YouTube, streaming video over the Web is commonplace and is supported by most browsers.

With the advent of HTML5 (§5.3), audio and video built-in applications are now a requisite part of the specificaton. In this chapter we make occasional use of markup formats, distinguished by tags; start tags begin and end with angle brackets as `<map>`; end tags have an additional backslash as in `</map>`. The full syntax and structure of markup for HTML and XML are covered in Chapter 5.

3.2   Server Software

There are many versions of software for Web servers. The most widely used is the Apache server, which is a freeware program, with versions available for all Unix operating systems. However, most of the major software houses have also developed server software. It is not feasible here to give a full account of server software and the issues involved in selecting, installing and maintaining it. Here we can only identify some of the major issues that Web managers need to be aware of. Security is a major concern for any server. Most packages make provision for restricting access to material via authorisation (access from privileged sites) and authentication (user name and password).

With heightened concern over the security of commercial operations, most server software now includes provision for encryption as well as other features to minimise the possibility of illegal access to sensitive information. The rapid rise of online commerce has led to a variant of HTTP, HTTPS, which provides encrypted data transfer, suitable for providing sensitive information such as credit card numbers. In terms of functionality, most server software today makes provision for standard services

---

[1] WiFi has become so widespread that new standards are continually appearing, each with higher bandwidth than the last. Similarly, telecom companies are offering increasing bandwidth over the mobile phone network.

such as initiating and running external programs, handling cookies (§4.5), allowing file uploads and making external referrals. Perhaps the most important questions regarding functionality are what versions of HTTP the server software is designed for and how easily upgrades can be obtained and installed. When installing a server, it is important to give careful attention to the structure of the two directory hierarchies. Source material used by HTTP servers normally falls into two main hierarchies:

- The *document hierarchy,* which contains all files that are delivered directly to the client.

- The CGI *hierarchy* (§3.3.1), which contains all the files and source data needed for processing information.

The above distinction is important because it separates freely accessible material, such as documents, from programs and other resources used in processing, which often have security implications. Care is needed too in the organisation of directories under each of these hierarchies. The names of directories normally form part of the URL for any item of information, so it is important that they have logical names, if the URLs are to be referenced directly. However, the widespread use of databases to serve Web pages has led to very long URLs, which are essentially machine readable only.

Moreover, they should reflect the sorts of queries that users will make. Many Web managers make the mistake of structuring directories and information in terms of system management, or in terms of the internal organisation of their institution or corporation. So, for example, users would normally prefer to look for tourist information organised under country or region, rather than (say) the names of individual travel companies or hotel chains. More importantly, the logical names of directories should never change once they are established. It is possible to circumvent this issue to some extent by using aliases to provide a logical hierarchy. In Unix, for instance, directories and files can be assigned logical names that are completely independent of their true storage location. For instance, the real file path

```
/documents/internal-data/file023.dat
```

might be assigned the much simpler logical path

```
hotel-list.
```

3.2.1 Practical Issues

Although not specifically related to online GIS, a number of general issues are important for any online information service. In maintaining a server, three important issues are system updates, backups and server logs. System updates consist of files and data that need to be changed at regular intervals. For example, a data file that is derived from another source may need to be downloaded at regular intervals. These sorts of updates can be automated by using appropriate system software. In the Unix operating system, for instance, the traditional method of automating updates is by setting *crontabs* (a system for setting automated actions against times) to run the necessary shell scripts at regular intervals.

Backups are copies of data on a server. Their function is to ensure that vital data is not lost in case of hardware or other failure. Backups are usually made regularly. For safety, in case of fire or flood, backup copies are best stored off site. Mirror images of data provide another form of backup. However, it may be unwise to rely on an outside organisation to provide the sole form of backup. As with server updates, backups can be automated. Any busy server would need daily backups, though if storage space is limited, these could be confined to copies of new or altered files. However, it is always wise to make a complete copy of the document and CGI hierarchies at regular intervals.

At the time of writing the second edition of this book (2015), the storage world is changing. The cost of storage has fallen dramatically, so much so that many organisations now provide large amounts of storage online, for free, or virtually for free. This has led to *cloud computing*, where data is stored and distributed across multiple servers. Cloud computing (§13.4.1) is still in a growth phase. There are trade-off issues in security, both for and against: the wide distribution of data means that no single catastrophic event will destroy any of the data if it is properly padded with redundancy, but the direct control or hold on the data is sacrificed.

Server log files provide important sources of information about usage of the system. The access log lists every call to the system. As well as recording the names of files or processes that are accessed, it also includes the time and address of the user. This information can be useful when trying to assess usage rates and patterns. The error log is useful in identifying faults in information services, as well as potential attempts to breach system security.

3.3 Server Processing

To be a complete range of media, the Web provides a methodology by which a Web server can carry out processing of various kinds in order to respond to a query. The processing falls into four main classes, which are listed below:

1. Allowing users to submit data to the server, usually via forms;

2. Uploading files to the server;

3. Data processing to derive the information required to answer a query;

4. Building and formatting documents and their elements.

The Common Gateway Interface (CGI) is a link between the Web server and processes that run on the host machine (Figure 3.1). CGI programs are accessed through the Web just as a normal HTML document is, through a URL. The only condition placed upon CGI programs is that they reside somewhere below the `cgi-bin` directory configured within the HTTP server. This is so the server knows whether to return the file as a document or execute it as a CGI program. The data from a form or query is passed from the client browser to the HTTP server:

1. The HTTP server forwards the information from the browser via CGI to the appropriate application program.

2. The program processes the input and may require access to certain data files residing on the server, such as a database.

3. The program writes either of the following to standard output: an HTML document or a pointer to an HTML document in the form of a Location Header that contains the URL of the document.

4. The HTTP server passes the output from the CGI process back to the client browser as the result of the form or query submission.

Programs to run CGI processes can be written in virtually any programming language. Early examples were PERL, C and shell scripts. PERL was the language of choice for many CGI programmers due to its powerful string manipulation and regular expression matching functionality. This made it well suited to handling CGI input from HTML forms, as well as producing dynamic HTML documents.

The trade-off among these is ease of use and power (such as PERL) and security (compiled languages, such as C). Other languages have now emerged, such as Python and PHP, which allow both client-side and server-side processing. Since this chapter is primarily about concepts and ideas, rather than an application cookbook, we shall stay mostly with a generic language, such as PERL. The basic structure of a CGI program is illustrated by Figure 3.2.

3.3.1 CGI and Form Handling

HTML forms are implemented in a way that produces key-value pairs for each of the input variables. These key-value pairs can be passed to the CGI program using one of two methods, GET and POST. In HTML5 there are two additional methods, PUT and DELETE. However, the uptake of these has been somewhat slow, and we shall not consider them further.

An important point to note is that with both GET and PUT methods, the data is encoded so as to remove spaces and various other characters from the data stream. A CGI program must decode the incoming data before processing.

3.3.1.1 GET Method

The GET method appends the key-value pairs to the URL. A question mark separates the URL proper from the parameters which are extracted by the HTTPD server and passed to the CGI program via the `QUERY_STRING` environment variable. The general format of a GET query is as follows:

```
http://server-url/path?query-string
```

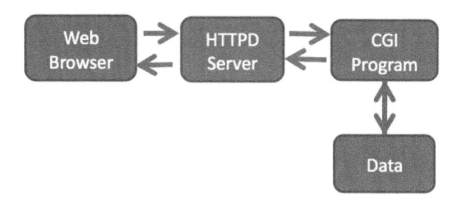

FIGURE 3.1: Role of the Common Gateway Interface (CGI), which mediates exchanges between a Web server and other programs and data.

In this syntax, the main terms are as follows:

server-url is the address of the server that receives the input string;

path is the name and location of the software on the server;

query-string is data to be sent to the server.

Below are some typical examples:

```
http://www.cityofdunedin.com/city/?page=searchtools_st
http://www.geo.ed.ac.uk/scotgaz/scotland.imagemap?30,29
http://www.linz.govt.nz/cgi-bin/place?P=13106
http://ukcc.uky.edu:80/~atlas/kyatlas?name=Main+&
                              county=21011
```

The GET method is generally used for processes where the amount of information to be passed to the process is relatively small, as in the above examples. Where larger amounts of data need to be transmitted, the POST method is used.

3.3.1.2   POST Method

In the POST method, the browser packs up the data to be passed to the Web server as a sequence of key-value pairs. When the server receives the data it passes it on to

CGI Program Structure

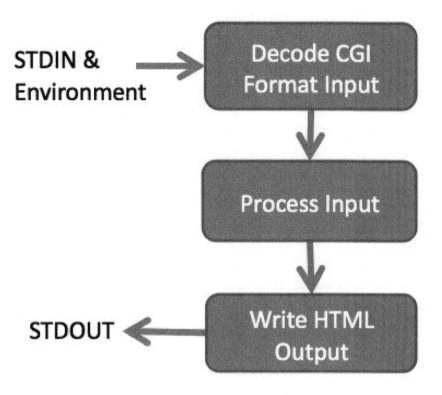

FIGURE 3.2: The structure of CGI programs, where any program must carry out the three tasks shown: first it decodes any form input, then carries out the required processing, and finally it must compile an output document to return to the user.

the CGI program as standard input. Here is a typical string that would be sent by the sample form, which is described in the section that follows.

```
register=tourism&country=Canada&attraction=LaketLouise&
description=&latdeg=Sl&latmin=26&longdeg=-116&
longmin=11&region=Alberta&hotel=YES&meals=YES&park=YBS&
history=YBS&website=http://www.banfflakelouise.com/&
email=info@banfflakelouise.com
```

The application program needs to unpack this string and carve it up into the required name-value pairs before it can use the data.

TABLE 3.1: Illustrative fields for a Web form where the number of attributes has doubled in HTML5

Field	HTML tag	Attribute	Function
Text	input	text	text entry
Textarea			box for longer text
Select	input	select	pull down list
Radio	input	radio	button selection
Check box	input	checkbox	box to tick
Hidden field	input	hidden	invisible data such as to indentify the form
Submit	input	submit	send form off for processing

3.4 Forms and Image Fields

An important aspect of the Web is that information access need not be passive. Users can send data to Web servers via forms. Forms are documents that include fields where data can be entered. These fields consist of text boxes and various other widgets (Table 3.1), which will be found in any basic book on HTML. Users normally submit form data by clicking on a SUBMIT button. The server receives the submitted data, processes it and sends a response back to the user (Figure 3.2).

The basic structure of an HTML form is as follows:

```
<form processing_options>
  input fields mixed with text
</form>
```

Here the *processing options* are attributes that describe the method that will be used to transmit the data (see GET or POST above (§3.3.1)) and indicate which program will receive and process the form data. Table 3.1 lists the types of input fields that can be included. The example below is the HTML source code for a simple form that might be used for (say) operators to register tourist attractions in an online database. The code contains two hidden fields, which provide technical data to the server. One, called "register" tells the server which register to use. This is essential if the same software handles several different services. The second hidden field tells the server which country the registered sites are located in (Canada in this case). This data would be necessary if the underlying database held information for many countries, but the form applied to one only. Note the use of the table syntax to arrange a clear layout of the fields. Web authoring systems usually make it possible to construct HTML documents and forms without seeing the underlying code at all. However, it is not a bad idea for authors to learn to manipulate HTML code directly. For instance, many automatic form builders are features that may not display well on all browsers.

```html
<html>
  <head>
  </head>
  <body>
    <h1>Tourist site register</h1>
    <form
      action="http://life.csu.edu.au/cgi-bin/geo/demo.pl"
      method="POST">
    <input type="hidden" name="register" value="tourism">
    <input type="hidden" name="country" value="Canada">

    <p>Attraction
      <input name="attraction" type="text" size="50"
          value="Enter its name here">

    <p>Description
      <textarea name="description" rows="2"
          cols="40"> Write a brief description here.
      </textarea>

    <h4>Location</h4>
    <table>
      <tr> <td><i>Latitude</i><br>
          <input name="latdeg" type="text" size="3"> deg
          <input name-"latmin" type="text" size="3"> min
             
        <td><i>Longitude</i><br>
          <input name-=longdeg= type="text" size="3"> deg
          <input name="longmin" type="text" size="3"> min

        <td><i>Province</i><br>
          <select name="region">
            <Option value="BC">British Columbia
            <option value="ALB">Alberta
            <option value="SAS">Saskatchewan
            <option value="MAN">Manitoba
            <option value="ONT">Ontario
            <Option value="QUE">Quebec
            <Option value="NB">New Brunswick
            <Option value="PEI">Prince Edward Island
            <option value="NS">Nova Scotia
            <option value="NFL">Newfoundland
          </select>

    </table>

    <h4>Facilities available</h4>

    <table>
```

```
    <tr><td valign="top"><i>Hotel</i> <td>
      <td><input type="radio" name="hotel" value="YES"> Yes
        <br><input type="radio" name="hotel" value="NO"> No
           
      <td valign="top"><i>Meals</i>
      <td><input type="radio" name="meals" value="YES"> Yes
        <br><input type="radio" name="meals" value="NO"> No
    </table>

    <h4>Features</h4>
    Park <input type="checkbox" name="park" value="TRUE">
    History <input type="checkbox" name="history"
              value="TRUE">

    <h4>Online addresses</h4>
    <table>
      <tr><td><i>Website</i> <br>
        <input name="website" type="text" size="30"
              value="http://">

      <td><i>Email</i> <br>
        <input name==email" type="text= size="20"
              value="Enter.name@location">

    </table>
    <input type="reset" value="Clear">
    <input type="submit" value="Submit">

  </form>
  </body>
</html>
```

Image input fields HTML provides online forms with the capability for images to be used as input fields. That is, if a user points at an image with a mouse and clicks the mouse button, then the location of the mouse pointer will be transmitted as image coordinates. This facility makes it possible to provide maps, within a form, that allow the user to select and submit geographic locations interactively. HTML provides for a user-defined widget called an image field. Although we restrict the discussion here to images that are maps, the image can potentially be anything at all. A typical entry for an image field might be as follows.

```
<input type="image" name="coord" src="world.gif">
```

In this example, the name of the field variable that is defined is coord and the image to be displayed is contained in the file called `world.gif`. The image field so defined has the following important properties.

- The browser transmits not one, but two values from this field. They are the x and y coordinates from the image and are denoted by `coord.x` and `coord.y`, respectively.

- These coordinates are image coordinates, not geographic coordinates. It is up to the processing software to make the conversion to latitude and longitude.

- The image field acts as a SUBMIT button. That is, when the user clicks on the image, the form data (including the image coordinates) are submitted to the server immediately.

To convert the image coordinates into geographic coordinates, we need to know the size of the image in pixels, and the latitude and longitude corresponding to the two corners. The simplest formula for converting the image coordinate 1 into the corresponding longitude 1 is then

$$L = \frac{x(L_1 - L_0)}{(x_{max} - 1)} + L_0 \qquad (3.1)$$

where L_0 and L_1 are the longitudes represented by the top and right sides of the map and x_{max} is the horizontal width of the image in pixels and x begins at zero. A similar formula would apply for extracting latitude from the y coordinate. Note that these formulae apply only on small scales. On a global scale it is necessary to take into account the map projection that is used.

3.4.1   Quality Assurance and Forms

The distributed nature of the Web means that hundreds or even thousands of individuals could potentially contribute data to a single site. This prospect raises the need to standardise inputs as much as possible and to guard against errors. One method, which can be used wherever the range of possible inputs is limited, is to supply the values for all alternatives. So, for instance, instead of inviting users to type in the text for (say) "New South Wales" (i.e., entering it as a text field), we can supply the name as one of several options and record the result in the desired format (e.g., "NSW"), by using a pull-down menu. This method avoids having to sort out the many different ways in which the name of the state could be written (e.g. "NSW", "N.S.W.").

3.5   Processing Scripts and Tools

The most basic operation on a server is to return a document to the user. However, many server operations need to include some form of processing as well. For instance, when a user submits a form, the server usually needs to interpret the data in the form and do something with the data (e.g. write it to a file, carry out a search), then write the results into a document that it can return to the user. The processing may include passing the data to various third party programs, such as databases or mapping packages. Some commercial publishing packages now provide facilities to install and manage the entire business. However, the processing itself is often managed using processing scripts. Scripts are short programs that are interpreted by the system on the fly. They are usually written in scripting languages, such as Perl (§3.5.1), Python (§3.5.2), Java, Shell Script (Unix/Linux) or Visual Basic (Windows).

3.5.1   Form Processing with Perl

Perl was very popular in the early days of the WWW. It is an extremely powerful text processor, with many arcane shortcuts and extensive code libraries. It was, however, very much a product of the heyday of Unix, with its laconic and at times slightly obfuscating, style. But it was the potential security risks which hastened its decline for WWW applications.

The following short example shows a simple Perl script that processes form data submitted to a Web server. The code consists of three parts. Part 1 decodes the form data (which is transmitted as an encoded string), and the data are stored in an array called list. This array, which is of a type known as an associative array, uses the names of the fields to index the values entered. So for the field called country in the form example earlier (which took the value Canada), we would have an array entry as follows: list{country}=Canada. Part 2 writes the data in tagged format to a file on the server. Part 3 writes a simple document (which echoes the submitted values) to acknowledge receipt to the user. A detailed explanation of the syntax is beyond the scope of our discussion. There are also many online libraries of useful scripts, tools and tutorials.

```perl
#!usr/local/bin/perl
# Simple form interpreter
# Author: David G. Green, Charles Sturt University
# Date  : 21/12/1994
# Copyright 1994 David G. Green
# Warning: This is a prototype of limited functionality.
#          Use at your own risk. No liability will be
#          accepted for errors/inappropriate use.
#
# PART 1 - Convert and store the form data
# Create an associative list to store the data
%list = ();
# Read the form input string from standard input
$command_string = <STDIN>;
chop($command_string);
# Convert codes back to original format
# ... pluses to spaces
$command_string =~ s/\+/ /g;
# ... HEX to alphanumeric
$command_string =~ s/%(..)/pack("c",hex($1))/ge;
# now identify the terms in the input string
$no_of_terms = split(/&/,$command_string);
@word = @_;
# Separate and store field values, indexed by names
for ($ii=0; $ii<$no_of_terms; $ii++)
{       @xxx = split(/=/,$word[$ii]);
```

```
        $list{$xxx[0]} = $xxx[1];
}
#
# PART 2 - Print the fields to a file in SGML format
$target_name = "formdata.sgl";
open(TARGET,">>$target_name");
# Use the tag <record> as a record delimiter
print TARGET "<record>\n";
# Cycle through all the fields
# Print format <fieldname>value</fieldname>
foreach $aaa (keys(%list))
{     print TARGET "<$aaa>$list{$aaa}<\/$aaa>\n";
}
print TARGET "<\/record>\n";
close(TARGET);
#
# PART 3 - Send a reply to the user
# Writes output to standard output
# The next line ensures that output is treated as HTML
print "Content-type: text/html\n\n";
# The following lines hard code an HTML document
print "<HTML>\n<HEAD>\n Form data
return\n<\/HEAD>\n<BODY>\n";
print "<H1>Form received.</H1>\n<P>Here is the data you
entered ...\n";
# Print the fields in the form FIELD = VALUE
foreach $aaa (keys(%list))
{     print "Field $aaa = $list{$aaa}\n";
}
print "<\/BODY><\/HTML>\n";
```

To understand how this script would be used in practice, suppose that it is stored in an executable file named *simple.pl* that is located in the *cgi-bin* hierarchy of a server whose address is *mapmoney.com*. Then the script would be called placing the following action command in the form:

```
<form action="http://mapmoney.com/cgi-bin/simple.pl"
      method="POST">
```

The purpose of the above example is to show the exact code that can be used to process a form. However, in general, it is not good practice to write scripts that hard-wire details such as the name of the storage file or the text to be used in the ac-knowledgment. Instead, the script can be made much more widely useful by reading in these details from a file. For instance, the return document can be built by taking a document template and substituting details supplied with the form for the blanks left in the template, or by the script itself. Likewise, the name of the output file could be supplied as a run-time argument to the script. To do this the script would need to replace the line:

```
    $target_name  "formdata.sgl";
```

with an assignment such as the following

```
$target_name =@ARGV;
```

The URL to call the script would use a "?" to indicate a run-time argument:

```
http://mapmoney.com/cgi-bin/simple.pl?formdata.sgl
```

This example still has problems. In particular, this example shows the extensions of both the script and the exact name of the storage file. For security reasons, it is advisable to avoid showing too many details.

3.5.2   Python Scripting Language

As the Perl star has waned, the Python (from Monty Python's Flying Circus) has risen. Python is an open-source scripting langage that is gaining in popularity over more traditional scripting languages, such as Perl. Python runs on all major operating systems and has implementations for C (CPython), Java (JPython) and C# (Iron-Python) programming languages. More importantly, Python is considered by many as an ideal programming language for beginners, because of its concise code. Unlike Perl, the Python language is small and must be written in a particular way, the lack of shortcuts making it easier for people to follow code examples when learning the language.

Python can be used in Web applications, scripts and standalone applications, and is a powerful language for both server-side processing and for client systems. Python functions have one or more arguements and a return value of type PyObject*.

The following Python code sample is the equivalent of the Forms Interpretor Perl script in §§3.5.1.

```python
#!/usr/bin/python
import cgi

# PART 1 - Get the form variables
form = cgi.FieldStorage()
print form;

# PART 2 - Write the form values to an SGML file
with open( '/tmp/formdata.sgl', 'a' ) as sgmlFile:
    sgmlFile.write( '<record>\n' )
    for key in sorted( form.keys() ):
        sgmlFile.write( '<%s>%s</%s>\n' % ( key, form.getvalue(
            key), key ) )
    sgmlFile.write( '</record>\n' )

# PART 3 - Send the results back to the browser depicting the
    input fields and their values
print 'Content-type: text/html\n\n'
print '<html>\n<head>\n<title>Form data returned</title>\n</
    head>\n<body>\n'
```

```
for key in sorted( form.keys() ):
   print 'Field %s = %s<br />' % ( key, form.getvalue(key) )
print '\n</body>\n</html>\n'
```

There is an active Python community online[2] where developers can get involved, with a lot of guides, downloads and installations guides all available online for programmers and non-programmers alike.

Python contains a repository of software tools that can be used on Linux, Mac OS X and Windows platforms. Known as the Python Package Index (PyPI), the index contained almost 60,000 packages from third-party development communities[3] at the time of writing. These packages are provided in a number of different stages across the standard development cycle (planning to retired (Inactive)) and can be used under 1 of 13 different licensing models from Creative Commons 1.0 to Proprietary License, Public Domain and freeware.

There is now a variety of spatial-specific Python projects, including projects that have implemented spatial joins and RTree indexing[4] available for use in other applications.

3.6   Online Map Building

To build a simple map across the Web, the following sequence of steps must take place:

1. The user needs to select or specify the details of the map, such as the limits of its borders and the projection to be used;

2. The user's browser (the client) needs to transmit these details to the server;

3. The server needs to interpret the request;

4. The server needs to access the relevant geographic data;

5. The server needs to build a map and turn it into an image (e.g., PNG format);

6. The server needs to build an HTML document and embed the above image in it;

7. The server needs to return the above document and image to the client;

8. The browser needs to display the document and image for the user.

This process is illustrated in Figure 3.3. In the above sequence, only Steps 2, 7 and 8 are standard operations. The rest need to be defined. In almost all cases,

---

[2] https://www.python.org/
[3] https://pip.pypa.io/en/latest/installing.html
[4] RTree is a tree data structure which groups and indexes nearby objects through holding a parent extent at the next higher level of the tree, which enables fast spatial operations.

Step 1 involves the use of a form, which the browser encodes and transmits. In Step 3, the server passes the form data to an application program, which must interpret the form data. The same program must also manage the next three steps: communicating with the geographic data (Step 4), arranging the map building (Step 5) and creating a document to return to the user (Step 6).

When building a map in the above example, what the system actually produces is a text document (which includes form fields) with the map inserted. The map itself is returned as a bit-map (pixel-based) image. The image is in some format that a Web browser can display. Until recently, this usually meant a GIF format, or even JPEG. Having to convert vector GIS data to a pixel image has been a severe drawback. The output loses precision. It cannot be scaled, and downloading a pixel image, even a compressed one, often requires an order of magnitude more bandwidth than the original data. Scalable Vector Graphics (SVG) are a much better solution.

But one new option is the canvas element in HTML5 which provides just such vector operations (§5.3).

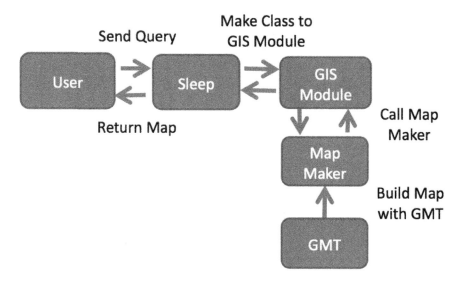

FIGURE 3.3: The flow of information from user to server and back that is involved in a typical system for building maps over the Web. The modules used here are as follows: SLEEP was a script interpreter with a module of GIS calls to the Mapmaker software (Steinke et al. 1996) which uses the GMT package of free map drawing tools (Wessel and Smith 1991, 1995).

3.6.1　The Use of High-Level Scripting Languages

Programming languages were invented to simplify the task of programming computers. High-level languages are computer languages that are designed to simplify programming in a particular context. It is far easier to write a program that carries out a specialised task if you can use terms and concepts that relate directly to the system concerned. The problem with general purpose languages is that the solution to a programming problem has to be expressed in terms that are very far removed from the problem's context. In particular, most automating of online services has involved writing programs in languages such as Perl, Java, C++ or shell script.

It has become commonplace in many computing packages to simplify the specification of processing steps by providing high-level scripting languages. The advantages are that most operations can be programmed far more concisely than general purpose languages. Also, because they are oriented towards a specific content area, they are usually easier to learn and to use. For example, to extract the contents of a Web form in the language Perl requires a program of at least a dozen lines. However, in a Web publishing language the entire process is encapsulated in a single command. High-level languages are desirable in developing server-side operations on the Web. The advantages (Green 1996, 2000; Green et al. 1998) include modularity, reusability and efficiency. As we shall see below and in later chapters, high-level languages can include GIS functions and operations. Most conventional GIS systems incorporate scripting languages to allow processes to be automated. In automating a website, particular scripts can be generalised to turn them into general purpose functions. To do this we start with a working script such as the following simple example:

```
SET BOUNDS 34.8S 140.1E 40.4S 145.2E
EXTRACT roads, topography,vegetation
PLOT roads
PLOT topography
PLOT vegetation
```

We then replace constant values by variables, which can be denoted by angle brackets.

```
SET BOUNDS <tlat> <tlong> <blat> <blong>
 EXTRACT roads, topography,vegetation
PLOT roads
PLOT topography
PLOT vegetation
```

This generalised script can now serve as a template for producing a plot of the same kind within any region that we care to select. If the user provides the boundaries from (say) a form, then we could generate a new script by using a Perl script to replace the variables in the template with the new values. Here's a simple example of a Perl script that does this.

```
#!/usr/bin/perl
# Build a simple script from a template
# The associative array markup contains
# replacement values
getvarsfromform;
```

```
filtertemplate;

sub filtertemplate {
while ($sourceline=<STDIN>)
{ chop($sourceline);
$targetline = $sourceline;
# Enter the input string into the template fields
for ($i=0; $i<$no_of_tags; $i++)
{ $work = $tag[$i];
$targetline =~ s/$work/$formvar{$work}/gi;
}
print "$targetline\n";
}
}
```

This script acts as a filter. Its function is similar to a merge operation in a word processor. We supply values for variables in a form. The function *getvarsfromform* (cf. the example in §3.5) retrieves these values as a table ($formvar). The Perl script then reads in the template as a filter and prints out the resulting script. To run this script we would use a call such as:

```
cat templatefile | filterfile > outputscript
```

where *templatefile* is the file containing the template, *filterfile* is the file containing the above Perl code and *outputscript* is the Perl resulting script.

Although the above procedure works fine, it is cumbersome to have to rewrite scripts for each new application. A more robust and efficient approach is to continue the generalisation process to include the Perl scripts themselves. This idea leads quickly to the notion of implementing Web operations via a high-level publishing language. The following example of output code shows what form data might take after processing by a script such as the above. The format used here is XML (§5) with tags such as country corresponding to form fields.

```
<country>
  <name>United State of America</name>
  <info>http://www.usia.gov/usa/usa.htm/</info>
  <www>http://vlib.stanford.edu/Servers.html</www>
  <government>http://www.fie.com/www/us_gov.htm</government>
  <chiefs>http://www.whitehouse.gov/WH/html/handbook.html
  </chiefs>
  <flag>http://www.worldofflags.com/</flag>
  <map>http://www.vtourist.com/webmap/na.htm</map>
  <spdom>
    <bounding>
      <northbc>49</northbc>
      <southbc>25</southbc>
      <eastbc>-68</eastbc>
      <westbc>-125</westbc>
```

```
     </bounding>
   </spdom>
   <tourist>http://www.vtourist.com/webmap/na.htm</tourist>
   <cities>http://city.net/countries/united_states/</cities>
   <facts>http://www.odci.gov/cia/publications/
   ...factbook/us.html</facts>
   <weather>http://www.awc-kc.noaa.gov/</weather>
   <creator>David G. Green</creator>
   <cid>na</cid>
   <cdate>1-07-1998</cdate>
</country>
```

A simple publishing script converts the above data from XML format (held in the file usa.xml) into an HTML document (stored in the file usa.html). In this case it would replace the tags with appropriate code. The reason for doing this will become clearer in Chapter 5.

Such a script would be easily prepared by a naive user, without understanding what the conversion operations actually are. Many browsers in 2015 are XML aware and they now use stylesheets to do these conversions; however, authoring a stylesheet is still a task for a web developer specialist.

3.7   Implementing Geographic Queries

A simple example of a geographic query online, is the following demonstration of computing great circle distances between points, selected from a map of the world.

The interface shown in Figure 3.4, is a simple HTML form containing a map image, descriptive text and several data fields. With this form, the user simply clicks on the image map to select a new location for position A or B. A radio button (SELECT POINT) defines which point is being selected. Following each selection, the server regenerates the form with the new values, and the great circle distance is displayed in the box at the right of the map.

To call the example, the URL addresses the interpreter for this service (here it is called `mapscript`). The interpreter links to a set of functions that perform relevant GIS functions. To enable the demonstration, we pass to the interpreter the name of the publishing script to be used (here it is in the file *circle.0*). The full call is therefore as follows:

```
http://lilliput.gov.lp/cgi-bin/gis/demos/mapscript?circle.0
```

The complete process needs two scripts. Firstly, a publishing script (`mapscript`) which defines initial locations for the two points involved and provides an entry into the *publishing service* shown in Figure 3.5. A second script (*circle.1*), processes subsequent calls made from the form. The complete list of files involved is:

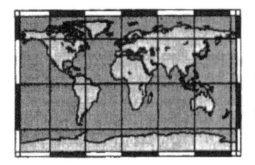

Great circle distances

This service calculates the distance
from the point A to point B.
*Click on the image to select a new
point.*
Lat/Longs are in degrees North.
Current distance 732.9   **Km**

Value	Point A	Point B	
SELECT POINT	○	⊙	Submit
Latitude (e.g. -34.5)	-37.45	-33.53	Reset
Longitude (e.g. 134.5)	144.5	151.1	

FIGURE 3.4: A simple form interface for computing great circle distance.

- An initialisation script (*circle.0*);

- A processing script (*circle.1*);

- The document template in Figure 3.6, which is essentially the form shown in Figure 3.4 but with variables in place of the locations of the two points;

- The map used, saved as a static JPEG image (*great-circle.jpg*);

- The interpreter (`mapscript`).

The HTML file that is displayed on the client browser does not exist as a stored file. It is generated on the fly by the server, and the key element in the form is the image input type, given by the element:

```
<input type="image" name="coord" height="255"
                    width="320" src="great-circle.jpg">
```

Clicking on the image generates a pair of values, here called *coord.x* and *coord.y*, which hold the coordinate values of the point on the image where the mouse is clicked. These values are passed off to the server when the user selects the "Submit" button on the form.

The publishing script combines input from the form (Figure 3.4) with a template in Figure 3.6 by replacing actual values (e.g., 147) for the corresponding XML variables (`along`). The resulting HTML code of the new form is shown in the right column of Figure 3.6 and is passed to the processing service as XML.

source	circle.xml	Value	(define the template file)
target	STDOUT		(send results to standard output)
var	alat	-34.5	(initial Latitude of Point A)
var	along	147	(initial Longitude of Point A)
var	blat	-33	(initial Latitude of Point B)
var	blong	149	(initial Longitude of Point B)
var	radius	6306	(Earths radius in kilometres)
form			(extract fields from the form)
circle			(calculate the distance)
sub			(place new values in the template)

FIGURE 3.5: The publishing script *(circle.1)* used in the great circle example.

XML Template		HTML Form	
(input circle.xml)		(standard output)	
<input	type="text" name="alat" value="<alat>"	<input	type="text" name="alat" value="-34.5">
<input	type="text" name="blat" value="<blat>"	<input	type="text" name="blat" value="-33">
<input	type="text" name="along" value="<along>"	<input	type="text" name="along" value="147">
<input	type="text" name="blong" value="<blong>"	<input	type="text" name="blong" value="149">

FIGURE 3.6: Use of a document template to replicate the form used in the great circle example.

Figure 3.7 shows the HTML source code that creates the form used in the great circle example. There are four key parts to this HTML that are worth hightlighting. These are:

- `<form action="http://mapcalc.gov.lp/cgi-bin/gis/gc-calc .pl method="POST">` which is the call to the form processing script;

- `<input type="image" name="coord" height="225" width="320" src="great-circle.jpg">` listing an image input field, which ensures that coordinates are read off the map;

- `<input type="hidden" name="radius" value="6366.19">` representing a hidden field, providing a value for the earth's radius;

- `<td><input type="text" name="alat" value="-37.45"> <td><input type="text" name="blat" value="-33.53">`

```html
<html>
  <head>
    <title>Great circle distance calculations</title>
  </head>
  <body>
    <form action="http://mapcalc.gov.lp/cgi-bin/gis/gc-calc.pl
        method="POST">
    <table>
      <tr>
       <td><input type="image" name="coord" height="225" width
          ="320" src="great-circle.jpg">
        <td><h2>Great circle distances</h2>
          This service calculates the distance from the point A
               to point B.
          <br><i>Click on the image to select a new point.</i>
          <br>Lat/Longs are in degrees North.
          <br><b>Current distance
        <input type="text" name="circle" value="732.9" size
          ="10"> Km
        <input type="hidden" name="radius" value="6366.19">
      </b>
      </table>
    <p>
     <table border="1" cellpadding="1">
       <tr><td><b>Value</b> <td><b>Point A</b>
       <td><b>Point B</b>
       <td rowspan="2"><input type="submit">
       <tr><td>SELECT POINT
       <td><input type="radio" name="site" value="source">
       <td><input type="radio" name="site" value="target"
          checked>
       <tr><td>Latitude (e.g. -34.5)
       <td><input type="text" name="alat" value="-37.45">
       <td><input type="text" name="blat" value="-33.53">
       <td rowspan="2"><input type="reset">
       <tr><td>Longitude (e.g. 134.5)
       <td><input type="text" name="along" value="144.5">
       <td><input type="text" name="blong" value="151.1">
     </table>
    </form>
  </body>
</html>
```

FIGURE 3.7: HTML code which generates the great circle distance Web form

```
... <td><input type="text" name="along" value="144.5">
<td><input type="text" name="blong" value="151.1"> being
```
the four lines of coordinate values processed by the publishing script.

This has been a simple implementation of a geographic query, where user actions on the client are processed by a server-based service. That is, mouse clicks are converted into coordinates and passed to server-based functions where they are processed. The server side code (not shown here) calculates the great circle distance and passes the result back to the client form where it is rendered.

With the increase in computing power on clients, including smartphones and tablets, geographic queries are now commonly carried out at the client, reducing the need for server based processing. These queries may include buffer searches and area calculations. More complicated calculations, such as processing vegetation statistics using large databases and answering spatial queries on proximity, would still be carried out on a server due to processing speeds and the amount of data involved.

3.8   Summary and Outlook

This chapter has introduced the basic concepts of server side processing:

- §3.1 introduced Web server technology, now part of our daily lives, and the HTTP protocol which underlies it. Some of the shortcomings of the first (and still widespread) version of HTTP were discussed.

- §3.2 introduced the software on the server side and §3.3 the sort of processing it carries out.

- Forms play a major role in much interactive Web processing and §3.4 gives a brief.

- Forms need scripts to process them on the server side, discussed in §3.5.

- The chapter ends with two examples of GIS processing, online map building in §3.6 and geographic queries in §3.7.

The examples in this chapter are indicative of the kind of scripting that developers of Online GIS are likely to encounter. Most commercial systems have in-built scripting features, which simplify the implementation of GIS online, much as we have demonstrated above.

Over the next few years there will be a significant transition from corporate owned data centres to cloud-based data centre infrastructure. As a result, there will be changes in the way server-side Online GIS operations will need to function.

There are already a number of cloud service providers that are offering cloud infrastructure that can dynamically expand or contract in terms of active server numbers. Running applications on these types of infrastructure requires significant ability for the server-side operations to also run in this ever changing *elastic* environment, bringing its own set of challenges for setting up server-side components of Online GIS systems.

4

Client-Side GIS Operations

CONTENTS

Of all the changes to Online GIS over the last decade, it has been the changes on the client side that have had the greatest influence on the popularity of Online GIS. The biggest changes have been the use of browser-based systems to access map content online and languages to enable client-side operations. Underpinning this has been the development of APIs to speed up development and extend ways in which users can render and query data. This chapter discusses the GIS operations which are performed at the client end. Clients used to be mostly desktop computers, but now include tablets and smartphones. Operating systems now include Android, new variants of Microsoft Windows and IOS (Apple) as well as the traditional big three, LINUX, Mac OS X and Windows.

4.1 Introduction

One of the fundamental differences between online and stand-alone GIS is the separation of the user interface from the main data and processing, as we saw in Chapters 2 and 3. To have to refer every operation, every choice back to the server is slow. It is also frustrating for the user. Therefore any steps that are used to move interactions to the client's machine are desirable. Many types of operations can be performed client side. The following list is indicative, but not exhaustive:

- Simple selection of geographic objects can often be implemented as client-side image maps and georeferenced image tiles.

- Some simple functions, such as elementary checks of validity for data entries, can be implemented in JavaScript, closely linked to ECMA Script,[1] which comes built-in to most Web browsers.

- Scalable Vector Graphics (SVG) (see §4.6.5).

- Java applets and JavaScript can be used to implement many interactive features, including:

  - context-dependent menus.

  - advanced geographic selection (e.g. rubber-banding).

  - Java has a full set of image processing classes enabling efficient operations to be carried out client side.

- Helper applications allow many processes to be off-loaded to other programs. This even includes some GIS operations. Other examples include:

  - Downloading tables as spreadsheets.

  - Passing images to suitable viewing programs.

Most of the above features can be employed to help implement an Online GIS. In this chapter, we explore some of the ways in which this can be done. For the most part we focus on how to use standard Web tools and facilities. However, it is important to bear in mind that commercial GIS packages are likely to provide many tools that will greatly expand (and simplify) the range of what can be achieved.

---

[1] http://www.ecma-international.org/

4.2 Image Maps

One of the first systems for implementing geographic queries was the **image map**. An **image map** was an interactive image. In a simple hypertext link, clicking on an image would request a single document. However, the image map called up different documents, depending on where on the image you clicked.

An early method in Online GIS for retrieving information that had a spatial context was to plot data on a diagram or map that shows how ideas, places etc., are related to one another. The image map construct allows us to use such diagrams as hypertext indexes. The source image may be a map, but it may also be the plans for a complex building site, sporting venue or ferry route.

For example, we could use a map of Australia to to serve as a simple geographic index to Australian states and cities. For simplicity, we could implement the index as a set of overlapping rectangles, one for each state. However, detailed polygons representing state borders could be used if precision was important (Figure 4.1).

4.2.1 Operation of an Image Map

An image map requires three elements:

1. An image (e.g. *file.html*) so that the map can be displayed.

2. An HTML file containing the image and linking the image with the map. To be active the image must be referenced via the ISMAP syntax.

3. A map table which defines regions on the image and lists what action to take for each.

In the example shown in Figure 4.1, the image is a map of Australia (as a raster image in jpg format). The associated map file (aus.map) containing the rectangle and polygon screen based coordinates is shown below:

```
default /links/ozerror.html
rect /links/wa.html 0,90 219,300
poly /links/nt.html
218,114 218,248 357,248 328,133 324,128
315,123 311,124 304,116 299,155 295,110
304,101 302,96 303,91 310,89 315,78
308,75 294,79 287,75 276,75 267,69
264,69 261,78 240,79 231,88 231,94
227,94 223,105 225,110 225,114 218,112
```

This first line defines the default action – what to return if no valid region is selected. The remaining lines define regions of the map and the action to take. For example, in the first line:

rect means that the region is a rectangle

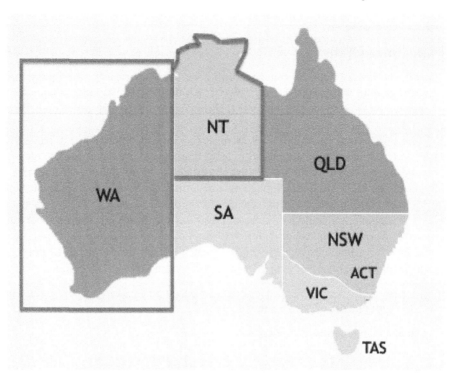

FIGURE 4.1: The image contained in the file ausmap.jpg, where the box around Western Australia and the polygon bordering the Northern Territory indicate the regions used in the image map example discussed in this section

/links/wa.html defines the file to retrieve when the region is selected; and

0,90 217,370 defines corners of a rectangle within the image.

Note that this example has been simplified to avoid printing pages of code! For some applications, the rectangle shape, as shown for Western Australia, would be suitable, but if precision is required, then the rectangles would be replaced by polygons, as shown here for the Northern Territory, that track the state borders more accurately. The syntax for calling an image map differs from ordinary hypertext. For example, to use the above image as a simple hypertext link, we would use the syntax:

```
<IMG SRC="ausmap.gif">  <A HREF="/links/ozweb.html">
```

However, to define the image map we use the following ISMAP syntax. Below, the attribute */cgi-bin/imagemap/ausmap* denotes the map file and *ausmap.gif* is the image.

```
<A HREF="/cgi-bin/imagemap/ausmap">
<IMG SRC="ausmap.gif" ISMAP></A>
```

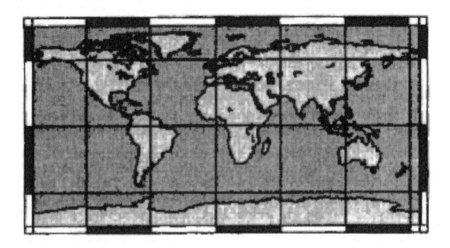

FIGURE 4.2: The world image map described in the text.

For more detailed instructions, see an online image map tutorial, such as w3schools.[2]

4.2.2   Client-Side Image Maps

Originally image maps were available only for server-side processing. That is, for every selection the coordinates were transmitted back to the server, which then performed the lookup of the map table. However, this procedure meant that image maps could not be used for stand-alone reference.

In client-side image maps, the browser itself processes selections on the image (there are many advantages to this approach).

The map in Figure 4.2 is a client-side image map. If you click on Australia, then a small map of Australia is loaded; if you click elsewhere a default file is loaded.

The above innovation requires two additions to HTML syntax. First, the parameter usemap tells the browser to look for the map table at the location indicated (usually within the same document) and the map element defines a block of code where the table is located. Here is the code used in the above example:

```
<img src="world.gif" usemap="#xxx" ISMAP>
<map name="xxx">
<area coords="232,87,266,117" href="aus.htm">
<area coords="0,0,300,154" href="error.htm">
</map>
```

In the above fragment of HTML code, the first line loads the image and activates the map. The remaining lines (which need not follow immediately, even though they do here) define the map table.

The use of image tiles from a pyramid image cache (§2.2.2) is essentially the same thing. These square map tiles of around 256 pixels x 256 pixels on the client side represent a known map area within the GIS data. As the geographic extents of these map tiles are known, pixel based coordinates (screen coordinates) collected through mouse clicks can be easily transformed into geographic coordinates and used in Online GIS processing.

4.2.3   Production of Image Maps

As we saw above, define hot regions on an image by listing the coordinates of the region's border. However, these are image coordinates, not geographic coordinates. So when creating an image map, it is necessary to convert borders so that they refer to the image, not the geography. There are two distinct ways of doing this. If the geographic coordinates are known, then they need to be converted using the appropriate transformation. However, if they are not, then the borders need to be digitised directly. Most image viewers allow the user to read coordinates directly off an image. However, several shareware programs allow the user to build a map file directly.

The production of image tiles for a pyramid map cache can be done directly from the GIS data warehouse (Chapter 10) and this process has largely been automated by a number of vendors. As there are likely to be millions of small map tiles needed to cover an average country at the various map scales, the maintenance of these modern day image maps becomes a whole lot more interesting.

4.3   Use of JavaScript in Client-Side Operations

JavaScript is a scripting language designed by Brendan Eich for use with Web documents. It began as a proprietary Netscape invention, but has now become commonplace on the Web and was originally tightly coupled to ECMA Script[3] in the early days of the Web. The code can be included within an HTML document to perform various functions. The main kinds of applications are

- To improve the user interface.

- To validate form data prior to submission.

- For animations and other client-side processing tools.

- To allow interaction and exploration of content (e.g. simulations and games).

---

[3] http://www.ecmascript.org/

Note that Java and JavaScript are quite different languages, although they share some syntactic structures. A simplified way to view the difference is to think of JavaScript as knowing what is on the Web page and being able to manipulate it. It is somewhat easier to use, being a scripted language embedded directly into the Web page and executed by the browser. Java, on the other hand, runs in an independent window or within the Web page but does not normally access its fields. In the following sections we examine two useful examples.

In the last decade, the use of JavaScript has grown enormously. It is still a slight security risk, but disabling it in a browser makes many websites unuseable. It is behind most Web pages, which dynamically adjust their content to match perceived user needs and desires.

4.3.1 Screening Data Input Fields

One of the greatest problems with any large scale data system is to ensure that all of the entries are valid and in a standard format. The best place to trap errors is at data entry. Web browsers provide two ways of reducing input errors:

- Controlling values in form fields; and

- Using JavaScript or other scripting functions to screen data input fields. There are now numerous scripting languages for client-side use, PHP being one very popular one. However, the principles are largely independent of the language, hence we consider only JavaScript.

The first of the above methods is the simplest. We ensure that values are entered in the correct format by controlling what is entered. To understand this, suppose that we set up a simple form in which every value is entered as a simple text field. The problem is that people can enter values in many different ways. For instance, the name "United States of America" can be written in many ways, such as

- United States

- USA

- U.S.A.

- US of A

Whereas humans can readily identify all of the above forms as variations of the same thing, they pose severe problems for computer searches, indexing and many other automated functions. To reduce this problem we can use use pull-down menus, radio buttons and other similar data fields that require the user to choose the value, rather than enter it. By adopting this method, we can ensure that the system enters the correct value.

For instance, in the example above, USA can be rendered as an abbreviation. Every time it is selected the machine enters it in the same way. Sample HTML code to generate such a form is listed below. Notice that, although country names are written on screen in full, the values returned for the variable "country" are all abbreviations. This can be seen in the `options` fields listed in the following source code.

```
<html>
  <body>
    <h1>Choose a country</h1>
    <form action="http://maps.gov.lp/cgi-bin/gis/opt1"
                                method="POST">
    <select names="country">
      <option value="USA">United States of America
      <option value="AUS">Australia
      <option value="UK">United Kingdom
    </select>
    <P><input type="submit"><input type="reset">
    </form>
  </body>
</html>
```

It is advisable to adopt the above approach as widely as possible in data entry forms. For instance, to enter dates we can provide a list of all days of the month, as well as a list of months of the year. In general, text entry can usually be avoided wherever there is a finite (and relatively small) number of possible values for a given field.

Another way to trap errors using selection lists is to provide redundant fields. For instance, suppose that the user is required to enter the name and latitude/longitude of a town in text fields within a data entry form. Then it is useful to include a field for correlated geographic units, such as country, state or province as well. This makes it possible to carry out redundancy checks and flag possible errors. For instance, the town name could be checked against a gazetteer of place names within the selected geographic unit. Likewise the latitude and longitude can be tested to confirm that they do lie within the selected unit.

4.3.2   The Need for JavaScript

The above method of providing choices during data entry applies only to fields that have a restricted range of possible values. However, it cannot trap all problems with data input. For example, the user may fail to *select* a value where one is required. And inevitably free text will be required in almost any data entry form. For text fields we can use JavaScript to carry out preliminary screening of input data before it is transmitted to a server. One example of a simple form for data entry could be a user entering the name of a town and the latitude and longitude of its location. The code for the form includes the required JavaScript, which is downloaded to the browser, along with the form. The field:

```
ONSUBMIT="return checkForm()"
```

within the FORM tag instructs the browser to run the JavaScript code just before submitting it to the server. The form data is transmitted only if the function check-Form returns a value of TRUE, otherwise an alert box could be presented to the user with details of the error detected. The code in the following sample shows how the JavaScript is integrated into the HTML code that produces the form. The JavaScript is enclosed in the `script` tag in the document `head`.

```html
<!DOCTYPE html>
<html>
  <head>

    <script src="javascript-text-checkingTH.js"
                  type="text/javascript"></script>
    <script>
      // -- Check the submitted form
      function checkForm()
      {
        if ( isLatLong() && isPlaceName() )
        {
          return true;
        }
        else
        {
          return false;
        }
      }

      <!-- code omitted for isLatLong and isPlaceName -->

    </script>
  </head>
  <body>
    <form action="form_processing_url" method="post"
        name="data_entry" onsubmit="return checkForm()">
      <h1>Add a town name and its location</h1>
      <input type=hidden namb="topic" value="">
      <p>Town Name: <input type="text" name="town"
                                    size="40">
      <p>Latitude: <input type="text" name="latitude"
                                    size="10">
      <p>Longitude: <input type="text" name="longitude"
                                    size="10">
      <input type="submit" value="Submit Details">
      <input type="reset" value="Reset">
    </form>
  </body>
</html>
```

In this example, we have applied three different kinds of checks.

- The first is to check that the name field is not empty. This is a useful way to ensure that required data fields are completed prior to submission. We can generalise the code used as follows, where angle brackets indicate variables that should be replaced with appropriate field names or code.

```
if (<field_name> ... "") { <action> }
```

- The second check ensures that the town name is of an acceptable form. For simplicity, the code used in the example given here

```
var ch = townstr.aubatriag(i, i + 1); if
  ((ch < "a" || "z" < ch) && (ch != " "))
```

is very restrictive. It ensures that each character in the name is either a letter, or a space. In real applications, a much greater variety of characters would be permissible.

- The final check is to ensure that values are given for latitude and longitude and that they lie within the permissible range. In this case we have used the convention that positive values for latitude denote degrees north and that negative values indicate degrees south.

```
if (townlat == "" || townlat < -90 || townlat > 90 )
```

The source code for carrying out the above tests is included in the JavaScript code that follows.

```
// Checks town field.

function isPlaceName ()
{
   var townstr = document.data_entry.town.value.toLowerCase();

   // Return false if field is blank.

   if (townstr == "")
   {
      alert("\nThe Town Name field is blank.")
      document.data_entry.town.select();
      document.data_entry.town.focus();
      return false;
   }

   // Return false if characters are not letters or spaces.
```

```
   for (var i = 0; i < str.length; i++)
   {
      var ch = townstr.substring(i, i + 1);
      for (var i = 0; i < townstr.length; i++)
      {
         var ch = townstr.substring(i, i + 1);
         if ((ch ="a" || "z" < ch) && (ch != " "))
         {
            alert("\nTown names may contain only letters and
               spaces.");
            document.data_entry.town.select();
            document.data_entry.town.focus();
            return false;
         }
         return true;
      }
   }
}

// Check Lat & Long fields

function isLatLong ()
{
   var townlat = document.data_entry.latitude.value;
   var townlong = document.data_entry.longitude.value;

   // Return false if latitude is outside the range -90, 90,
   // or if longitude is outside the range -180,180.

   if (townlat == "" || townlat < -90 || townlat > 90)
   {
      alert("\nLatitude must be in the range -90 to 90");
      return false;
   }
   if (townlong == "" || townlong < -180 || townlong > 180)
   {
      alert("\nLongitude must be in the range -180 to 180");
      return false;
   }
   return true;
}
```

4.3.3 Geographic Indexes

JavaScript can also be used for many other purposes. For instance, we can enhance image maps by eliminating the need to refer back to the server. In this example, we replace calls to the server with calls to a JavaScript function. Whenever a mouse click is made on the image map, this function generates a fresh document and inserts the

```
<HTML>
  <HEAD>
    <SCRIPT src="javascript-imagemap.js">
    </SCRIPT>
  </HEAD>
  <BODY>
    <h1>Information Map</h1>
    <img src="world-image-map.jpg" usemap="#xxx" ISMAP>
    <map name="xxx">
      <area coords="232,87,266,117"
          href="JavaScript:DataDisplay(1)"
          onmouseover="self.status='Australia';return true"
          onmouseout="self.status='';return true">
      <area coords="0,0,300,154"
          href="JavaScript:DataDisplay(99)"
          onmouseover="self.status='Try again';return true"
          onmouseout="self.status='';return true">
    </map>
    <P>
  </BODY>
</HTML>
```

FIGURE 4.3: JavaScript combined with an image map.

appropriate text into it. The JavaScript code in Figure 4.3 provides the functionality for this example.

An important implication of the above is that a number of indexing and other operations can readily be transferred from the server to the client. This is desirable where practical, unless it would compromise proprietary or other concerns. Given that websites often contain images that total several hundred kilobytes in size, it is not impractical to download data tables of hundreds, and perhaps even thousands of entries. JavaScript can be used to make complex mapping queries self-contained. For example, Environment Australia's online "Australian Atlas" uses JavaScript to allow users to select layers and other options in the course of making customised environmental maps.

4.3.4   Other Applications of JavaScript

JavaScript provides a means to transfer a very wide range of operations from server to browser. For example, continuity in an interactive session means that the form or dialog to be displayed depends on the choices and inputs made by the user. This process is slow if the documents must be repeatedly downloaded from the server.

4.4 AJAX: Dynamic Web Page Updating

One of the functions frequently required when working with map data is zooming in. Downloading the whole map at full resolution requires too much bandwidth and the map itself may take up too much space on the client machine. At the same time we don't want to keep downloading the whole page every time we zoom in somewhere. Thus there is a need to be able to fetch pieces of a Web page as desired. Ideally, too, we would want to be able to fetch additional things asynchronously, so that ongoing interactivity with the Web page is not impaired.

The technology introduced for asynchronous piecewise Web page update was AJAX (Asynchronous JavaScript and XML). It requires an additional HTTP header, XMLHttpRequest, a W3C Working Draft as of January 2014.[4] In some ways this is a bit of a misnomer, in that the request can fetch other things besides XML and it can also be used over HTTPS in some implementations.

The way it works is that a Web page will call a JavaScript function which first creates an XMLHttpRequest object. In this simple example, the request may be for information about a restaurant pointed to on a map.

```
var cafeInfo
cafeInfo=new XMLHttpRequest();
```

A request for a file, *curryhouse.txt* on the server now has to be generated:

```
cafeInfo.open("GET","cuuryhouse.txt",true);
xmlhttp.send();
```

The GET option is analogous to the GET/POST options in HTML. The second argument is the filename on the server and the last argument specifies that asynchronous communication is to be used.

We now need to set up a handler for when the data arrives, poll for arrival of the data, and then use JavaScript to update the current Web page in the appropriate way.

The handler is a custom defined function, implementing the XMLHttpRequest method *onreadstatechange.* When the request is complete, the readyState is set to 4 and the status is set to 200 if the request was successful, otherwise the ubiquitous 404 to indicate file not found. The data arrives as responseText (a string) or responseXML (XML data).

```
xmlhttp.onreadystatechange=function()
{
  if (xmlhttp.readyState==4 && xmlhttp.status==200)
  {
    document.getElementById("CafeData").innerHTML=
      cafeInfo.responseText;
  }
}
```

---

[4] http://www.w3.org/TR/XMLHttpRequest/

The final stage is the updating of an element on the Web page. The JavaScript method *document.getElementsById* finds the element in the page with the (unique) id `CafeData` and sets the content of this element to the string which came back.

As with all the XML standards we have looked at in the book, the full specification is lengthy, with many options and constraints. This is a simple bird's eye view of an important facility for processing online spatial data.

4.5   Cookies

A *cookie* is an HTTP header that passes data between a server and a client, introduced by Lou Montulli at Netscape, but his original proposal is still available.[5] The main function of cookies is to help a server to maintain continuity in its interactions with users. As we have seen, HTTP is memoryless (§2.2.1). Every interaction is independent of previous interactions, even if a user is browsing through a site during a single interactive session. One of the motivations for cookies was to be able to facilitate interactive sessions by providing a means for a server to keep track of a user's previous activity.

The core of a cookie is the name of a variable and its value. Usually this variable provides a user identification which allows the server to look up relevant background details on the history of the user's previous interactions. However, the value could be anything at all. For instance, in the context of a GIS session, it could encode details of the type of operation that the user is currently undertaking, although this information is often more simply passed as hidden form fields.

There are two cookie headers: `Set-Cookie` is used by the server when a browser makes a request for a URL. When the browser has a cookie set for a site, it will send this back to the server using `Cookie`. Although the cookie header is a text string, usually quite short, modern browsers such as Firefox and Safari store the cookies as binary files. Thus, seeing the content of a cookie can only be done through the browser itself.

A minimal cookie looks something like this:

```
Set-Cookie: name=Puccini;
            expires  Sat, 10-Apr-2012 12:15:00 GMT;
            domain=maps.gov.lp;
            path=/coastal;
            secure;
            httpOnly
```

with the name/value keywords as the main content, along with the expiry date attribute. This defaults to `session`, meaning that the cookie only survives until the browser is closed down.

---

[5] `http://curl.haxx.se/rfc/cookie_spec.html`

TABLE 4.1: Examples of entries for a cookie

Field	Meaning
name=value	The variable name and its value (here name = Puccini)
domain	The website domain (maps.gov.lp)
path	The path to which the cookie applies (/coastal)
expires	When the cookie ceases to be valid (ex-presssed here as Greenwich Mean Time)
secure	indicates that the cookie is only to be sent along a secure link (such as HTTPS)
HttpOnly	was introduced after the original cookie specification to block cookies being modified by JavaScript

A typical format for a cookie header is given in Table 4.1, which also sets out the meaning of the fields used in this header, as well as other fields assumed in the process.

The domain field specifies the *server(s)* to which the cookie applies and defines whether all machines in the given domain can access it. The secure field defines whether the server needs to provide a secure connection before it can access the entry.

Cookies can be created and retrieved using JavaScript (as well as other commonly used languages such as Perl and VBScript). In JavaScript there is a default object *document.cookie*, which handles interactions with cookies. Passing a cookie to this object causes it to be created and stored (Whalen 1999).

Thus the cookie example above would appear as (without the HttpOnly attribute)

```
document.cookie="name=Puccini;
          expires  Sat, 10-Apr-2012 12:15:00 GMT;
          domain=maps.gov.lp;
          path=/coastal;
          secure"
```

4.6 Interactive Graphics

There are now several ways of getting interactive graphics for drawing and interacting with map data. We mention the new canvas element for HTML5 in §5.3 and cover Scabelable Vector Graphics, briefly, in §4.6.5. But one of the most important technologies is the general purpose programming language Java. It was developed by SUN Microsystems to meet the need for a secure way of introducing processing elements into HTML pages. It is object oriented and includes a wide variety of graphics and other functions.

Like JavaScript, the Java language can be used to provide client-side functionality. Many of the functions described above for JavaScript (e.g., context-sensitive menus) can also be implemented using Java. However, another important use for Java is its drawing ability. Here we look at three examples of interactive operations that are fundamental in GIS which can be easily implemented in Java: rubber banding, tracing polygons and drawing maps.

4.6.1   Drawing Geospatial Data at the Client

An important use of Java is to reduce the processing load on a Web server by passing the task to the client. Another is to reduce the volume of data (especially large images) that need to be passed across the network. One of the heaviest processing loads (when repeated on a large scale) is simply drawing maps. While some of this rendering of map data has now been replaced by passing pre-existing map tiles to the client for rendering, there are still circumstances where geospatial data in the form of vector data are needed for Online GIS processing and must be processed in preparation for rendering. When done on the server, there is both the time cost of processing and a network cost in the form of maps presented as large images that need to be transferred from server to client. In many instances it is both faster and involves less data transer to pass raw data to the client, together with the Java code needed to turn it into a map. Other ways of doing this include the new `canvas` element in HTML5 (§5.3).

The following fragment of Java code is a simple example of a function that draws a map as an image. In this case it also draws a rectangle on the map to indicate a selected area. As we shall see later, SVG now provides an alternative.

```
public void paint ( Graphics g )
{
  g.drawimage( mapimage, 0, 0, mapWidth, mapHeight, this);
  width = Math.abs( topX - bottomX );
  height = Math.abs( topY - bottomY );
  upperX = Math.min( topX, bottomX );
  upperY = Math.min( topY, bottomY );
  g.drawRect( upperX, upperY, width, height);
}
```

4.6.2   Selecting Regions

Earlier we saw two ways of making geographic selections via a standard Web browser. The first was to use an image within a form as a data entry field. This method allows the user to enter the image coordinates of a single point within a map.

The second method was to define an image map within a standard HTML document. The image map construct allows users to select a pre-defined region from a map.

The above types of selection omit several important GIS operations:

- Although in principle the form method could be used to select a variety of objects (e.g., a road), in practice the user could not confirm that the correct object had been selected without reference back to the server, which is time consuming and clumsy.

- Selecting arbitrary, user-defined regions requires the user to be able to draw a polygon by selecting a series of points.

- Single-point clicks suffice for defining single points. However, to define a set of points, or to digitise (say) a line, such as a road or river, or a region, the user must again be able to select a set of points and have lines joining them drawn in.

- Dynamic movement is an interactive process in which a user selects an object and moves it. This is useful to denote the changing position of a car on a road. It is closely related to rubber banding (§4.6.3).

- Zooming and panning involve redrawing a map interactively.

In the following sections we look at some of the methods needed for carrying out these functions.

4.6.3 Rubber Banding

Rubber banding is the process of selecting a region (usually a rectangle, circle or some other regular shape) by choosing a point and holding down the mouse button whilst simultaneously "pulling" the shape out until it reaches the desired size.

In practice, rubber banding is an example of interactive animation. That is, the map's image, with the shape overlaid on it, is repeatedly redrawn and displayed in response to the user's mouse movements. Conceptually this procedure involves the following steps:

```
Get mouse location
Calculate shape coordinates
Copy image of base map
Draw shape image over base map
Redisplay map image
```

The following Java source code illustrates some of the functions required in the above example. Note that the listing includes functions for dealing with three distinct events involving the mouse: that is, the mouse being pressed, dragged and released.

```
public void mousePressed( MouseEvent e )
{
  xPos = e.getX();
  yPos = e.getY();
  mousePosition = myMap.checkPosition(xPos, yPos);

  if(mousePosition)
  {
    setTopX( xPos );
    setTopY( yPos );
    tempUpperLong = myMap.calcLong(xPos);
    tempUpperLat = myMap.calcLong(xPos);
  }
  else
  {
    showStatus("Mouse is outside the map area");
  }
}

public void mouseReleased ( MouseEvent e )
{
  xPos = e.getX();
  yPos = e.getY();

  if(rodentDrag == 1 && rodentRelease == 0)
  {
    setBottomX( e.getX() );
    setBottomY( e.getY() );
    mousePosition = myMap.checkPosition(xPos, yPos);

    if(mousePosition)
    {
      lowerLat = myMap.calcLat(yPos);
      lowerLong = myMap.calcLong(xPos);
      upperLong = tempUpperLong;
      upperLat = tempUpperLat;
      myMap.convertToPixels(upperLong, lowerLong, upperLat,
          lowerLat);
      zoom.setVisible(true);
      rodentRelease = 1;
      mousePosition = false;
    }
    else
    {
      countryinfo.setText("Outside map. Try again\n");
      mousePosition = false;
    }
  }
}
```

```
public void mouseDragged( MouseEvent e )
{
  if(mousePosition)
  {
    setBottomX( e.getX() );
    setBottomY( e.getY() );
    rodentDrag = 1;
    repaint();
  }
  else
  {
    showStatus("Mouse is outside the map area");
  }
}
```

4.6.4 Drawing and Plotting Maps, Graphs and Diagrams

The final application we consider here is the use of Java to draw entire maps on the client machine. The ability to draw polygons interactively is crucial in GIS. It is used in both drawing and selecting geographic objects. The ability to draw and interact with maps on the client machine has several potential advantages:

• For vector maps, it usually requires less data to download the vector data that defines a map than to download an image of the map. It also reduces server-side processing.

• Client-side drawing cuts down the amount of processing required of the server.

• It speeds up processes such as zooming and panning, which otherwise require the constant transfer of requests and responses between client and server.

The need to develop Java applets to deal with the above issues may decrease with the advent of SVG, described in the next section (§4.6.5).

4.6.5 Scalable Vector Graphics (SVG)

Methods of drawing and plotting maps and other online figures are likely to change dramatically with the introduction of new standards and languages for plotting vector graphics on the Web. One of the problems with online graphics with early Online GIS had been that Web browsers could display only raster images, usually in GIF or JPEG formats. However, pixel-based images suffer from the problem that they cannot be rescaled, which is a problem when printing screen images. They also tend to lose resolution from the original. Diagonal lines, for instance, often exhibit a staircase effect. Another huge problem is that pixel-based images can produce very large files. This was one of the biggest factors in slowing delivery of online information, especially for users linked via modems. Given that a lot of the GIS information available online is vector based, the necessity of delivering this data as pixel images has been a great nuisance.

In August 2011, the World Wide Web Consortium released the recommendation for Scalable Vector Graphics (SVG) 1.1.[6] These examples use some of the XML markup features described more fully in Chapter 5. The language includes syntax for describing all the common constructs and features of vector graphics. It also recognises the need for integrating graphics with hypermedia elements and resources. This ability is essential, for example, to create image map hot spots in an image, or to pass coordinates to a GIS query.

Because polygons defining regions are so common in GIS, perhaps the most appropriate example is the following, in which the code defines a simple filled polygon drawn inside a box, much as a simple map might be. Example *map01* shown in Figure 4.4 specifies a closed path that is shaded grey. The following abbreviations are used for pen commands within the path command:

```
M   moveto,
L   lineto,
z   closepath
```

Using these pen commands, the SVG code for Figure 4.4 is as follows.

```
<?xml version="1.0" standalone="no"?>
<!DOCTYPE svg PUBLIC "-//W3C//DTD SVG 20001102//EN"
    "http://www.w3.org/TR/2000/CR-SVG-20001102/
                              DTD/svg 20001102.dtd">
<svg width="4cm" height="4cm" viewBox="0 0 400 400">
<title>
Example map01 - a closed path with a border.
</title>
<desc>A rectangular bounding box</desc>
<rect x="1" Y="1" width="500" height="400"
                  style="fill:none; stroke:black"/>
<path
d="M 200 100
   L 300 100
   L 450 300
   L 350 300
   L 300 200
   L 300 300
   L 100 300
   L 100 200
   L 200 200
z "
```

---

[6] http://www.w3.org/Graphics/SVG/About.html

FIGURE 4.4: Screen capture of the figure produced by the SVG code given in the text.

```
style="fill:grey; stroke:black; stroke-width:3"/>
</svg>
```

Pasting this code into a file and opening it in a browser will render the simple eight sided polygon.

Although we have deliberately kept this example brief, it is easy to see how the method could be used to draw, say, the coastline of North America. Note that we can readily integrate SVG code into the object-oriented method. For instance, each object, say North America, would have associated with it methods to draw a graph of itself, either as a list of coordinates or as calls to sub-objects (e.g. Canada, USA). The method might include the following steps for each object:

- Convert the list of lat-long bounding coordinates into SVG path commands.

- Print the name of the object by writing an SVG text command.

Creating an entire map from a description of the data objects is simply a matter of working recursively through the hierarchy of objects in the map. At each stage, you convert each object into its SVG representation. The process may sound cumbersome, but it is the kind of processing that computers thrive on.

Because SVG is based on XML (see Chapter 5), there is ample scope to bundle up different map elements as objects wrapped in the corresponding XML elements. So, for instance, we could render the above example as a hierarchy of map objects, with the box indicating the border of the entire map, and the path indicating one feature in the map.

The compact nature of vector graphics means that downloading even complex maps would still require less data than an image of the same area. We anticipate that the introduction of SVG is likely to have several effects. One is that the need for Java code as a rendering tool for maps will be reduced. SVG viewers are likely to include the operations (e.g. zooming and panning) described earlier. So, in many instances, a map can be downloaded from a server as an SVG source and manipulated directly.

Third party packages are likely to include scripting commands and other high level features to simplify the generation of an SVG source. One important step would be converters for turning files in proprietary GIS formats into SVG code. Many SVG related tools are already available. Most basic are programs and plug-ins for viewing SVG data. Other basic tools include filters for converting translating data between SVG and several GIS and CAD/CAM formats. They also include converters for SVG into standard image and output formats, such as GIF, Postscript and PDF.

4.7   Application Programming Interfaces

There has been a significant shift in how developers are now coding client-side applications for the Web. Many of the websites now available on the Internet are using libraries of functions and datatypes defined in APIs. Chapter 2 gave a bird's eye view of some spatial examples in §2.3.4.

An API is essentially an index into a set of reusable functions and data types (data structures) that can be imported into an application and used. Functions and data types in an API are usually related to a common theme or requirement, such as GIS functions. They might include simple drawing tools for adding features, or be a set of functions that can be used for page layouts or Graphic User Interface (GUI) components to organise elements of a Web page.

APIs are not binary libraries, but human readable source code. Traditional libraries were often provided as binaries, where developers could use the functionality, but could not see what the code was doing. With APIs the specification is for the functionality to be provided, but *not* the actual implementation, which may be done by more than one individual or group in different ways for different hardware. The existence of an API does not guarantee that the implementation even exists.

APIs can be included in an application by using a few lines of simple JavaScript (§4.3). An example of code that imports the Dojo API is

```
<script src="http://ajax.googleapis.com/ajax/libs/dojo/
                    1.10.4/dojo/dojo.js">
</script>
```

This JavaScript tells the client application where to go and get the API source code (in this case, hosted on the Google CDN website[7]), to import into the client session. It also includes some other details such as the API version number or release version of the API (say version 1.10.4). Developers would develop their application directly against this release version of the API, and as such, can be confident that their application will still function as expected.

But will the API always be available online? To ensure this, the developer could copy the API and host it locally on their own Web server, redirecting the client application to that instance of the API. The open source approach used in APIs has created communities of developers that contribute and maintain the libraries for others to use. Some software vendors still provide APIs, such as Google and Esri. There are also a lot or non-vendor organisations offering robust APIs which are free to access.

One example of a quality API from a non-profit organisation is the Dojo Toolkit, from the Dojo Foundation,[8] discussed in more detail below.

There are a lot of APIs now freely available on the Web. Some of the popular ones which are commonly found in Online GIS websites include:

WebGL API This API is a JavaScript library that provides functions and data structures for the online rendering of 2D and 3D data on the Web. Originally developed by Mozilla, the WebGL (Web Graphics Library) API is now maintained by a non-profit organisation, Khronos organisation.[9] Based on OpenGL, WebGL uses the HTML5 canvas element (§5.3) and Document Object Model (DOM (§1.2.1)), a cross-platform API for accessing HTML, XHTML and XML objects. WebGL supports a large number of browsers on both desktop environments and mobile devices.

DoJo API Dojo is a JavaScript library which enables quick development of cross-platform JavaScript or Ajax (§4.4) applications and websites. It contains around three thousand JavaScript components including tools for the optimisation and unit testing of JavaScript, HTML and CSS. It also supports the generation of documentation, utility classes and widgets. Dojo provides a capability for client-side and server-side storage, even running on the server side. It can be accessed from the Dojo Toolkit Web page.[10]

Python/C API Another popular example of an API is the Python/C API. This API gives C and C++ programmers the ability to use Python code in one of two ways. They can use the API to *extend* Python by embedding functionality written in the C/C++ programming languages. The API can also be used to provide access to the Python interpreter from C and C++ applications so that Python code can be run within C and C++ applications. Python was discussed in Chapter 3 in §3.5.2.

There are a number of popular geospatial APIs available from vendors, including

---

[7] Google CDN is a globally available hosting and distribution network for open source JavaScript libraries (APIs).

[8] http://dojotoolkit.org

[9] http://www.khronos.org/webgl

[10] http://dojotoolkit.org/download/

the Google Maps API from Google and the Esri JavaScript and REST (§3.1.1) APIs, which are now commonly used in many Online GIS applications.

APIs offer a number of advantages for developers. They

- Save a considerable amount of time, when developing an Online GIS website. Why reinvent the wheel when there are freely available APIs which provide the required functionality?

- Are externally managed, maintained and updated to support new technology. These changes can be easily inherited by client applications, without the need to maintain this code.

- Are easy to integrate using simple import statements in a client application.

- Often run across many platforms, supporting all popular browsers. Developers using APIs can get this cross-platform support without the need to develop code.

- Give developers in organisations access to communities of developers, all working towards creating friendly, feature-rich and robust libraries of well-tested source code that is freely available.

Because of these advantages, APIs are replacing large development projects, where developers no longer redevelop their own functionality, but are choosing to implement open source functions developed and maintained by other groups or organisations. This trend is likely to continue and as the functionality of these APIs becomes richer, so will the quality of Online GIS websites.

4.8   Examples

4.8.1   Example of an Interactive Geographic Query System

A simple demonstration of the use of Java applets is an online query system for Australian towns. In this system the user downloads an applet that displays a map of Australia together with dots for towns (Figure 4.5). To obtain information about a particular town, the user selects a town by mouse click (Figure 4.5a). When a town is selected in this way, the details are displayed in boxes on the screen. The user can zoom into particular regions by rubber banding (Figure 4.5b).

4.8.2   Asian Financial Crisis

The company `Internetgis.com` has developed a system, called ActiveMaps, which provides a class library designed for Java developers who want to add GIS/mapping functions to their applets or applications. The service provides geographically based information about the Asian financial crisis. It combines a map of the region with data about the economy of each country. The user interface consists

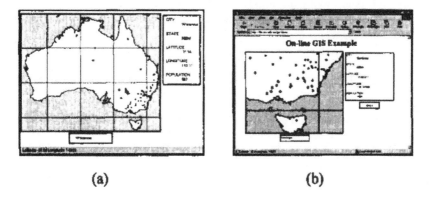

(a) **(b)**

FIGURE 4.5: A demonstration of an online geographic query system using Java. The Australian towns query system allows users access to (a) the online user interface, which is invoked when the system is started. (b) An example of zooming (using rubber banding) and displaying selected data.

of a map, together with a variety of tools for customisation and queries. The mapping features includes panning and zooming, using rubber banding, plus the facility to insert data on the map.

4.8.3 NGDC Palaeogeographic Data

The US National Geophysical Data Centre provides large, public repositories for many kinds of palaeoclimatic data, such as tree ring records and the Global Pollen Database (NGDC 2000). To assist users, NGDC developed a Java applet called WebMapper for finding palaeoclimate data held at the World Data Center for Paleoclimatology. The service includes zooming and site selection.

4.9 Summary and Outlook

- Building on the previous discussion of server-side processing (Chapter 3), §4.1 introduced the basic Web technology that can be adapted to create an Online GIS. The techniques described here are based on free software and can be used as an integral part of a website without incurring the start up and maintenance costs of a commercial GIS.

- §4.2 described image maps, which are the cornerstone of Online GIS on the client side.

- Many Web pages today are highly dynamic, making extensive use of JavaScript and other scripting languages. §4.3 introduced the use of JavaScript and §4.4 described AJAX, a technology for allowing this dynamic Web page updating to go on asynchronously in the background.

- §4.5 described cookies, which are widely used for preserving state information during a Web session and maybe beyond.

- Online GIS has benefitted enormously from interactive graphics described in §4.6. The chapter concludes with some examples in §4.7 and §4.8.

An important area of interest for minimalist client-side operations is of course the mobile market (Chapter 11) using smartphones and tables, and Location-Based services (Chapter 12) that use these devices.

At the time of writing, this market is growing rapidly. In the chapters that follow, we turn to look at ways in which individual Online GIS can be integrated into more comprehensive geographically based information systems.

5

Introduction to Markup

CONTENTS

This chapter contains one of the most fundamental ideas of the book: embedding information about structure and content within the text, using the same characters as the text itself. In many of the word-processing packages in use today, information about format and style is relegated to special character symbols, making the file look to all intents and purposes like a binary file. The idea of text markup has a long history and has become of major importance with the advent of the World Wide Web.

5.1 Markup Languages

In the early days of computing, formatting data was pretty much up to the idiosyncrasies of individual programmers. If you wanted to be able to plot the boundary of a country, then you might throw the data into a file as line after line of data, each line containing pairs of coordinates. The file would contain numbers, just numbers. As likely as not, the coordinate pairs would be jammed up against one another. Clarity didn't matter because the data were set up for just one program to read and you told your program what the exact format was. If it ever occurred to you that others might one day want to use that data, then you just shrugged your shoulders and said *Hey, that's ok. If they need to know what the format is, then just come and ask me.*

Today's computing scene is completely different. Data and processing are distributed across networks. An Online GIS is likely to need to draw on data from many different sites. Likewise, any data set is likely to be accessed and used by many different programs. Given this environment, there is no time to go and ask the author what the format is. The content, structure and format of a data file need to be immediately transparent to any program that accesses it. This need creates a demand for ways of indicating the structure of data. But ad hoc explanations are not enough. They need to be universal standards that everyone can understand and apply. And so markup languages were born.

In the very early days of computer document processing, full screen displays of a document as it would look when printed, so-called what you see is what you get (WYSIWYG), was far from practical. Not only were machines not powerful enough, either in processor speed or in memory, to manipulate documents on the fly, but computer monitors were hardly up to the task. Windowing systems, universal today, were still to come. Yet there was a need to go beyond plain text, just an endless list of words with simple punctuation, to text that had different fonts or structure (numbered paragraphs, chapters, indexing and so on).

Thus began the idea of embedding special characters, called *markup tags,* in the text. Markup tags introduced, not more text, but control sequences. Early examples of such systems were the utilities *nroff* and *troff* on Unix and *runoff* on DEC machines.[1]

In runoff, for example, a dot at the beginning of the line signified that the line was not part of the text, but contained commands to the processing system. Once a document had been written up with the associated commands, or markup, a program would read it, put the commands into action and produce a printable form.

A major breakthrough in computer typesetting occurred in the late 1970s, early 1980s, when Donald Knuth developed the language TEX (Knuth 1984).[2] Designed to facilitate the typesetting of mathematics (and improve the accuracy in the process), TEX introduced a whole range of new concepts and methods for representing and pro-

---

[1] Although UNIX is still very much with us, DEC, the Digital Equipment Corporation, disappeared long ago. It was, in its heyday, a major innovator in operating systems (VMS), hardware (VAX) and utilities, such as MAIL, the precursor to today's email.

[2] TEXis pronounced "tech", the third letter being the Greek letter χ. At the time it was a virtuoso piece of typesetting.

cessing typographical data. TEX is still a markup language with a very, very powerful macro facility (i.e., a system for storing and re-using sequences of commands), but it is concerned primarily with layout and presentation. Moving a full stop a fraction of a millimetre is a lot of fun in TEX. TEX is, however, complex! Users frequently relied on collections of macros or ad hoc templates from elsewhere. The derivative language LATEX, which followed in around 1986 (Lamport 1986), introduced an integrated macro framework using pre-written TEX style files. It was much easier for the non-specialist to use, but its really important contribution was its emphasis on document structure.

A LATEXdocument consists of a range of nested environments, nested within one another. From the document environment downwards they indicate the structure of the document as sections, lists, paragraphs, etc. For research papers in the quantitative sciences, there is still no significantly better typesetting system. Although nroff and troff are still often found on Unix machines, most of these systems are now obsolete. But one, the Geography Markup Language from IBM, evolved into SGML, the Standard Generalised Markup Language. which subsequently gave birth to XML. Originating around the same time as TEX, the emphasis in SGML was, right from the beginning, on structure rather than presentation.

SGML was a well-kept secret for many years. In fact, one book was even called the Billion Dollar Secret (Ensign 1997). It had some big customers: you don't get much larger than the US military. SGML, the Standard Generalised Markup Language, is not in itself a markup language, but rather a meta-language: it enables precise document specifications, or *Document Type Definitions* (DTDs), to be written. The merit of such a scheme is that it becomes possible to precisely control the structure of a document. The military found this particularly desirable, having vast quantities of paperwork specifications and a need for powerful document control techniques.

SGML really became a household tool with the beginning of the World Wide Web, although, with apologies to Voltaire, using SGML was like writing prose for many people. What everybody saw, of course, was not SGML but one DTD written within it: HTML, the Hypertext Markup Language. Early exponents of the Web followed HTML quite closely, but the Web grew at a pace hardly anybody dreamed of. HTML 1.0 was followed rapidly by ever more complex versions. It is currently at level 5, which is at least four times the size of the original HTML specification! Up to HTML 4, new features were gradually added and the specification tightened in various ways. But HTML allows structures which are not syntactically correct XML (such as omitting end tags). This led to the development of XHTML, which is XML compliant.

HTML 5 is a major advance on HTML 4, with a number of features which make it suitable for cross-platform games. However, from the mapping point of view, the drawing tools are potentially very useful. We discuss this further in §5.3.

The World Wide Web now covers almost everything imaginable: from academic papers to advertising; from pornography to space science; from the trivial and ephemeral to ground-breaking new research. In 1999 the number of web-sites hit somewhere around 50 million, with over 200 million forecast for the year 2000. As the second edition of this book is completed, the size is now 4.3 billion pages (Anon 2014). With this explosion has come a truly enormous problem: finding things. Because the Web grows organically, without any central control, it is inherently disordered. Hence the need for metadata, and along with this need came the latest markup innovation, XML. Originally SGML was concerned primarily with document structure. SGML would ensure that all the pieces of the document were present and in a correct order. It's easy to see how this might be important. So, imagine competing tender documents for a new aircraft. Important items such as delivery date, price structure, and lots of other details need to be present. Why make a (fallible) human proofreader check that everything is there if it can be done by the document structuring package?

But along with the disordered growth of the Web another aspect started to assume more importance: the use of markup tags to convey semantic information as well as structure. Let's see how this works. Imagine that we want to find a recipe for the old English dish of jugged hare. If we look for hare we find zoological data; we find children's stories about mad March hares; we find polemics against hare coursing, one of England's less animal welfare-oriented pursuits; we might even find references to hares as food somewhere too. Now imagine that we have tagged the occurrence of jugged hare with the label dish:

```
<dish>jugged hare</dish>
```

The word `dish` in angled brackets is the tag; we'll cover the syntax in more detail below. Now if we restrict our search to hare inside the dish tag, we get rid of most of the spurious hits. But we can do even better. We still might hit the biography of the Duke of Upper Ernest, whose favourite dish was roast hare. Suppose now that all recipes for hare are embedded within an additional tag like this:

```
<recipe>
        <dish>juggedhare</dish>
</recipe>
```

Now we can search for hare inside food inside recipe and we should hit just what we want. When we study the Geographic Markup Language (GML), which is an extension of XML, we shall see how easy it is to incorporate geographic constraints, such as recipes from a particular region. The beauty of all this, the beauty of XML, is that we don't need anything special here. We don't need an elaborately configured database; any particular metadata standard will do. But there could be one complication. Different people might use tags in different ways. Dish could be used to tag pottery. Thus we have come to the idea of namespaces, which allow one to register the use of a tag. We'll cover this in §5.4. SGML is pretty arcane in these days of desktop publishing. XML removes a lot of the complexity (and a bit of the rigour too). But it has a free-form nature which makes it much easier to use. It was the big growth event of the years 1999–2000. As a standard it is pretty stable, but new derivatives based on it keep appearing. We shall look at those which are germane to spatial data.

5.1.1 Spatial Tags

We are not concerned here with all the many aspects of Web publishing. We need just enough knowledge to work through the spatial metadata standards, which are strongly dependent on SGML/XML. Now, although SGML came first and is in some respects the more powerful of the two, we will start with the more straightforward XML. XML does have one advantage over SGML for spatial metadata: it allows much more tightly controlled data types. Thus we can specify that a number shall be between 0 and 180 (such as we would require for a spatial coordinate).

Firstly, we will pick our way through the syntax, watching the functionality and application rather than the fine syntactic detail. Then we look at two important related standards. We consider the important issue of attaching unique interpretations or meaning to XML tags through the namespace recommendation in §5.4. There is now a firm recommendation for XML namespaces. The other key topic is the area of XML schema, which we consider in §5.5.

5.2 XML: Structural Ideas

SGML is one of the most flexible and general standards imaginable. It is composed of markup concepts and principles, but it does not specify exactly how these get represented in the source document. It does, however, have an example (commonly used) set of syntactic elements, known as the *reference concrete syntax*. We shall skip over all these details and work with the syntactic specification chosen for XML.

Conceptually, an XML document is made up of a sequence of one or more elements, which may be nested inside one another. The resulting document structure we refer to as a *document tree*. Each *element* has a single parent element right on up to the top or outermost element, which we refer to as the *document root*. Elements may be qualified by one or more *attributes*. Marked up documents can get quite complicated and so there is a sophisticated abbreviation mechanism, through the use of entities.

Perceptive readers may already have noticed a similarity between the idea of elements in an XML tree and the object hierarchies in Object Oriented (OO) technology. XML is highly suited to representing information objects, and conversely, XML elements can rightly be thought of as information objects. We briefly discussed OO concepts in §1.2 and will examine these concepts and their associated syntax in a little bit more detail in §5.2.1.

5.2.1 Elements and Attributes

An element begins with its name in angled brackets, as in `<dish>` above. It ends with the same structure but including a forward slash immediately after the first bracket, as in `</dish>`. The content of the element goes in between and can in-

TABLE 5.1: Correct nesting of elements

Correct	Wrong
<recipe>	<recipe>
<dish>	<dish>
recipe text	recipe text
</dish>	</recipe>
</recipe>	</dish>

clude other elements to an arbitrary extent. The content can be as simple as a word or phrase, as in our dish example. But it can be as complicated as an entire document, maybe the recipe example above, or even an entire cookery book (which would, of course, be the document root).

Sometimes an element may have no content (for example, an element indicating the insertion of an object such as a graphic). In this case the closing tag can be rolled into the start tag as in

```
<image src="mypic.jpg"/>
```

In this example, the trailing forward slash "/" acts as a closing tag. An element may have attributes which qualify the element in some way. So a recipe might have attributes of foodtype and usage. It might look like this:

```
<recipe foodtype="game" usage="maincourse">
```

The attributes, which are essentially keyword-value pairs, go inside the start tag. Now, in SGML, elements (along with everything else) are defined beforehand in the DTD. The DTD specifies the order that elements go in, what elements may be nested inside others, the sorts of characters elements can contain and a variety of special features which relate to how an element is processed. A *parser* can then check whether any given document conforms to this DTD.

In XML we do not have a DTD as such. But simple rules still have to be obeyed. The most important of these is that elements should not be interleaved. In other words, each element must finish entirely within the parent element, as shown in Table 5.1. XML has an arguably more powerful way of describing document structure, the XML *Schema*, which we discuss in §5.5.

5.2.2   What's in a Name?

SGML defines what each name can contain quite precisely. Sometimes the reference concrete syntax is quite restrictive, e.g., names are restricted to eight characters. XML is more fluid than SGML, and common sense will normally suffice. More or less any mixture of letters and characters with limited punctuation is acceptable. Why should we limit punctuation? Well, we do not want it to interfere with the control syntax. So, since angle brackets mark the beginning and end of tags, using them in a name presents complications! We would need a mechanism to quote them, to say that they were to be interpreted literally and not as indicating the start or end of a tag. We do this with entities, which we consider next.

5.2.3 Entities

In writing documents in standard ASCII text, we occasionally come across characters we don't have a representation for, particularly in non-English words. They might be accents over letters, or different letters entirely. Another situation which causes difficulty is that of the special punctuation character, such as the non-breaking space. We represent these cases using character entities. The format is simple. The entity begins with the ampersand symbol, is followed by a character string and terminated by a semi-colon. So, for the non-breaking space the character entity is ` `. It is not surprising to find that `&lt;` and `&gt;` are the entities for the angled brackets while `&amps;` is the ampersand itself.

5.2.4 The Document Type Definition

As we have already indicated, the DTD provides a very tight definition of document structure. We do not have the time and space to go into the fine details here, particularly since the future, especially for online documents, will be XML.

In addition to the definitions of the elements and attributes we have just discussed, the DTD contains also a certain amount of front matter. First, it contains information about the version of SGML and the locations of any files of public entities, such as special character sets and so on. Then, quite different from XML, it contains definitions of the syntax used for the document. SGML is truly self-describing. Various organisations, including ANZLIC and the FGDC, have created DTDs for spatial metadata; however, these DTDs have since been replaced with other XML schemas, so they have been removed from this edition of the book.

For the purposes of the Web at any rate, SGML has been largely superceded by XML. But it will take some time for all the DTDs out there to be revisited and converted to XML. There is one DTD which is proving a bit intransigent, HTML itself. At the time of writing, HTML has just entered version 5, with a whole lot of new features. HTML is does *not* require HTML conformance. There is a parallel effort, XHTML, to produce an XML-conformant HTML. But with the billions of Web pages already out there, servers are not going to switch over to XHTML only any time soon.

Although few people these days would develop Web pages by editing the HTML directly, behind the scenes it has retained backwards compatibility with the earliest version of HTML. One of the consequences of this is that end tags can be omitted, where so allowed by the DTD. In fact, this applies to most HTML tags, but this is not allowed in XML. HTML 5 has become a key format for the mobile world, with it becoming more and more the medium of choice for mobile computer game development. It is similarly important for the Online GIS world, although in not quite so obvious a way. For this reason we look at the relevant new features in the next section (§5.3).

5.3   HTML 5

HTML 5 looks, of course, a lot like earlier versions of HTML. The features which make it a big step forward for the mobile games and Online GIS developers are

- Native multimedia: videos, audio files and so on are now part of HTML with their own tags, `video` and `audio`. The Web browser and server have to provide native support for these without additional plug-ins.

- The `canvas` element takes web graphics to a new level. A simple use of the `canvas` element is simply to allocate a piece of screen real estate in which a graphic or game will play. But it goes further than this: all the common graphics commands, such as for drawing lines, polygons and colouring regions, are available as sub-elements of the `canvas` element. Thus the browser has to support the direct drawing of graphics as described by the procedural statements embedded within `canvas`.

5.4   XML Namespaces

As we saw earlier, XML namespaces (Clark 1999) are devices that allow us to attach meaning to XML tags. In SGML it is possible to refer to external entity sets or DTDs, but there is no precise network link. With XML, which was driven by the needs of the Web, indices to elements and attributes are, naturally enough, available as URLs.[3]

XML namespaces are defined within the XML schema, which we come to in §5.5, but for present purposes we need to know just two things about their use. It is important to realise though that the URI defining a namespace is essentially a descriptive document rather than the individual names themselves. It's almost like a metadata document. So if we look at the namespace document for XML schema,[4] we find the opening statement

> This document describes the XML Schema namespace. It also contains a directory of links to these related resources, using Resource Directory Description Language.

---

[3] Strictly speaking, all these documents are now written in terms of the more general concept of URIs (Uniform Resource Identifier) but the difference is unimportant here.

[4] `http://www.w3.org/2001/XMLSchema`

It is a short document without any of the elements we find in schemas (simpleType, complexType etc.).

1. The *xmlns* tag is used to define the URL where the namespace is to be found;

2. Tags may be now indexed according to the namespace using the form *gis:bridge* to indicate a tag called bridge in the *gis* namespace.

Tags without a namespace prefix inherit the namespace from the next outermost namespace tag. These facilities are important for the Resource Description Framework, which we look at in detail in §8.5. To guide us through RDF we don't really need to know very much more, but it is worth looking at an example:

```
<section xmlns:gis="http://myserver.com.zz/smdogis/xml">
    In the bottom left corner of the image is the
    <gis:bridge>Sydney Harbour Bridge</gis:bridge>
    and not far away is one of Australia's finest buildings
    the <building>Sydney Opera House </building>
</section>
```

There has been a certain amount of confusion over the fine details of the XML namespace definitions, which have arisen through the need to maintain backward compatibility with XML 1.0. The note by James Clark (Clark 1999) explains this in more detail, but we do not need to probe further in the present context.

XML allows a pretty unrestricted use of names for elements. This is fine when one author is handling relatively small documents. But for larger scale systems, websites, etc., we could run into name conflict. Names may get reused in different contexts, creating considerable confusion. The solution is the use of namespaces, defined as specific URIs. Each tag may then be qualified by a namespace indicator to make its significance precise.

Thus the Resource Description Framework (§8.5) has a namespace prefix, often denoted by *rdf*, hence the *creator* tag can be prefixed as

```
<rdf:creator>Garfield</rdf:creator>
```

It is important to realise, though, that this prefix is arbitrary, something we demonstrate with numerous examples in Chapter 6 and Chapter 9. We get the prefix from the *target namespace*, discussed below in §5.4.1, This format enables a fairly common word, such as creator, to be given a precise reference. This nomenclature could get fairly complex, but there are rules, which we do not have space to discuss here, that allow developers to set up assumptions and hierarchies of namespace references. The SGML DTD provides a precise description of document structure: which elements go where; what element is nested within what. But a DTD says nothing about the meaning or representation of the terms. Let's eavesdrop on a conversation at the bridge table:

If the palooka sitting East had two bullets he would have doubled. So, he must have a stiff in one of the minors, we can throw him in to suicide squeeze West.

If you've read the right books (on the card game bridge), then you might recognise some of this. The rest is jargon or slang. You'd need a glossary or thesaurus to make sense of it. In a gangster movie, stiff would probably refer to a corpse, bullets would of course be the real thing and a squeeze could be a gangster's moll or pressure to be brought on a victim.

XML recognises that the various meanings of terms can change in different contexts and sets out to define them, through the concept of namespaces. A document may use more than one externally defined namespace (somewhere on the Web) and there are various defaults defined for names which are not prefixed with a namespace. We won't go into the exact details of the specification, but just look at a couple of simple examples to get an idea of what happens.

First, imagine we have a namespace which defines various sailing terms which we might label as sail. So the tag, crew, which could refer to a type of haircut, is appropriately labelled:

```
<sail:crew>Fred Bloggs</sail:crew>
```

This tells us that the term crew is an element in the sail namespace. So how do we locate the definitions of sailing terms? We use the attribute, defined in XML, xmlns to provide a URI (Uniform Resource Indicator) for where the information is found. But it is more complicated than this. Firstly, the processing software has to locate the schemas. Secondly, it may do so in different ways. The *xmllint* program, distributed on Unix and Mac OS X, has a schema parameter, with the basic syntax

```
xmllint myxmlfile.xml -schema myschema.xsd
```

xmllint has the XMLSchema built in, so it does not need to fetch this from anywhere. The *XInclude* schema is also built in, but this needs to be accessed with a separate parameter, as

```
xmllint --xinclude myxmlfile.xml -schema myschema.xsd
```

There are two other ideas we need to grasp. The first is how we nest elements and infer default namespaces.

The second is just to recognise that we are using at least one namespace without referring to it: the XML namespace itself. One of the beauties of XML is it is incredibly self-referential. Most concepts are defined within XML, as we shall see in the remainder of this and the following chapter.

Although XML design does not espouse terseness, continual addition of prefixes would start to make a document hard to read. Hence we use the concept of inheritance. A tag without a prefix inherits its prefix from the parent element, or grandparent element and so on up the document tree. Thus the following fragment:

```
<sail xmlns="http://sailing.vir/terms">
        <dinghy>Laser</dinghy>
        <cat>Stingray</cat>
</sail>
```

is equivalent to

```
<sail xmlns="http://sailing.vir/terms">
        <sail:dinghy>Laser</sail:dinghy>
        <sail:cat>Stingray</sail:cat>
</sail>
```

Note that dinghy is pretty specific, but cat is not. In fact, a Web search for cat would have many hits which have absolutely nothing to do with boats, hence the importance of the namespace. There is an alternative syntax where we spell out the attribute more explicitly.

```
<sail:boat xmlns:sail="http://sailing.vir/terms">
        <dinghy>Laser</dinghy>
        <cat>Stingray</cat>
</sail:boat>
```

5.4.1   The Target Namespace

The `targetNamespace` element is the glue which binds schemas and XML documents. The schema begins with a declaration such as

```
<?xml version="1.0" encoding="UTF-8"?>
<xs:schema
    targetNamespace="http://www.opengis.net/giraffe"
    xmlns:tgmd="http://www.opengis.net/giraffe"
    xmlns:xs="http://www.w3.org/2001/XMLSchema"
    elementFormDefault="qualified" version="3.2.1.2">
```

The namespace xs denotes the XML schema itself and `targetNamespace` is one of the attributes of the `schema` element. The URI specified here *has to match* that specified in the XML document. The namespace itself in this schema document would have the prefix tgmd, which matches the URI of the `targetNamespace`. The attribute `elementFormDefault` specifies that whenever an element of the target namespace is used in the schema, it has to be qualified with the tgmd prefix.

In the XML document the namespace must match, but the prefix is arbitrary. So, in §9.3, we have the declaration

```
<quoll:MD_Metadata
    xmlns:quoll="http://www.opengis.net/giraffe"
    xmlns="http://www.w3.org/2001/XMLSchema"
    xmlns:xi="http://www.w3.org/2001/XInclude">
```

where the attribute xmlns:quoll matches the namespace denoted by tgmd above. But it could be anything, parrot, pumpkin, puddle, pronk, prurgle....

5.4.2 Include and Import Tags

With any long and complex document, be it a book, a piece of software or some online document set, it is often advantageous to break it up into separate files which cross reference each other. XML has such mechanisms, but they are slightly different for schemas(§5.4.2.1) and other XML documents(§5.4.2.2). We deal with them each in turn, dealing with only sufficient detail to suffice for the discussions in the book.

5.4.2.1 Inclusion in Schemas

The schema specification[5] contains two elements for including other schema documents:

include which takes the `schemaLocation` attribute. The contents of this attribute are any URI (URL or filename) and they tell the processor where to find the schema. The `include` element requires that the target namespace is the same for the parent and the included document.

import has the same function and `schemaLocation` as `include` but there is one crucial difference. The included document now must have a *different* namespace from the parent document.

5.4.2.2 Inclusion in XML Documents

Now, although the names are the same, the `include` tag used in XML documents, such as we find in §9.3, is actually different. It comes from the *XInclude* schema.[6] Practically it behaves very similarly, but the attribute is different, `href` as in HTML, as in

```
<xi:include href="md-citeinfo.xml"/>
```

Note that it might be necessary to tell the XML processor to use the *XInclude* schema for this to be interpreted correctly, as we saw in §5.4.

5.5 XML Schema

In SGML the Document Type Definition served to exactly encode the rules of document structure. In XML we have much looser rules, which merely control things like the embedding of tags inside one another. The need for stronger control is satisfied by XML schemas. They describe the tags and their possible values in considerable detail. As we hinted above, control over the element context may include things like the range of possible numerical values which might be taken.

---

[5] http://www.w3.org/TR/xmlschema-1
[6] http://www.w3.org/2001/XInclude

XML began life as a simplified version of SGML. It has now come to dominate current development on the Web. With its spread into areas such as metadata, digital signatures, generic document structures and so on, the need for an additional structuring mechanism became necessary. This mechanism is the XML schema. In some ways it is a bit of a reinvention of the wheel, in that a lot of what it does is similar to an SGML DTD. But in other ways it has gone beyond the DTD framework to allow more precise specification of structure and content.

The XML schema requirements were slightly different from SGML, as discussed by Malhotra and Maloney (Malhotra and Maloney 1999), such as being both more expressive than XML DTDs, yet relatively simple to implement. Key requirements were that they have to be self-describing and expressed in XML.

The root element for an XML schema is xs:schema.

There are many bells and whistles to the schema documents. We cannot possibly include all of them here, but the W3C website has the URLs for tutorial documents and the full reference specifications. What we propose to do here is to look at a metadata DTD (in fact a subset of the ANZLIC DTD) and see how the XML schema document would express the same concepts.

We shall actually start in the middle, rather than at the beginning, to focus on the definitions of elements and their attributes. Right at the root of the document tree is the definition of the anzmeta element. This element does not contain any actual text but merely other elements. The DTD entry to define such an element is

```
<!ELEMENT anzmeta - - (citeinfo, descript, timeperd,
       distinfo?, cntinfo+)>
```

The two hyphens indicate whether it is possible to omit the start or end tags (in this case no; possible omission would be indicated by a letter o instead of the hyphen). In XML this situation does not arise, as start and end tags are obligatory. In brackets we then have a list of elements which make up the anzmeta element. The comma separating them has a specific meaning here: the elements must appear in this order only; the ampersand symbol would be used to indicate that several elements are required but may occur in any order. Two other symbols appear: ? indicates that an element is optional; + indicates that the element must occur one or more times. This syntax is of course very much like common regular expression syntax.

In the XML schema this looks a lot more complicated. The definition of the element is fairly simple:

```
<xsd:complexType name="anzmeta" type="anzmetaType"/>
```

where xsd denotes the schema namespace and the element is empty (denoted by the closing symbol />. What makes it more complicated is the type attribute. Any element may have either a simple or complex type; to users of programming languages, such as C or C++, this is very similar to the difference between a string or integer and a class or structure. We shall see other programming analogies later. anzmetaType is complex because it is made up of other simple or complex types.

Incidentally, the name of the type could be anything. We've followed a practice common in object-oriented programming of appending a descriptor (Type) to

the name of the instance. This is merely a convention. So here is the definition of
`anzmetaType`:

```
<xsd:complexType name="anzmetaType">
    <xsd:element name="citeinfo" type="citeinfoType"/>
    <xsd:element name="descript" type="descript"/>
    <xsd:element name="timeperd" type="timeperd"/>
    <xsd:element name="distinfo" type="distinfo"
                minOccurs=0/>
    <xsd:element name="cntinfo" type="cntinfo"
                minOccurs=1 maxOccurs="unbounded"/>
</xsd:complexType>
```

So we move recursively down through the definition of each element and its
type. Note that for `distinfo` and `cntinfo` we have specified an attribute which
indicates the number of times the element may occur, equivalent to the SGML "?"
and "+" operators, respectively. We haven't specified them for the other elements,
because the defaults are adequate. The default for `minOccurs` is 1 and the default
for `maxOccurs` is `minOccurs`, i.e. the element occurs once and only once.

In the `cntinfo` case we have specified that the maximum number of occur-
rences is unbounded. We could also specify a finite number (say we going to allow
up to one such element for each state or territory which in Australia would make
`maxOccurs` 8). This gives added flexibility over SGML, which cannot express such
a precise range.

All the sub-elements, `citeinfo` etc., will have their associated type definitions
and we do not need to go through all of them. But there are a few more features we
would like to illustrate. Here is the cntinfo type (not strictly according to the ANZLIC
definition):

```
<xsd:complexType name="cntinfoType">
        <xsd:element name="cntorg" type="xsd:string"/>
        <xsd:element name="cntpos" type="xsd:string"/>
        <xsd:element name="address" type="addressType"
                minOccurs=0
                maxOccurs=1/>
        <xsd:element name="city" type="xsd:string"/>
        <xsd:element name="state" type="stateType"/>
      <xsd:element name="postcode" type="postcodeType"/>
        <xsd:element name="cntvoice" type="telNumType"
                minOccurs=0
                maxOccurs=1/>
</xsd:complexType>
```

The first point to note is that, although we have defined `city`, etc. as just plain
strings (built-in simple types), we have chosen to define a special type for `state`.
Here it is:

```
<xsd:simpleType name="stateType" base="xsd:string">
        <xsd:pattern value="[A-Z]{2}"/>
</xsd:simpleType>
```

The states in the USA all have two-letter upper-case abbreviations. Thus we define a sub-type of string which restricts strings to precisely this form. Any other string will generate an error message.

Such restrictions on XML elements are called *facets*.

Australia has fewer states than the USA. In this case we might want to tighten things up even more and specify only the allowed abbreviations. We do this with enumeration:

```
<xsd:simpleType name="stateType" base="xsd:string">
        <xsd:enumeration value="ACT"/>
        <xsd:enumeration value="NSW"/>
        <xsd:enumeration value="NT"/>
        <xsd:enumeration value="QLD"/>
        <xsd:enumeration value="SA"/>
        <xsd:enumeration value="TAS"/>
        <xsd:enumeration value="VIC"/>
        <xsd:enumeration value="WA"/>
</xsd:simpleType>
```

Another explicit example occurs in the definition of `jurisdic` which is a sub-element of `citeinfo`. Here the jurisdictions are spelt out explicitly:

```
<xsd:simpleType name="jurisdicType" base="xsd:string">
        <xsd:enumeration value="Australia"/>
        <xsd:enumeration value="Australian Capital
                                            Territory"/>
        <xsd:enumeration value="New South Wales"/>
        <xsd:enumeration value="New Zealand"/>
        <xsd:enumeration value="Northern Territory"/>
        <xsd:enumeration value="Queensland"/>
        <xsd:enumeration value="South Australia"/>
        <xsd:enumeration value="Tasmania"/>
        <xsd:enumeration value="Victoria"/>
        <xsd:enumeration value="Western Australia"/>
        <xsd:enumeration value="Other"/>
</xsd:simpleType>
```

The first thing we can do with a schema is to specify what elements and other things belong inside an element:

```
<xsd:element name="distinto" type="distinfoType">
        <xsd:complextype name="distinfoType"/>
</xsd:element>
```

Latitude values, for example, range from 0 to 90 degrees. So what we need to do is to take a simple type, integer, and restrict its application.

```
<xsd:simpleType name="latitude" base="xsd:integer">
        <xsd:mininclusive="0"/>
        <xsd:maxinclusive="90"/>
</xsd:simpleType>
```

Sometimes we might want to mix element content with some basic text data. This might work as follows:

```
<distinfo>
        The following options are available:
        <ol>
        <li> mapiinfo</li>
        <li> arcinfo</li>
        <li> OpenGiS</li>
        </ol>
        These data are also available
        in a variety of other non-standard formats.
</distinfo>
```

The schema that would represent this is

```
<xsd:element name="distinfo" content="mixed">
        <xsd:element name="ol"/>
</xsd:element>
```

Note that we cannot just mix text and elements at random. We still have to specify the order in which the elements appear even though we might intermix plain text amongst them. Suppose we want to restrict the particular strings which might be used. There is a regular expression syntax to do something just like this. The details are complex, so let's just look at a simple example.

A common problem with much legacy data is that the origin is uncertain, i.e., we do not know the start date. We could just leave this out, or add some not-known tag. An alternative would be to make the element explicitly null. In the schema this would take the form

```
<xsd:element name="begindate"
        type = "date" nullable = "true"/>
```

and the date element itself would look like this:

```
        <begindate xsi:null="true"></begindate>
```

We can of course make the element an empty element, which we do simply by including the attribute `content="empty"`.

There are two things to note about this format. First, we have applied a specific namespace (an instance of an XML schema) to the null attribute. Second, the tag is not an empty tag, but a tag with nothing in it (which has a conventional close tag). We might want to specify that an element is made up of a collection of elements. We have several ways of doing this: choice, sequence and all. The element choice allows just one element from a selection:

```
<xsd:choice>
        <xsd:element>
        <xsd:element name="type"/>
        <xsd:group ref="junk"/>
        <xsd:element>
</xsd:choice>

<xsd:group name="junk">
        <xsd:sequence>
        ...
        </xsd:sequence>
</xsd:group>
```

Note that sequence is the default anyway. With the element all, we have to use all of the elements, but they can be in any order, e.g. minOccurs and maxOccurs.

Finally, it is worth noting that XML schema are somewhat more precise with mixed content models than DTDs often are.

5.5.1 Application Schemas and Profiles

As specifications grow in complexity, to take in more and more different cases, their overall size may become quite daunting. The OGC metadata standard has over 70 component schema documents. To make this complexity tractable, schemas are both simplified and extended to form shorter more focused schemas relevant to a particular domain. These are referred to as *Application Schemas*.

A *Profile* is a restricted subset of a schema particularly in the GIS domain[7].

5.5.2 Where to Find DTDs and Other Specifications

We have seen in the preceding sections a wide variety of specification, for document structure and semantics. There are still a few more to come. So, how does some given document know where to find these specifications? There are two mechanisms: a generic header at the top of the document and embedded URIs throughout.

---

[7] http://www.ogcnetwork.net/gmlprofiles

5.5.2.1   Document Headers

In SGML we begin with quite a complicated header block specifying a great deal of things to do with syntax used for defining SGML constructs, etc. We needn't worry too much about this, since XML has tended to lock in defaults for many of these options. The key component comes right at the beginning, the doctype declaration in which the DTD is given a name:

```
<!DOCTYPE myDTD [
<!ENTITY t ISOpub PUBLIC "ISO 8879-1986//ENTITIES
                         Publishing//EN">)>
```

The embedded declaration defines the public entities, the expressions such as   to represent a non-breaking space. In this definition we have reference to an International Standards Organisation (ISO) definition, a public text class (ENTI-TIES) and a public text description (Publishing) and finally after the second set of //, a public text language code (EN for English) (Bryan 1988).

In XML the situation is a little simpler:

```
<?xml version="1.0" encoding="UTF-8"?>
<?xml:stylesheet href="annrep99.css" type="text/css"
                         charset="UTF-8"?>
```

First we have the declaration of the version number of XML and an encoding specification. The second line provides something different from the SGML frame-work: a specific style sheet for presenting the document. A processing system may not need to make use of this: a query agent would be interested in content rather than presentation, for example.

5.6   XQuery: The XML Query Language

Two query languages appear in this book: **XQuery**, the subject of this section, which is used for querying XML documents in general; and **SPARQL** (§8.6), which is used for querying RDF documents, as we discuss in Chapter 8.

XQL was a generic language for querying XML documents, implemented in a number of software packages. It was essentially defined in a proposal to a W3C Query Language workshop in 1998 by Joe Lapp of webMethods and David Schach of Microsoft (Lapp and Schach 2014). The W3C has a working group on XML query languages and launched a recommendation in 2010 **XQuery**.

XQL enables selections of subsets of a document based on XML elements, along with pattern matching for the contents of elements themselves. XQuery has a number of characteristics in common with SQL. Apart from the sort of operations which can be performed, it is declarative, rather than procedural. Thus XQuery implementations might use a range of algorithms or techniques for efficient query processing: they

TABLE 5.2: The predefined namespaces for XQuery

	URL Reference
1	xml = http://www.w3.org/XML/1998/namespace
2	xs = http://www.w3.org/2001/XMLSchema
3	xsi = http://www.w3.org/2001/XMLSchema-instance
4	fn = http://www.w3.org/2005/xpath-functions
5	local = http://www.w3.org/2005/xquery-local-functions

have nothing to do with the language. The result of an XQuery query is itself an XML document.

XQuery uses the idea of expanded names, or **Qualified Names**, **QNames**, which are basically strings with an optional namespace prefix followed by a colon, as in the examples we've seen earlier in this chapter. Some namespaces are predefined, as shown in Table 5.2. It relies also on the notion of paths, through an XML document tree. A path recommendation from the W3C was released in 1999, and we consider this next in §5.6.1.

It's useful to us in the metadata context because we can use it to extract different parts of the metadata for particular purposes. Consider the fragment in Figure 5.1, an illustrative document based loosely on the ANZLIC DTD.

Suppose we want to check the data quality of the dataset described by this metadata. The DTD provides an element, `<dataqual>` for precisely this, containing sub-elements describing, for example, the accuracy, completeness and logical consistency. We get all of these as a sub-document with the XQuery query `anzmeta\dataqual` and we can now check that the data meets our quality requirements. We wouldn't necessarily write these queries explicitly ourselves. They can be generated automatically by a specialised metadata query program or they can be part of a program for extracting the data itself. Let's take a more detailed look. First we will look at the way we drill down through the elements of an XML document, then at how to put in specific pattern matches.

5.6.1 XPATH: Locating Elements and Attributes

The intense wave of activity in Web language specifications at the end of the 1990s has had the considerable benefit of using similar syntax wherever possible. So the syntax for locating elements in XQuery is very closely linked to finding components in an XML document in the **XPath** specification (W3C 2014b). The most widely implemented XPath version is 1.0, which dates back to 1999. The specification grew in size to version 2.0 and the latest version is XPath 3.0, which became a W3C recommendation in April 2014 (W3C 2014b). The big change going from 2 to 3 is the addition of a lot more function capacity for performing transformations and tests

```
<lillimap>
    <descript id="Bathurst">
        <abstract>
            Covers Bathurst and surrounding area
        </abstract>
        <theme>
            <keyword>NSW</keyword>
            <keyword>city</keyword>
        </theme>
    </descript>
    <distinfo>
        <!-- we do the digital version first-->
            <digform>
                <formname res="full">
            Unsigned 8 bit generic binary
                </formname>
            </digform>
                <acct>
                    Available online to the
                                    general public.
                </acct>
                <nondig>
                    <formname res="600dpi">
                Hardcopy map
                    </formname>
                </nondig>
                <acct>
                    Available for purchase
                        from NSW Government
                            OneStop shops.
                </acct>
    </distinfo>
</lillimap>
```

FIGURE 5.1: A sample document for the query examples.

on sub-trees. By 3.0, XQuery has become pretty complicated. The EBNF grammar[8] now has 196 clauses. Obviously, this chapter can cover only a tiny subset!

There are seven document components: root nodes, elements, attributes, text nodes, namespace nodes, processing instruction nodes and comment nodes. The XML document is a tree and paths are routes through this tree. Query results then select nodes or sub-trees of the contents thereof. Figure 5.1 shows a sample document and Figure 5.2 shows the corresponding tree, where we have omitted some of the text nodes to avoid making the diagram too cluttered. Each element has associated with it a namespace node, and again we haven't shown these in Figure 5.2 for clarity. The root node does not have an expanded name.

There are definite rules for the **Document Order** of the nodes in the tree (W3C 2014b):

1. The root node is the first node and each node occurs before all of its children;

2. Namespace nodes immediately follow the element node and are followed by the attribute notes;

3. Children and descendents come before siblings, but the relative order of siblings is the order of the `children` property of the parent node.

There are two ways of getting at the pieces of an XML tree, be those pieces sub-trees, elements, whatever:

- A Unix-like syntax, including some regular expression operators (§5.6.1.1);

- A syntax using logical descriptors for tree components – parent, child, sibling and so on (§5.6.1.3).

5.6.1.1 The UNIX Syntax

The syntax is almost identical to the pathname syntax defined for Uniform Resource Indicators, which is of course itself very similar to UNIX. So to get the positional accuracy inside the data quality element we have simply:

```
anzmeta/dataqual/posacc
```

with forward slashes (/) separating each element. As per UNIX, the root node is / and the current and parent nodes are . and .., respectively. We might also want to access, not the immediate children of a parent node, but simply the descendants. So we might want to pick the theme nodes in Figure 5.1, which are quite deep in the DTD structure. Now we use a double slash operator (//):

```
lilliput/description/theme:
```

---

[8] EBNF, Extended Backus-Naur Form, is a common standard for expressing the grammar of computer languages.

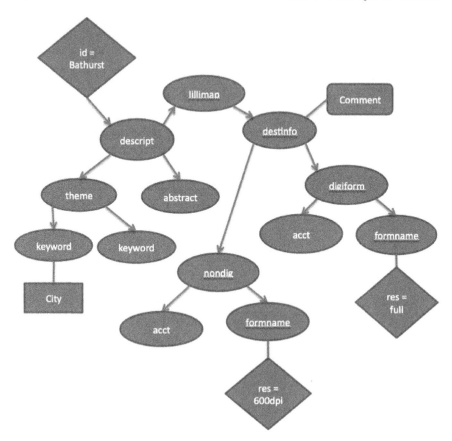

FIGURE 5.2: The XML tree corresponding to the document in Figure 5.1, where elements are shown as ovals, attributes as diamonds, textboxes as rectangles and comments as rectangles with rounded edges.

```
<theme>
        <keyword>NSW</keyword>
        <keyword>city</keyword>
</descript>
</theme>
```

The `//` operator gets all the nodes matching the node name or node path. Alternative paths can be selected using the `|` operator, familiar from UNIX as an `or` operator. These examples all get elements. We can also get attributes using the `@` symbol; `//@res` picks out all the resolution attributes

```
full
600dpi
```

Similarly a few UNIX ideas from regular and logical expressions transfer across. One useful concept is the asterisk (⁎) to represent a wild card. Thus the query

```
descript/*/keyword
```

finds all keywords that are precisely the grandchildren of the descript element. There are numerous other functions and operators such as element () for any element node or node () ⁎ for a sequence of zero or more nodes, where the asterisk is again the regular expression operator.

5.6.1.2   Predicates: Adding Constraints to Paths

After the node locator we have been discussing, we can add a **Predicate**, enclosed in square brackets [ ].

These are mostly numerical conditions or string matches. The following examples all refer to Figure 5.1.

We can count nodes just by enclosing numbers in brackets, as [1]. Thus

```
/lilliput/distinfo/formname[2]
```

selects the formname with res=600dpi. This could also be written as

```
/lilliput/distinfo/formname[last()]
```

since it is the last element and in this case

```
/lilliput/distinfo/formname[1]
```

could be writtten

```
/lilliput/distinfo/formname[last()-1]
```

We could also select this as a condition on the attribute

```
//formname[@res="600dpi"]
```

If we omitted the value string "600dpi" we would find all the formname elements with a res attribute set.

In Figure 5.1 res is set equal to a string. If it had been defined as a numeric value, we could have a condition such as @res=600[9].

---

[9] Most of the numeric comparisons are UNIX like, but this one is not – as one might have anticipated.

5.6.1.3   XPATH: Family Locations

The family locations use the idea of an `axis` and take the form

```
axis::node[predicate]
```

where the axis is taken from the current node, obtained by a previous path selection. The `node` component conditions the nodes to be selected from the axis.

The axes comprise family terms: ancestor/descendent; parent/child; and siblings. Thus `ancestor::*` in Figure 5.1 refers to `lillimap` if the current node is `descript`.

We can also add `-or-self` to `ancestor` and `descendent` to include the node itself. So `ancestor-or-self::descript` includes the `descript` node itself.

There are also the more general terms, `following/preceding`, which refer to everything before or after. They can also be restricted with `-siblings` to refer to the preceding or following siblings of a node. Thus, if the current node is `descript`,

```
\verb|following-siblings::distinfo
```

refers to the two `distinfo` nodes.

Finally, before we leave this section, it is very important to stress that the recommendations are extremely complicated. There are many final details as to what characters can be used where, what constitutes a match, how sequences are processed and many more. The purpose here is simply to give the general flavour of how XPATH works. The same caveat applies equally strongly to XQUERY, which we come to in the next section (§5.7).

5.7   XQuery Grammar

As we noted above, the XQuery grammar is very sophisticated, but it shares many features of XPath. We will thus discuss just one principal construct, the **FLWOR** statement. Some things can be done in XPath without FLWOR, but the latter is more powerful.

5.7.1   FLWOR: XQuery Expressions

The processing of an XQuery has its own special syntax denoted **FLWOR**, standing for **F**or; **L**et; **W**here; **O**rder by; and **R**eturn.

- **For** as you might expect is an iterator. It will iterate over all the nodes of a sub-tree, or over some other expression, such as a numerical loop.

- **Let** defines a local variable.

- **Where** selects a subset of the range of the **for** loop.

- **Order** determines the order of the results.

- **Return** does just as it says.

FLWOR, itself, occupies 29 of the grammar clauses of XQuery 3.0.

The **for** statement behaves just like any loop construct, but with its own syntactic sugar. So,

```
for $x in (100,200,300)
```

the loop variable $x takes the values 100, 200 and 300.

But usually the **for** statement will be selecting nodes from a tree. Using the pre-defined function **fn:doc**, we can use a complete document from a URL.

```
for $x in fn:doc("lilliput-map.xml")
```

using the data from Figure 5.1, which we have given an arbitrary file name.

XQuery behaves in a very similar way to XPath. So, here is a query which just extracts the distinfo element.

```
for $x in doc("map-sales.xml")/lillimap/distinfo
return
$x
```

resulting in

```
<?xml version="1.0" encoding="UTF-8"?>
<distinfo>
    <!-- we do the digital version first-->
    <digform>
        <formname res="full">
            Unsigned 8 bit generic binary
        </formname>
    </digform>
    <acct>
        Available online to the general public.
    </acct>
    <nondig>
        <formname res="600dpi">
            Hardcopy map
        </formname>
    </nondig>
    <acct>
        Available for purchase from NSW Government
                            OneStop shops.
    </acct>
</distinfo>
```

```
<?xml version="1.0" encoding="UTF-8"?>
<distinfo>
    <!-- we do the digital version first-->
    <digform>
        <formname res="full">
    Unsigned 8  bit generic binary
        </formname>
    </digform>
    <acct>
        Available online to the general public.
    </acct>
    <nondig>
        <formname res="600dpi">
    Hardcopy map
        </formname>
    </nondig>
    <acct>
        Available for purchase from NSW Government
                            Onestop shops.
    </acct>
</distinfo>
```

Note that this query returns the distinfo element as part of the return document. Multiple elements are returned as a list, but this may not be a valid XML
document. Thus

```
for $x in doc("map-sales.xml")/lillimap/descript
                               /theme/keyword
return
$x
```

returns

```
<?xml version="1.0" encoding="UTF-8"?>
<keyword>NSW</keyword>
<keyword>City</keyword>
```

The **let** clause defines local variables, but could introduce another file. We might,
for example, want to use data from a file containing a list of government shops (Figure 5.3):

```
let $e := fn:doc("lilliput-gov-shops.xml")/town
```

To begin with, let's get the towns with postcodes greater than 300 which have
some form of retail outlet. The code is

```
<ul>
{
    for $x in doc("gov-shops.xml")/govshops/category/town
    where $x/address/postcode/@value>300
    order by   $x/@name
    return
    <li>
        {data($x/@name)}
    </li>
}
</ul>
```

The **for** statement selects towns from the source document.

The **where** statement selects the postcodes of interest.

Curly brackets are used to enclose processing instructions. Everything else goes straight to the output. The condition is on the attribute of the postcode element and thus is preceded by the @ symbol.

The output is sorted alphabetically according to the name of the town. Finally, we get to the **return** clause. This is usually an XML fragment, but it could be a chunk of HTML to send back to a browser, as in this case. We have used the **data** function here, which extracts the data from the attribute and does not report the tag as well.

The result is

```
<?xml version="1.0" encoding="UTF-8"?>
<ul>
  <li>Belfaborac</li>
  <li>Swifttown</li>
</ul>
```

The language is a lot more powerful than these simple examples suggest. Many of the constructs can be nested to any level (FLWOR with FLWOR, for example). Among the numerous features of FLWOR we have omitted, there are conditional **if, then, else** and **switch**, **try/catch** expressions common to many languages, plus numerous others which condition on variable type or quantified expressions, such as **some, every**, which return true or false depending upon whether a list of some sort **satisfies** some condition. For the reader who would like to dig deeper, the **saxon** package provides a comprehensive set of java classes for processing XQuery (Kay 2014).

A more complicated example is to combine XML files. Here we want to find the towns which sell the 1200 dpi maps. These are only available at the full mapping shops, not the subsidiary outlets in places such as post offices. The code is

```
<ul>
{
    for $shops in doc("gov-shops.xml")/govshops/category
    let $maps := doc("map-sales.xml")/lillimap/distinfo
```

```
<govshops>
    <category type="full">
        <town name="Swifttown">
            <address>
        <street number="5">McLochlan</street>
        <postcode value="303"/>
            </address>
        </town>
    </category>
    <category type="sub">
        <town name="Belfaborac">
            <address>
        <venue>Post Office</venue>
        <street number="20">Clarence</street>
        <postcode value="321"/>
            </address>
        </town>
        <town name="Gulliver Point">
            <address>
        <venue>Bank of  Lilliput</venue>
        <street number="5">Fortress</street>
        <postcode value="249"/>
            </address>
        </town>
    </category>
</govshops>
```

FIGURE 5.3: Retail outlets for Lilliput Maps. They may be full government shops or counters within some other shop or building

```
                                             /nondig/formname
    where $maps/@res="1200dpi"
        and $shops/@type = $maps/@category
    return
    <li>
        {data($shops/town/@name)}
    </li>
}
</ul>
```

which results in just one town:

```
<?xml version="1.0" encoding="UTF-8"?>
    <ul>
        <li>Swifttown</li>
    </ul>
```

5.8 The Future of XML

SGML originated in the 1960s and is a much older standard than XML. As such, SGML is the basis upon which many markup languages have been written and there are a number of good books available. The book by Bryan (1988) is formal and thorough and an excellent reference text. Two of the earlier books which discuss the broader implications of structured documentation are by Alschuler (1995) and Ensign (1997). Charles Goldfarb has been a pioneer of SGML and subsequently XML technologies and his XML handbook (Goldfarb and Prescod 1998) is a definitive reference at the time of writing the first edition of *Online GIS and Spatial Metadata*.

XML has been used by many organisations and standards groups over the last decade. It now underpins the languages of many formal standards and it now provides human readable specifications of all kinds of systems, being used across a wide range of applications including a number of Online GIS. XML continues to have a strong influence on the creation of open standards for GIS and this is unlikely to change in the near future.

SGML, although it will always be around and is mostly backwards compatible with XML, is on the decline as organisations use XML to describe business data. Figure 5.4 shows the current situation.

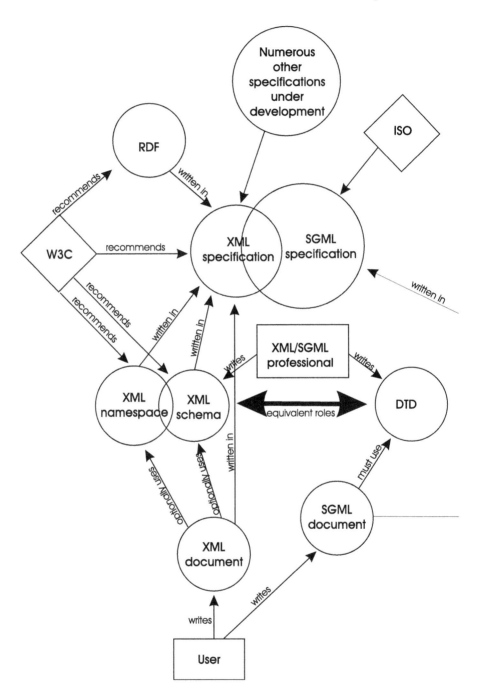

FIGURE 5.4: Comparison of XML and SGML.

5.9   Summary and Outlook

This has been a long chapter, but it has covered the markup standards, which are the foundation of the metadata discussions to follow.

- §5.1 introduced the idea of markup, how document structure and content can be separated from presentation.

- §5.2 briefly described XML, a descendent of SGML, which now features in literally thousands of pages of standards and specifications. Here we saw how, in the last decade, many standards have become very large and comprehensive and how we can only sample some of the key ideas.

- §5.3 introduced HTML 5, which is particularly important for Online GIS. It introduces its own graphics canvas element and a set of drawing primitives, which any supporting browser should implement.

- Digging deeper into the structuring nature of XML, §5.4 considered XML namespaces, the specification of vocabularies for particular purposes.

- But the vocabularies themselves need a formal specification, particularly in how they relate to one another and embed, and how the type of data may be restricted. This happens in the XML Schema, discussed in §5.5. Where to find DTDs and other specifications is discussed in §5.5.2.

- The structuring aspect of XML allows it to be queried in a precise way with its own query language, XQuery, introduced in §5.6, with a more detailed look at the grammar in §5.7.

- §5.8 looks towards the future of XML.

6

XML and Online GIS

CONTENTS

Since the first edition of this book, organisations and individuals have been using XML to solve a wide range of information management tasks. These tasks include schemas to aid interoperability across systems as well as simply improving the exchange of data in common human readable form. There is now a signficant number of XML implementations that can be applied to a wide range of GIS-related disciplines and we now have XML schemas specific to Web delivery to support different forms of geography and geometry. XML schemas now exist that define water, buildings, city data and civil engineering. This chapter provides an overview of four common XML schemas used in Online GIS.

6.1 XML Implementations

This chapter discusses four commonly used XML schemas in Online GIS: GPX (§6.2), KML (§6.3), GML (§6.4), and LandXML (§6.5). They range from the light weight GPX used to translate Global Positioning System (GPS) data to and from devices, through the more complex civil engineering data described using LandXML and the standards-based geographic data elements that are formalised using GML. There is another example, PMML, in Chapter 10 (§10.10.1).

But why have yet more markup languages for geographical information? XML has the following advantages:

- *Implicit metadata.* By having geographically meaningful tags we can do some spatial related searches based on text only. It is still difficult, although a fast moving research frontier, to do image searches based on image queries. So, for example, it is not easy at present to write search queries, such as find a fish in the given set of images. Auxiliary text markup is a way around such problems. As we have seen already, markup carries implicit metadata.

- *A fast way of delivering vector-based geographical data to the Web browser.* As we saw in Chapter 4, there are now standards for vector graphics on the Web. So, we could mark up our map data, say, in SVG directly. Unfortunately, this was not flexible enough to meet all user needs. Browsers vary in resolution; they may be used by people with sensory disabilities; networks vary in speed; in fact, there are many reasons for tailoring a Web page on delivery. A spatial language written in XML can be easily transformed to other XML schemas. Thus, SVG can still be used for displaying GML at the client by simply transforming the geospatial language into SVG.

This chapter will provide a solid background to the ideas introduced in Chapter 5 using a practitioner's view, covering a brief history of each as well as an overview of schemas and discussions around recent applications. Its also worth noting that each of the schemas contains metadata elements as part of their taxonomy, highlighting the importance of metadata creation at the time of data capture to support downstream usage.

As with many of the Online GIS technologies, there are plenty of trusted websites showing in-depth detail of each standard, and we have included some of these hypertext links throughout Chapter 6 as guides to further online reading.

6.2 GPX Schema

GPX is an XML schema that can be used for the storage, manipulation and translation of GPS data from a device to an application. The schema is an open standard that was first released in 2002 and has been a stable format (1.1) since 2004. [1] Understanding and using GPX is a perfect place to start using XML, as the usage is simple and it can be applied to many real-life situations.

The schema is included in the XML Document by using the following schema component:

```
<?xml version="1.0" encoding="utf-8"?>
<echidna:gpx xmlns:xsd="http://www.w3.org/2001/XMLSchema"
xmlns:echidna="http://www.topografix.com/GPX/1/1"
              version="1.1"
              creator="TH">
```

Note that the prefix we use for the GPX schema is completely arbitrary, discussed further in §6.4 – here we've used a spikey Australian icon, the echidna.

The basic element of GPX is the `point` feature. The tags of these point elements store location, as well as optional elevation and time information for each point. The mandatory attributes of any point feature are the point coordinates. These coordinates are always stored as latitutde and longitude values, representing the observed location of the *point*. The coordinate system used by GPX is WGS84 datum[2] and all measurements are in metric units. This is a common *intermediary* global coordinate system used when converting between coordinate systems. This means that to go from coordinate system A to B, the coordinates are first transformed from coordinate system A to WGS84, and then converted from WGS84 to coordinate system B.

All other elements in GPX are optional (excluding point). Some vendors have extended GPX by adding additional elements specific to the sensors on their devices, such as air temperature and depth of water. Other common extensions to GPX by vendors include adding attributes and elements for the GPS collection of location data such as address points, addressing, business names and street asset types, which can all be tied to the location using the coordinates of each element. These extensions to the GPX schema are generally for the field collection of data.

6.2.1 GPX Schema Details

The essential elements of GPX data are Waypoints, Routes and Tracks. These three elements are all made up of data points and are often described as (schema tags are shown in brackets after the name):

WayPoint (wpt) The meaning of waypoint is very much as it sounds. Waypoints are points along a route, or points along the *way*. They are important decision points that point the way. For example, they provide important destination information at junctions in the road, a natural break point in the travel or a general change in direction, to name a few.

---

[1] http://www.topographix.com/gpx.asp
[2] http://www.ga.gov.au/scientific-topics/positioning-navigation/geodesy/geodetic-datums/other/wgs84

Track (`trk`) A track is made up of a number of track points that are normally captured at a set time interval relevant to the speed of the device. Users should set a
time interval so that they capture the *shape* or the track, without capturing too many
track points. For example, you would need a longer time interval if capturing the
track taken during a camel ride than you would in a motor bike ride. A track in GPX
may also be made up of a number of Track Segments, each segment representing a
natural grouping of points.

Route (`rte`) A route object is a thinned track that retains changes in direction such
as turns and stops, therefore retaining shape with minimal track points. It would
also show decision points including left or right turns, roundabouts and the names
of roads right down to what road lane to be in and how far away the next turn is.
Anyone who has used a navigation system such as TomTom or NavMan will be
familar with these types of directional information.

The following XML code fragment shows how a point feature in GPX may be
used. The point is the Sydney Opera House. In the summer cruise ships and ocean
liners sail past almost every day, so we have coded this as a way point, where the
longitude and latitude are included as attributes of `wpt`. The elevation is very close,
since the bulding is on the waters edge.

```
<echidna:gpx xmlns:xsd="http://www.w3.org/2001/XMLSchema"
        xmlns:echidna="http://www.topografix.com/GPX/1/1"
             version = "1.1"
             creator = "TH">

 <echidna:wpt lat="33.8587" lon="151.2140">
  <echidna:ele>10</echidna:ele>
  <echidna:time>2002-10-10T12:00:00-05:00</echidna:time>
  <echidna:magvar>
    <!--Magnetic variation (in degrees) at the -->
    20
  </echidna:magvar>
  <echidna:geoidheight>
    <!-- Height (in meters) of
geoid (mean sea level) above
WGS84 earth ellipsoid.  As
defined in NMEA GGA message. -->
    0
  </echidna:geoidheight>
 </echidna:wpt>
</echidna:gpx>
```

In all, there are 20 global schema components defined in GPX 1.1. These global
elements are all listed in Table 6.1.

By way of example, two of the element schema representations and XML instances representations have been included here. These are `gpxType` and `ptType`.

TABLE 6.1: GPX Schema 1.1 elements

GPX Element	Element Description
gpx	the root element
gpxType	represents the metadata followed by waypoints, routes, tracks and extensions
metadataType	containing metadata about the GPX file
wptType	representing a waypoint, point of interest or named feature
rteType	represents a route, a set of waypoints leading to a destination
trkType	represents a track, an ordered set of points describing a path
extensionsType	the element that can contain elements from other namespaces
trksegType	can be used to show continuation of a track from the previous track segments
copyrightType	holds ownership details of the GPX file
linkType	holds links to external resources such as an image or Web page URL
emailType	holds an email address, as two seperate attributes, id and domain
personType	a person or organisation
ptType	a geographical point, with optional elevation and time
ptsegType	an ordered sequence of points used for polygons or polylines
boundsType	two lat/long pairs defining an extent of an element
latitudeType	latitude of a point in decimal degrees on WGS84 datum
longitudeType	longitude of a point in decimal degrees on WGS84 datum
degreesType	holds a bearing, heading or course
fixType	holds a GPS fix type
dGPSStationType	represents a differential GSP station

These samples are taken directly from the GPX 1.1 Schema Definition and the notation [0..*] represents zero to many instances of these tags.

A gpxType Schema component Representation:

```
<?xml version="1.0" encoding="utf-8"?>
<xsd:schema
    xmlns:xsd="http://www.w3.org/2001/XMLSchema"
    xmlns="http://www.topografix.com/GPX/1/1"
    targetNamespace="http://www.topografix.com/GPX/1/1"
    elementFormDefault="qualified">
 <xsd:complexType name="gpxType">
  <xsd:sequence>
   <xsd:element name="metadata" type="metadataType"
                        minOccurs="0"/>
```

```
    <xsd:element name="wpt" type="wptType" minOccurs="0"
                    maxOccurs="unbounded"/>
    <xsd:element name="rte" type="rteType" minOccurs="0"
                    maxOccurs="unbounded"/>
    <xsd:element name="trk" type="trkType"
             minOccurs="0" maxOccurs="unbounded"/>
    <xsd:element name="extensions" type="extensionsType"
                        minOccurs="0"/>
  </xsd:sequence>
  <xsd:attribute name="version" type="xsd:string"
                    use="required" fixed="1.1"/>
  <xsd:attribute name="creator" type="xsd:string"
                        use="required"/>
 </xsd:complexType>
</xsd:schema>
```

A GPX document consists of the route element, some metadata followed by way-points, tracks and routes. GPX also has extension hooks and elements which can be reused in other schemas, such as the point example below.

The ptType XML example contains data from a trail walk around an abandoned mining site in Lilliput. It contains a single track segment (trkseg) with multiple track points (trkpt). Each track point (trkpt) contains the optional elements of (ele) and (time):

```
<gpx>
<...
<trk><name>Lilliput Native Walk</name>
<cmt>Mount Creek Mine</cmt>
<desc>
    A walking track around the abandoned Mount Creek
    gold mining site last active in the late 1500's.
</desc>
<lilliput:layer>Trailmap</lilliput:layer>
<lilliput:type>Simple Walking Track</lilliput:type>
<trkseg>
    <trkpt lat="13.636661" lon="-36.169123">
            <ele>1160.729980</ele>
            <time>1599-03-25T16:32:31Z</time>
    </trkpt>
    <trkpt lat="13.636663" lon="-36.169138">
            <ele>1161.691406</ele>
            <time>1599-03-25T16:32:32Z</time>
    </trkpt>

    <trkpt ... </trkpt>
```

```
<trkpt lat="13.630696" lon="-36.167094">
        <ele>1055.946533</ele>
        <time>1599-03-25T16:47:44Z</time>
</trkpt>
<trkpt lat="13.630697" lon="-36.167106">
        <ele>1055.465820</ele>
        <time>1599-03-25T16:48:01Z</time>
</trkpt>
</trkseg>
</trk>
</gpx>
```

6.2.2   GPX Application

As stated, GPX is an XML schema that can be used for the translation of GPS data from a device to an application. It stores location information in its tags, including elevation and time and is used to interchange data between GPS devices and software packages. Users can use GPX applications to view their location information, annotate maps with the outputs from GPS devices, tag photographs and locate photographs with the geolocation information in the metadata. There are a number of applications on the market, including websites that can read and render these GPX data directly. While this form of Online GIS is still in its infancy, standards like GPX provide a simple cabability to share data online.

A simple example of the use of GPX would be for the planning and recording of a camel ride adventure. GPX can be used to load topographic data from another GPX compliant software onto the GPS device for use during the ride. The tracks and stops being planned for the camel ride can be loaded into the device as vector features and these can be used during the ride to orient the camel team against other location features and can be used by the device owner to check their location relative to the planned tracks, and to record progress and metadata along the way. Finally, when the camel ride is over, users can export a full history of the trip, where they went and how long it took as GPX data, and share this data using the simple GPX Exchange Format used for interchange of GPS data.[3]

Another simple example may be to strap GPS devices to players on a football team and use the devices to track where and how much the players travel during a game. This application has been used successfully now for many seasons and GPX is a specification that can provide the data back to monitoring systems in real time.

---

[3] http://www.topografix.com/gpx.asp

6.3 KML (Keyhole Markup Language)

KML is an XML grammar and file format used to notate 2D and 3D spatial information for rendering over the Internet in geobrowsers and client applications. KML was developed by Keyhole Inc. (thus the K in KML), and its initial release was in the late 1990s. It was then purchased by Google in 2004 and optimised as a Web delivery standard. Version 2.2 was then handed to the OGC by Google and it became an offical OGC standard in 2008, making it a fully supported international open standard. Google has retained an active interest in its ongoing development at OGC with representation on the OGC KML 2.3 Standards Working Group (SWG).[4]

An important distinction of KML from other standards such as LandXML is that it is encoded using the Geography Markup Language (GML), another OGC supported open standard discussed in §6.4, later in this chapter. The use of GML means that KML is now a powerful interoperability standard for the geographic data used in many forms of online 2D and 3D GIS.

To optimise the delivery of KML files over the Internet, they are often distributed as zipped files. These compressed files use ZIP 2.0[5] and have an file extension of *.kmz. These files are uncompressed automatically at the client. KMZ file contents include a single root KML document and a combination of overlays, images, icons or COLLADA 3D models. The root KML document in the zip file is a file named `doc.kml`, referencing the remaining documents in the compressed KML.

6.3.1 KML Schema Details

KML shares some of the same structural grammar as GML 2.1.2,[6] although not all KML data can be viewed in some Online GIS as GML can. The KML file specifies a set of GIS features including points, line strings, linear rings, polygons, place markers, images, 3D models and textual descriptions for display in any geospatial software implementing the KML encoding. In addition, there are a number of other attributes in KML that are common to 3D gaming software, including altitude, tilt and heading, which together define a camera view along with a timestamp or timespan.

The Header elements for the KML schema which can be viewed online[7] are

```
<?xml version="1.0" encoding="UTF-8"?>
<schema
    xmlns="http://www.w3.org/2001/XMLSchema"
    xmlns:kml="http://www.opengis.net/kml/2.2"
    xmlns:atom="http://www.w3.org/2005/Atom"
    xmlns:xal="urn:oasis:names:tc:ciq:xsdschema:xAL:2.0"
```

---

[4] http://www.opengeospatial.org/standards/gml
[5] http://zip-2-0.winsite.com/
[6] http://www.opengeospatial.org/search/node/GML
[7] http://schemas.opengis.net/kml/2.2.0/ogckml22.xsd

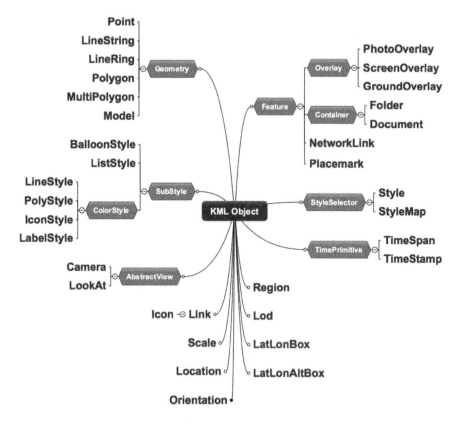

FIGURE 6.1: The complete element tree for KML 2.2.

```
targetNamespace="http://www.opengis.net/kml/2.2"
elementFormDefault="qualified" version="2.2.0">
  ...
</schema>
```

The hexagonally boxed elements in the tree shown in Figure 6.1 and Figure 6.2, such as Geometry, are analogous to the abstract elements in an Object-Oriented model (§1.2). They are not used in KML, but are a parent object of similar related objects. In the case of Geometry, its children elements (or objects) include `Point`, `Linestring`, `LinearRing`, `Polygon`. The full schema can be viewed online[8].

There are a lot of elements that make up the KML 2.2 schema, many more than with GPX and too many to list here. To give an example, however, one KML element that is commonly used is the `LookAt` element. This element is associated with other elements derived from the `abstractFeature` element. The purpose

---

[8] http://developers.google.com/kml/documentation/kmlreference

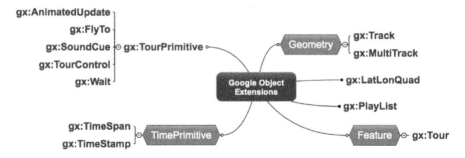

FIGURE 6.2: The Google extension elements in KML 2.2, all shown with the gx: namespace.

of the LookAt element is to position the 3D viewer in relation to the object being viewed. The geobrowser or client application understands the KML when it reads it, and based on the LookAt element attribute values for latitude and longitude coordinates, altitude and tilt, the viewer can orient the viewing position at that location when the user selects the object from the panel.

To do this, the LookAt element uses altitude, which refers to the distance above the earth's surface or the sea floor where gx:altitudeMode is used.

The Tilt element provides an angle that the LookAt position is from being vertically above the altitude point on the earth's surface, and range is the distance the viewing point (camera) is from this altitude point along the tilt line. A heading angle is finally used to rotate the camera point around the altitude point a height and distance fixed by tilt and range.

The gx characters shown in the gx:altitudeMode are the name of a namespace extension to the KML standard. The gx tags are associated here with the second namespace in the example below, but the prefix is, of course, arbitrary.

```
<kml xmlns="http://www.opengis.net/kml/2.2"
    xmlns:gx="http://www.google.com/kml/ext/2.2">
```

The LookAt element description from the KML schema file showing all of the possible attributes and abstract types is

```
<?xml version="1.0" encoding="UTF-8"?>
<schema xmlns="http://www.w3.org/2001/XMLSchema"
   xmlns:kml="http://www.opengis.net/kml/2.2"
   xmlns:atom="http://www.w3.org/2005/Atom"
   xmlns:xal="urn:oasis:names:tc:ciq:xsdschema:xAL:2.0"
   targetNamespace="http://www.opengis.net/kml/2.2"
   elementFormDefault="qualified"
   version="2.2.0">
```

```
<element name="LookAt" type="kml:LookAtType"
  substitutionGroup="kml:AbstractViewGroup"/>
<complexType name="LookAtType" final="#all">
  <complexContent>
    <extension base="kml:AbstractViewType">
      <sequence>
        <element ref="kml:longitude" minOccurs="0"/>
        <element ref="kml:latitude" minOccurs="0"/>
        <element ref="kml:altitude" minOccurs="0"/>
        <element ref="kml:heading" minOccurs="0"/>
        <element ref="kml:tilt" minOccurs="0"/>
        <element ref="kml:range" minOccurs="0"/>
        <element ref="kml:altitudeModeGroup"
                        minOccurs="0"/>
        <element ref="kml:LookAtSimpleExtensionGroup"
                minOccurs="0" maxOccurs="unbounded"/>
        <element ref="kml:LookAtObjectExtensionGroup"
                minOccurs="0" maxOccurs="unbounded"/>
      </sequence>
    </extension>
  </complexContent>
</complexType>
<element name="LookAtSimpleExtensionGroup"
              abstract="true" type="anySimpleType"/>
<element name="LookAtObjectExtensionGroup"
              abstract="true"
        substitutionGroup="kml:AbstractObjectGroup"/>
</schema>
```

Here, we have included KML LookAt element samples for three famous national parks from around the globe. This element specifies a virtual camera looking at a point defined by a latitude and longitude. If there is an additional Point element, Google Earth places an icon at that point. Range specifies the distance of the camera from the point in metres.

```
<?xml version="1.0" encoding="utf-8"?>
<kml xmlns="http://www.opengis.net/kml/2.2">
  <Placemark>
    <name>Yellowstone National Park</name>
    <description>First National Park in the world,
    famous for geothermal features</description>
    <LookAt>
      <longitude>44.6000</longitude>
      <latitude>20.5000</latitude>
      <tilt>0</tilt>
      <range>25000</range>
```

```
    </LookAt>
    <Point>
      <coordinates>
        44.6000,20.5000,-33.933300,0
      </coordinates>
    </Point>
  </Placemark>
</kml>
```

In the next example we include the `AltitudeMode`, here using `clampToGround`, which is actually the default. This identifies the point as on the ground surface regardless of its altitude.

```
<?xml version="1.0" encoding="utf-8"?>
<kml xmlns="http://www.opengis.net/kml/2.2">
  <Placemark>
    <name>Fjordland National Park</name>
    <!--    <visibility>0</visibility> -->
    <description>
      Fjordland is the largest National Park
                in New Zealand.
    </description>
    <LookAt>
      <longitude>-45.4167</longitude>
      <latitude>77.7167</latitude>
      <heading>90</heading>
      <tilt>20</tilt>
      <range>100</range>
    </LookAt>
    <Point>
      <altitudeMode>clampToGround</altitudeMode>
      <coordinates>-45.4167,77.7167,0</coordinates>
    </Point>
  </Placemark>
</kml>
```

In the last example, we change the `AltitudeMode` to `absolute`, where the actual elevation is used. KML uses 3D geographic coordinates in its reference system. These are longitude, latitude and altitude values in that order, representing x, y and z. It uses negative values for west of the international dateline, south of the equator and below mean sea level. As such, the values for latitude and longitude are N = +ve latitude; S = -ve latitude; E = +ve longitude; W = -ve longitude.

As with GPX, all coordinates are held using the World Geodetic System of 1984 (WGS84) datum for latitude and longitude, while WGS84 EGM96 `Geoid` is used as the vertical datum. It is possible to still display data with only x, y values. With these data, KML will assign an altitude (height) value of 0 to each coordinate pair, placing the data at sea level.

KML, like many other XML standards, allows extensions to the standard elements. Version 5.0 of Google Earth has a number of extensions that can be used when the `http://www.google.com/kml/ext/2.2` namespace is added to the KML element, where here we have assigned the (arbitrary) prefix `gx`:

```
xmlns:gx="http://www.google.com/kml/ext/2.2"
```

These elements are all prefixed with `gx` and are described in detail on Google development websites.[9] It's worth noting that not all geobrowsers and clients may be able to handle these KML extension elements.

The official KML schema can be freely downloaded from the OGC website[10] and the complete specification for KML.[11] There is also a developer's guide available from Google which can be found online.[12]

6.3.2 KML Application

One of the popular Online GIS viewers using KML is the Google Earth[13] application, a viewing globe created by Google. This application is a free download and runs on a number of platforms, including PC, Mac and mobile devices. It supports the rendering of vector and image (raster) data, and uses a worldwide terrain dataset to represent the earth's surface, providing an online 3D representation of the world.

State governments in Australia are implementing a number of Online GIS Globes using Google technologies and KML. The Queensland (QLD) globe was released in 2013 as part of the Governments Business and Industry Portal,[14] providing easy access to a wide range of themed globes covering such data topics as vegetation management, coal seam gas, mining and land valuations. Datasets for the QLD Globe are provided through the Open Data Strategy of Queensland Government and the application provided significant value to Queensland communities during natural disasters, such as cyclones and flooding.

To use the QLD Globe, as with other Globes, the user simply downloads the Google Earth application and installs it. Once installed, the KML file format is associated with Google Earth in the same way a Word document is associated with Microsoft Word, and whenever a KML file is opened, the device uses Google Earth as the default application to render the file. The KML or KMZ (compressed KML) file provides a URL link to the Google Earth Engine (GEE) or Google Maps Engine (GME) server where the data and services are hosted. A simple KML file for the QLD Globe vegetation management theme looks like

---

[9] http://developers.google.com/kml/schema/kml22gx.xsd
[10] http://schemas.opengis.net/kml/
[11] http://www.opengeospatial.org/standards/kml/
[12] http://developers.google.com/kml/documentation/kmlreference
[13] http://www.google.earth.com
[14] https://www.business.qld.gov.au/business/support-tools-grants/services/mapping-data-imagery/queensland-globe

```
<?xml version='1.0' encoding='UTF-8'?>
<kml xmlns="http://www.opengis.net/kml/2.2"
     xmlns:gx="http://www.google.com/kml/ext/2.2">
  <gx:GoogleMapsEngineLink>
    <href>http://globe.information.qld.gov.au/
                  vegmanagement</href>
  </gx:GoogleMapsEngineLink>
</kml>
```

The first line in this KML code sample:

```
<?xml version='1.0' encoding='UTF-8'?>
```

defines the XML version number and the encoding (character set) as UTF-8.

The next element is the `<kml ...` tag that defines two namespace variables, `xmlns` and `xmlns:gx`, and sets them to the KML version 2.2 and KML extention 2.2. Elements defined in the second namespace `xmlns:gx` can then be referenced by using the `gx` characters.

The next three lines set the `<gx:GoogleMapsEngineLink>`, and Google Earth extension to KML, to point to the QLD globe server and the `vegmanagement` KML service published at

```
<href>http://globe.information.qld.gov.au
                  /vegmanagement</href>
```

This KML example is all that is needed to start viewing the vegetation globe from QLD government. All additional KML needed (vector data, imagery, table of contents, query and search tools, etc.) are then sourced from the KML service, online.

Alternatively, a user can open up the Google Earth application and import the KML file from their file system, linking Google Earth to the GME server, reading the map extents, zooming the globe viewer to the location of the data and then rendering it. Geospatial data can also be rendered into Google Earth using *flat files*, a proprietory Google format optimised for performance. However, KML provides an easily customisable and human readable form of data which can be downloaded, rendered and used in querying.

Another example KML file below contains a bounding box to represent the location of an event:

```
<?xml version="1.0" encoding="UTF-8"?>
<kml xmlns="http://www.opengis.net/kml/2.2" >
  <Document>
    <name>Anchoring Location of Gullivers ship
                  near Lilliput</name>
    <Placemark>
      <styleUrl>#purpleInRedPoly</styleUrl>
      <MultiGeometry>
```

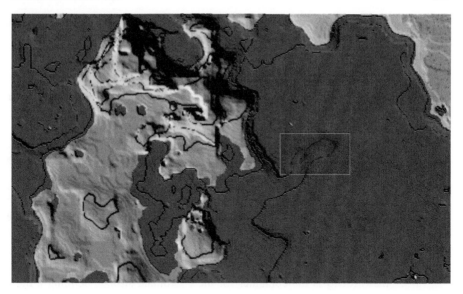

FIGURE 6.3: The location where Gulliver's ship, the Antelope, was anchored just off the island of Lilliput to take shelter from a storm; before their rowboat heading to shore capsized and all but Gulliver were drowned.

```
<Polygon>
  <outerBoundaryIs>
    <LinearRing>
      <coordinates>
145.10016,-42.17326,0
145.10069,-42.17326,0
145.10069,-42.17310,0
145.10016,-42.17310,0
145.10016,-42.17326,0
      </coordinates>
    </LinearRing>
  </outerBoundaryIs>
</Polygon>
      </MultiGeometry>
    </Placemark>
  </Document>
</kml>
```

This KML data shows the location that Gulliver's ship, the Antelope, was anchored as it took shelter from a storm off the coast of Lilliput. The data is a four sided polygon in geographic coordinates, longitude, latitude and height (in this case, a height of 0) and is shown in Figure 6.3.

This KML `document` contains a `name` element and a `Placemark` element.

The name element's contents are self-explanatory. The Placemark element consists of a number of other elements. The first element:

```
<styleUrl>#purpleInRedPoly</styleUrl>
```

sets the symbology of the object defined by the `MultiGeometry` element, in this case, a red ouline with purple infill. There are a number of predefined symbology files avaible for use with the styleUrl element. The roles of the other elements are

- `MultiGeometry` contains the definition of a multipart feature, which can be made up of many spatial geometries.

- `Polygon` contains the polygon data itself. It can contain a number of different elements, including inner rings (holes) and outer rings (boundaries). In this case, the polygon consists of just one feature, described by three elements:

  - `outerBoundaryIs` defines an outer ring or boundary of the feature.

  - `LinearRing` is the container for the coordinate strings, defining a linear feature that loops back to its starting point.

  - `coordinates` contains the list of the coordinates. Here, the coordinates represent a rectangular polgyon; however, the coordinate element could also be a point feature (one coordinate pair) or a linear feature. Five coordinate points are required for this polygon as the polygon loops in a clockwise direction, starting at the top right.

The other element tags (prefaced with the / character) are the end tags for each KML element.

The resulting polygon, shown in Figure 6.3, is located off the west coast of Van Diemans Land (now called Tasmania) in the southern seas around Australia, as indicated by the negative (-ve) latitudes of around -42 degrees. It is also located at sea level, with a height coordinate of 0 for all coordinates.

There are a couple of considerations for organisations implementing an enterprise application with Google Earth. Google Earth requires a client-side installation of the Google Earth application. Also, Google has recently announced the deprication of their GEE and GME products in 2016. These products host and publish data and services. At the time of writing, it was unclear what server options there will be for publication of data; however, there appears to be a strategic alliance between Esri and Google which may provide a migration path for existing customers of GEE and GME.

6.4   GML: Geographic Markup Language

GML is an XML schema set, which can be used to describe geographical features, introduced and copyrighted by the Open Geospatial Consortium (OGC)[15] current ISO standard (19136:2007).[16] The Geographical Markup Language has been an important initiative of the OGC for Online GIS. One of its key capabilities was a standard that fully supported GIS vector objects for distribution over the Web. Some of this focus on GIS vector objects should be attributed to Dr Carl Reed, who was involved in the development of DeltaMap, one early GIS systems and who has been the CIO of OGC up until his retirement this year (2015). DeltaMap, which later became known as GenaMap, used a powerful topology-based data structure of points, lines and polygons to define its GIS vector objects. The strengths of this early GIS is in many ways reflected in GML.

GML is both a *modelling language* and an *open interchange format* for geographic data. Unlike the other schemas we have discussed, it ships with an RDF (§8.5) file, `gml_32_geometries.rdf`, as well as the schema documents, all combined in a single zip file.[17]

GML supports complex data types consistent with GIS data structures and geographical features. Another key advantage of GML is its usefulness to the GIS industry in that it supports all forms of geographic information, not only vector GIS data, but also cell data such as imagery and data from digital sensors such as LiDAR.[18] These additional data types are important components of current and future Online GIS systems.

GML support for *Profiles* simplifies the use of GML. A profile is a limited set of GML specific to an application domain. Anyone can create a GML profile to suit their application needs, and, provided it is a true profile (sub-set of GML), the profile data can be mapped to any other GML-based information system and used to integrate geographic information between these systems. GML itself defines a number of profiles, including Point, GML Simple Features, GML2JP2 (where JP2 is the raster image format JPEG2000 [19]) and GeoRSS.[20]

Users can create their own GML Application Schema (§5.5.1). They can use either Profiles (§5.5.1) or the full GML schema. This use of GML is similar to the inheritance and polymorphism of OO technologies (§1.2) where communities can substitute terms such as roads, tracks, expressways instead of using generic GML terms such as lines. Such derived schemas give users the flexibility to apply GML to many situations and systems with local definitions and jargon, while still providing guarantees that data can be sucessfully integrated across systems that are GML

---

[15] http://www.opengeospatial.org/legal/

[16] http://www.iso.org/iso/iso_catalogue/catalogue_tc/catalogue_detail.htm?csnumber=32554

[17] http://schemas.opengis.net/gml/gml-3_2_1.zip

[18] http://oceanservice.noaa.gov/facts/lidar.html

[19] http://www.jpeg.org/jpeg2000/

[20] http://www.georss.org/

compliant. Examples of Application Schema include CityGML,[21] GeoSciML,[22] and INSPIRE.[23]

To see how this works, the following example shows the header for the building schema within CityGML. It provides the building namespace (§5.4), but it inherits the GML and CityGML namespaces, all defined as attributes in the top level, schema element. The GML namespace is prefixed by gml and the CityGML space by core, while xs refers to the XML schema namespace itself. It then imports the schema definitons from GML and CityGML. The complex type, BuildingInstallationType, extends AbstractCityObjectType of CityGML (core prefix), using the xs:extension with the base attribute to denote the type being extended.

```
<?xml version="1.0" encoding="UTF-8"?>
<!-- CityGML Version No. 2.0, February 2012 -->
<!-- CityGML - GML 3.1.1 application schema for 3D city
models -->
<!-- International encoding standard of the Open
Geospatial Consortium   see
http://www.opengeospatial.org/standards/citygml
-->
<!-- Jointly developed by the Special Interest Group 3D
(SIG 3D)  of GDI-DE, see http://www.sig3d.org -->
<!-- For further information see:
http://www.citygml.org -->
<xs:schema
    xmlns="http://www.opengis.net/citygml/building/2.0"
    xmlns:core="http://www.opengis.net/citygml/2.0"
    xmlns:xs="http://www.w3.org/2001/XMLSchema"
    xmlns:gml="http://www.opengis.net/gml"
    targetNamespace=
    "http://www.opengis.net/citygml/building/2.0"
    elementFormDefault="qualified"
    attributeFormDefault="unqualified" version="2.0.0">
  <xs:annotation>
    <xs:documentation>
      CityGML is an OGC Standard.
      Copyright (c) 2012 Open Geospatial Consortium.
      To obtain additional rights of use,
      visit http://www.opengeospatial.org/legal/ .
    </xs:documentation>
  </xs:annotation>
  <xs:import namespace="http://www.opengis.net/gml"
          schemaLocation=
          "http://schemas.opengis.net/gml/3.1.1/
                              base/gml.xsd"/>
```

---

[21] http://www.opengeospatial.org/standards/citygml
[22] http://www.ogcnetwork.net/geosciml
[23] http://inspire.ec.europa.eu/index.cfm/pageid/2/list/datamodels

```
<xs:import
    namespace="http://www.opengis.net/citygml/2.0"
    schemaLocation=
    "http://schemas.opengis.net/citygml/2.0/
                                cityGMLBase.xsd"/>

<!-- Definition of complex and simple types -->
<xs:complexType name="BuildingInstallationType">
  <xs:annotation>
    <xs:documentation>A BuildingInstallation is a
    part of a Building which has not the significance
    of a BuildingPart. Examples   are stairs,
    antennas, balconies....
    </xs:documentation>
  </xs:annotation>
  <xs:complexContent>
    <xs:extension base="core:AbstractCityObjectType">
      <xs:sequence>
        <!-- sequence elements -->
        <xs:element name="usage" type="gml:CodeType"
                    minOccurs="0" maxOccurs="unbounded"/>
        <!-- more sequence elements -->
        <xs:element
            name="lod2ImplicitRepresentation"
            type="core:ImplicitRepresentationPropertyType"
            minOccurs="0"/>
        <!-- more sequence elements -->
      </xs:sequence>
    </xs:extension>
  </xs:complexContent>
</xs:complexType>
<!-- more schema content -->
</xs:schema>
```

6.4.1 GML Schema Overview

The GML schema framework is somewhat larger than the other schemas in this chapter. At the time of writing, GML 3.2.1 contains 29 separate schema documents, which cross import each other. There is no one root element. Instead the schemas work as a set of building blocks for profiles and Application Schema.

As discussed, GML uses a vector model to encode the shape of a `geometry`. The key GML geometry types include `Point`, `Linestring` and `Polygon`, consistent with GIS systems. GML 3.0 extended these to include *coverages* and *raster* models. GML also defines a *feature* to represent an instance of a physical object. These features do not always have geometric, or location, attribution whereas a geometry instance must define location or region attribution to be valid in the vector model.

Unlike GPX, the GML schema is extensive in terms of support for geographic information and GIS. For example, sub-schemas describe the mathmatical operations to support topology, the properties of objects that are invariant under continuous deformation. In GML, `gml:AbstractTopoPrimitive` acts as the base type for all topological primitives which can be extended to query the relationships between objects using simple algorithms. GML also defines `Temporal` schemas to describe the temporal characteristics of geographic information. Moreover, GML describes a spatio-temporal model for both feature and attribute time stamping. As users become more sophisticated in what they expect to be able to access through Web services, these qualities of geographic information can be described using GML and can be integrated with any system which is GML compliant.

The GML schema is split across a number of seperate schema files to describe the GML elements and how they can be used. These files include a number of geometry types including zero dimension, 1 dimension and 2 dimension as well as complex types and primitive types. There are a number of other elements in the standard to cover datums, coordinates, coordinate operations, measures and observations, as well as formal definitions for dictionary, direction, grids, value objects, coverages, style and data quality.

GML is too extensive to document in full here. However, a simple sample of GML is the use of the Point element to hold the instance of a point feature. An XML instance will look like

```
<gml:Point gml:id="p21" srsName="http://www.opengis.net
                        /def/crs/EPSG/0/3308">
    <gml:coordinates>151.3445,-32.4455</gml:coordinates>
</gml:Point>
```

In this example, the GML point feature has an ID allocated to the object, stating that the coordinate system is the one defined by EPSG:3308 (Lamberts Project with 2 standard parallels)[24] and point is located at longitude 151.3445 and latitude 32.4455.

GML does not have a default coordinate system as do KML and other XMLs supporting GIS information. This means that all data in GML must have a coordinate sytem explicitly defined. The GML standard enforces this coordinate system to be defined using a Coordinate Reference System bound in the EPSG Geodetic Parameter Registry.[25] For the OGC Web Feature Service discussed (§10.8.2), this standard serves up GML as `gml:coordinates` elements. This holds true for GML 1.0 and GML 2.0; however, for GML 3.0 documents `<gml:pos>` and `<gml:posList>` are the preferred element types to use to hold coordinates.

Features in GML can share geometries using a *remote property reference* which formalised the spatial relationships between objects in the real world where these relationships can be defined through the sharing of location, boundary shapes and dimensions.

---

[24] http://www.opengis.net/def/crs/EPSG/0/3308
[25] http://www.epsg-registry.org.

6.4.2 GML Example

This example, which is somewhat more complex than many in the book, illustrates several general aspects of schema processing as well as the way GML is used. Firstly, it uses both a Profile and an Application Schema: the Point Profile,[26] and an Application Schema written for this example makes use of several of the GML schemas and is included as an appendix to this chapter.

The main river scene example is a stub file, which includes all the features as separate files. The XML processor needs to recognise the **XInclude** schema.[27] Sometimes it will be necessary to specifically request this. For example, the utility **xmllint**, which comes native on Mac OS X, will validate the picnic scene with the following command:

```
xmllint --xinclude river-build.xml -schema picnic.xsd
```

There is one aspect of the namespace use, which can be very confusing to a beginner. The namespace prefix introduced by `xmlns:` is *just a name, even though it looks like a URI/URL*. The URL is **not** looked up by processors such as *xmllint*. Furthermore, the prefix is entirely arbitrary, which we have emphasised by using animal names, which have no obvious connection to the schema names or instances. The *only thing which matters* is that the `targetnamespace` name has to match the `xmlns` name in the schema or instance using it. So, we have in *tgml.xsd* (§6.8)

```
targetNamespace="http://www.opengis.net/giraffe"
```

where this URL certainly does not exist. In *picnic.xsd* (§6.7) we import this namespace and then use it as

```
xmlns:dingo="http://www.opengis.net/giraffe"
```

where the animals have switched continents!

A full GML specification consists of a set of features in which we have two components:

1. A *spatial reference system*, which uses the *CoordinatedReferenceSystem* and related schemas.

2. A *collection of features*, each having spatial and non-spatial elements.

We will start the discussion by building the `geometry elements`, then bind them into features and finally a `feature collection`. The result of these XML elements will be a set of objects that can be distributed over the Web and used by GML aware systems for rendering and querying.

Referring to the map in Figure 6.4, we start with an elementary construct, the point:

---

[26] http://www.ogcnetwork.net/gml-point
[27] //http://www.w3.org/TR/xinclude/

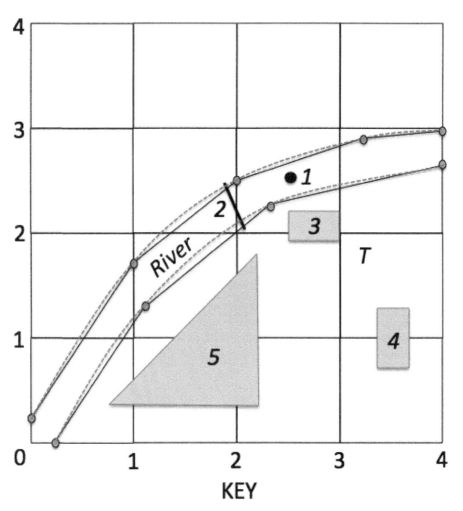

FIGURE 6.4: The river map defined using GML in the text

```
<?xml version="1.0" encoding="utf-8"?>
<!-- <Geometry xmlns:pig="http://www.opengis.net/gml" id
    ="location" srsName="swift1701"> -->
 <goat:picnic xmlns:goat="http://www.opengis.net/gorilla
    " xmlns:pig="http://www.opengis.net/junk">
 <pig:Point xmlns:pig="http://www.opengis.net/junk">
  <pig:pos>3.2 1.8 </pig:pos>
 </pig:Point>
<!-- </Geometry> -->
</goat:picnic>
```

The point is itself made up of the fundamental element, the coordinate list, posList, in this case having just one tuple. The coordinates are real-world co-ordinates, with a reference described in the spatial reference system, swift1701.

We could make this into a feature by wrapping it in a feature element:

```
<?xml version="1.0" encoding="utf-8"?>
<!-- <gml xmlns:gml="http://www.opengis.net/gml/3.2">
    -->
<gml:Feature featureType="telephone" fid="1" name="
    Riverside Phone">
 <Description> The phone by the river</Description>
 <Property name="number" type="integer"> 633901013 </
    Property>
 <Geometry id="location" srsName="swift1701">
  <Point>
   <pos>3.2, 1.8 </pos>
  </Point>
 </Geometry>
</gml:Feature>
<!-- </gml> -->
```

The feature has added a description of the feature and one of an arbitrary number of properties, in this case the number of the phone. The *fid* is simply a unique identifier. We have one more point element on the map, the flag in the river. Here it is as a feature:

```
<Feature featureType="flag" fid="2" name="flag02">
  <Description>Shallow water flag</Description>
  <Property name="flagCode" type="string">W</Property>
  <Geometry name="location" srsName="swift1701">
    <Point>
      <pos>2.5, 2.5 </pos>
    </Point>
  </Geometry>
</Feature>
```

Moving up now in complexity, we come to the rope bridge, which at the scale of the map, we just represent as a line. There is no single line construct, just a string of line segments, as a coordinate list, separated by white space:

```
<Feature featureType="structure" fid="8" name="Gulliver
    bridge">
 <Description>Rope bridge across the river</Description>
 <Geometry name="centreline" srsName="swift1701">
  <LineString>
   <posList>2.1, 2.1 1.9 2.4</posList>
  </LineString>
 </Geometry>
</Feature>
```

We keep repeating the srsName; it could in fact be different for each geometry. When we come to the river, we have each bank encoded as a line string, both banks together forming a geometry collection.

```
<Feature featureType="flag" fid="3" name="River Lilli">
  <Description>Delineates the river Lilli</Description>
  <Feature featureType="riverBank" fid="4" name="North
      Bank">
    <Description>North bank of the river</Description>
    <Geometry name="boundary" srsName="swift1701">
      <LineString>
        <posList> 0.0, 0.2 1.0, 1.7 2.0, 2.4 3.2, 2.8 4.0,
            3.0 </posList>
      </LineString>
    </Geometry>
  </Feature>
  <Feature featureType="riverBank" fid="5" name="South
      Bank">
    <Description> South bank of the river</Description>
    <Geometry name="boundary" srsName="swift1701">
      <LineString>
        <CList> 0.2, 0.0 1.1, 1.3 2.3, 2.2 4.0, 2.7 </
            CList>
      </LineString>
    </Geometry>
  </Feature>
</Feature>
```

We now come to the three features of the map, which all occupy areas and we represent as polygons. A polygon is simply a closed list of points, where each (x,y) coordinate pair follows one after the other without separators.

```
<Feature featureType="carpark" fid="6" name="Car Park
    1">
  <Description> Car Park 1 of the recreational area </
      Description>
  <Geometry name="extent" srsName="swift1701">
    <Polygon>
      <posList>0.7, 0.3 2.3, 1.8 2.3, 0.3</posList>
    </Polygon>
  </Geometry>
</Feature>
```

The buildings are covered in a similar manner:

```
<Feature featureType="building" fid="7" name="Park
    buildings">
  <Description> Buildings owned by the Park authority </
      Description>
  <Feature featureType="building" fid="4" name="Boat
      Shed">
    <Description>Boat shed for canoes </Description>
    <Geometry name="extent" srsName="swift1701">
      <Polygon>
        <posList> 2.5, 2.0 3.0, 2.0 3.0, 2.2 2.5, 2.2 </
            posList>
      </Polygon>
    </Geometry>
  </Feature>
  <Feature featureType="building" fid="4" name="Kiosk">
    <Description>Refreshment kiosk</Description>
    <Geometry name="extent" srsName="swift1701">
      <Polygon>
        <posList> 3.4, 0.7 3.7, 0.7 3.7, 1.2 3.4, 1.2 </
            posList>
      </Polygon>
    </Geometry>
  </Feature>
</Feature>
```

The final XML document, including files for each of the elements we have discussed, follows.

```
<?xml version="1.0" encoding="utf-8"?>
 <platypus:picnic xmlns:platypus="http://www.opengis.net
    /gorilla"
            xmlns:roo="http://www.opengis.net/junk"
            xmlns:possum ="http://www.opengis.net/
                giraffe"
```

```
xmlns:xi="http://www.w3.org/2001/XInclude">
 <roo:Point xmlns:roo="http://www.opengis.net/junk">
  <roo:pos>3.2 1.8 </roo:pos>
 </roo:Point>
 <xi:include href="bounding-box.xml"/>
<xi:include href="telephone-feature.xml"/>
<xi:include href="bridge-feature.xml"/>
<xi:include href="flag-feature.xml"/>
<xi:include href="northbank-feature.xml"/>
<xi:include href="southbank-feature.xml"/>
<xi:include href="carpark-feature.xml"/>
<xi:include href="building-feature.xml"/>
</platypus:picnic>
```

6.5   LandXML

LandXML is an XML schema[28] that describes a data structure that can be used for the exchange of data between land development organisations, civil engineering and surveying communities. LandXML is based on the following XML standards: XML 1.1, XML Schema 1.1, XML Namespaces 1.1 and xpath 2.0 and has its own namespace.[29] It was originally proposed as an XML formatted design standard that grew from the proposed AASHTO E-ASE ASCII file format development[30] in the late 1990s. With the digital nature of instrumentation used across these sectors and the availability of digital outputs needed to be shared between systems, there is a fast growing need for these systems to be interoperable. LandXML provides this interoperability, albeit in a loosly defined and flexible way.

There is a significant global community supporting LandXML through a volunteer organisation called LandXML.org. This organisation reports that it has 664 organisations across 41 countries and that many of these organisations are actively using the LandXML schema and documents. At the time of writing, 70 registered software products support LandXML, and many organisations around the world rely on LandXML as a data exchange format for mission critical activities.

Unlike GML, LandXML is a non-proprietary data standard. It is free to use and there are no direct costs to join LandXML.org, with an industry consortium of partners driving the development and ongoing support of LandXML. It has three primary goals,[31] as published on the LandXML.org Website.

---

[28] http://www.landxml.org/schema/LandXML-1.2/LandXML-1.2.xsd
[29] http://www.landxml.org/schema/LandXML-1.2
[30] http://www.landxml.org/landxml-gml.aspx
[31] http://www.landxml.org/landxml-gml.aspx

- To transfer civil engineering / survey design data between producers and consumers;

- To provide a data format suitable for long-term data archives;

- To provide a standard format for official electronic design submission.

LandXML was first released in July 2000 as version 0.88 and its current release is the LandXML-1.2 schema which was ratified on August 15, 2008. Work is currently underway on a LandXML 2.0 working draft which expands the previous use of LandXML to include the LandXML Project Delivery File Specification, to include the entire project life-cycle, including electronic signatures, data file versioning and portability between office and field. There are also plans for LandXML 2.0 to provide suitable grammar for support of Point Cloud and LiDAR data formats, support for cross-sections as used in road design, additions to `As-built` data such as human constructed objects like buildings, walls, fences, etc. Schema details of LandXML are covered briefly in the next subsection.

6.5.1   LandXML Schema Details

LandXML 1.2 has many elements defined in its schema, as well as nearly 300 complex types and over 150 simple types. The first element in LandXML is the LandXML root element. All other elements are contained inside this element.

These other elements can belong to any of the following categories:

Overall Schema General Features, Attributes, Metadata, Inheritance.

Geometrics Points, Lines, Curves, Spirals, Chains.

Survey Raw observations, GPS, Staking.

Surfaces TIN, DEM, Surface Source Data.

Road and Parcels Roadway and parcel related data structures.

Spatial Reference Systems Coodinate Systems.

Hydrology and Hydrolics Storm water, Sewer, Supply Water.

To demonstrate how a LandXML schema is implemented, let's go through some of the element types that make up a typical LandXML document.

The first two elements of a LandXML document are the XML `prolog` element, followed by the `LandXML` root element. All other LandXML document elements must be contained inside this LandXML root element:

```
<?xml version="1.0" encoding="UTF-8"?>
<LandXML version="1.0" date="2015-02-12" time="11:44:26"
 xmlns="http://www.landxml.org/schema/LandXML-1.2"
 xmlns:xsi="http://www.w3.org/2001/XMLSchema-instance"
```

```
xsi:schemaLocation="http://www.landxml.org/schema/
                                    LandXML-1.2
http://www.landxml.org/schema/LandXML-1.2/
                                    LandXML-1.2.xsd">
```

The attributes of this element include mandatory elements such as version, date and time, and also include namespaces and schema location information (including the location of the xsd file).

LandXML also contains measurement units of the LandXML data in the `Unit` element. This element looks like

```
<Units>
  <Metric linearUnit="meter" temperatureUnit="celsius"
          volumeUnit="cubicMeter" areaUnit="squareMeter"
          pressureUnit="milliBars"
          angularUnit="decimal dd.mm.ss"
          directionUnit="decimal dd.mm.ss" />
</Units>
```

Coordinate system information is defined in the `CoordinateSystem` element. In this example, the coordinate system is a made up local coordinate system for the island of Lilliput.

```
<CoordinateSystem datum="LLMD"
        horizontalDatum="Local" verticalDatum="LLHD" />
```

The `FeatureDictionary` element lists the enumerations and reference data versions used in the LandXML document. This includes, in this example, a link to the reference data xml file. In this instance, the `DocFileRef` element contains the name, location and attributes of the files that make up the feature dictionary:

```
<FeatureDictionary name="ReferenceDataContext"
                        version="LL-101" >
  <DocFileRef name="ll-gov-enumerated-types.xsd"
   location="http://www.lilliputmappingauthority.gov.ll/
                        xml_file/0010/13768/
                        xml-gov-ll-refdata-101.xml"/>
</FeatureDictionary>
```

The `CgPoints` element is a container for all of the CgPoint elements which contain point feature attributes and coordinates for each point.

```
<CgPoints zoneNumber="56" >
  <CgPoint state="proposed"
          pntSurv="boundary"
          name="1">145.10016 -42.17326</CgPoint>
  ... more points ...
</CgPoints>
```

The LandXML `parcel` element contains a definition of each land parcel, comprised of links back to the CgPoint elements in the LandXML file. These parcel elements also contain attributes of each parcel, as shown in the following partial parcel example:

```
<Parcels>
  <Parcel name="1" class="Lot" state="proposed"
          parcelType="Single" parcelFormat="Standard"
                            area="321.4">
    <Center pntRef="222"/>
    <CoordGeom name="647564">
      <Line>
        <Start pntRef="1"/>
        <End pntRef="2"/>
      </Line>
      ... more points ...
    </CoordGeom>
  </Parcel>
</Parcels>
```

There are a number of other elements common to many LandXML implementations, including information about the survey itself. Monuments are also included in LandXML to capture related survey information such as control points and reference points, and these are always linked to CgPoint elements in the LandXML using the pntRef attribute (such as pntRef="4") in the following LandXML data:

```
<Survey>
  <SurveyHeader name="12345" jurisdiction="Newfoundland"
        desc="Plan 324455" surveyorReference="1234-5678"
              surveyFormat="Standard" type="surveyed">
    <AdministrativeDate adminDateType="Date Of Survey"
                              adminDate="2015-02-05" />
    ... additional metadata about the survey ...
    <AdministrativeArea adminAreaType="Survey Region"
                              adminAreaName="Urban"/>
  </SurveyHeader>
  ... instrument setup and observationGroup ...
</Survey>
<Monuments>
  <Monument name="1" pntRef="4" type="Tree Stump"
                state="Found" originSurvey="DP1441"/>
</Monuments>
```

The LandXML file is then ended with the following element:

```
</LandXML>
```

The LandXML file for the Lilliput survey data (excluding instrument setup, observation groups, reduced observations, etc., which can for part of a complete LandXML document) is listed below.

```xml
<?xml version="1.0" encoding="UTF-8"?>
<LandXML version="1.0" date="2015-02-12" time="11:44:26"
   xmlns="http://www.landxml.org/schema/LandXML-1.2"
   xmlns:xsi="http://www.w3.org/2001/XMLSchema-instance"
   xsi:schemaLocation="http://www.landxml.org/schema/
                                       LandXML-1.2
                      http://www.landxml.org/schema/
                            LandXML-1.2/LandXML-1.2.xsd">
  <Units>
    <Metric linearUnit="meter" temperatureUnit="celsius"
          volumeUnit="cubicMeter"
          areaUnit="squareMeter"
          pressureUnit="milliBars"
          angularUnit="decimal dd.mm.ss"
          directionUnit="decimal dd.mm.ss" />
  </Units>
  <CoordinateSystem datum="LLMD" horizontalDatum="Local"
          verticalDatum="LLHD" />
  <FeatureDictionary name="ReferenceDataContext"
          version="LL-101" >
    <DocFileRef name="ll-gov-enumerated-types.xsd"
location="http://www.lilliputmappingauthority.gov.ll/
      xml_file/0010/13768/xml-gov-ll-refdata-101.xml"/>
  </FeatureDictionary>
  <CgPoints zoneNumber="56" >
    <CgPoint state="proposed"
     pntSurv="boundary"
     name="1">145.10016 -42.17326</CgPoint>
    <CgPoint state="proposed"
     pntSurv="boundary"
     name="2">145.10069 -42.17326</CgPoint>
    <CgPoint state="proposed"
     pntSurv="boundary"
     name="3">145.10069 -42.17310</CgPoint>
    <CgPoint state="proposed"
     pntSurv="boundary"
     name="4">145.10016 -42.17310</CgPoint>
    <CgPoint state="proposed"
     pntSurv="sideshot"
     name="222">145.10034 -42.17318</CgPoint>
    <CgPoint state="proposed"
     pntSurv="reference"
```

```
      name="6">145.10138 -42.17307</CgPoint>
  </CgPoints>
  <Parcels>
    <Parcel name="1" class="Lot" state="proposed"
            parcelType="Single"
            parcelFormat="Standard"
            area="321.4">
      <Center pntRef="222"/>
      <CoordGeom name="647564">
<Line>
  <Start pntRef="1"/>
  <End pntRef="2"/>
</Line>
<Line>
  <Start pntRef="2"/>
  <End pntRef="3"/>
</Line>
<Line>
  <Start pntRef="3"/>
  <End pntRef="4"/>
</Line>
<Line>
  <Start pntRef="4"/>
  <End pntRef="1"/>
</Line>
      </CoordGeom>
    </Parcel>
  </Parcels>
  <Survey>
    <SurveyHeader name="12345"
                  jurisdiction="Newfoundland"
                  desc="Plan 324455"
                  surveyorReference="1234-5678"
                  surveyFormat="Standard"
                  type="surveyed">
<AdministrativeDate adminDateType="Date Of Survey"
                  adminDate="2015-02-05"/>
    ... additional metadata about the survey ...
    ... including purpose, AdminArea ...
      <AdministrativeArea adminAreaType="Survey Region"
                          adminAreaName="Urban"/>
    </SurveyHeader>
    ... instrument setup and observationGroup ...
  </Survey>
  <Monuments>
```

```
    <Monument name="1" pntRef="4" type="Tree Stump"
            state="Found" originSurvey="DP1441"/>
  </Monuments>
</LandXML>
```

It's worth noting that the CgPoint is one of the most common elements found in a LandXML document. This element can be used for any spatial point data, including setting the extents of the survey and locating other geospatial objects.

There are estimates that tens of thousands LandXML files are being used every year to design, survey and build the world's infrastructure. There are many private and government processes that rely on a stable LandXML standard as do many software vendors working in these industries, each with substantial investments made.

6.5.2   LandXML Application

The Intergovernmental Committee of Surveying and Mapping (ICSM)[32] in Australia and New Zealand ratified a LandXML schema as the national standard for lodgement of digital plans of survey. These LandXML documents contain the bearing and distance information derived from survey observations as well as additional land parcel information, administration metadata and other information required to define and support land tenure (ownership) in Australia. The national-level document from ICSM that defines LandXML is titled "ePlan Protocol LandXML Structured Requirements" and is available from their website. [33]

The Land and Property Information (LPI),[34] a division of the Department of Finance, Services and Innovation, is the land registration authority in New South Wales (NSW), Australia. As such, the LPI is responsible for the creation and management of land title in NSW and it underwrites land transactions as part of the NSW government. Any streamlining of registration processes will has significant benefits for NSW, thus its keen interest in the ongoing development of LandXML.

ICSM and the NSW LPI are active partners in the LandXML.org standards development, including LandXML 1.2. In particular, NSW LPI have developed a localised implementation of LandXML 1.2. known as the *NSW Land XML Recipe*. This recipe is a *profile* of LandXML that extends LandXML 1.2 to be specific to the requirements of NSW land registration. The element tree showing all LandXML elements defined in the NSW Land XML Recipe is available online.[35] This NSW profile of LandXML contains elements for all relevant survey information used in the Deposited Plan (DP) to establish *legal definition* of a registered title. The XML schema contains survey geometry, dimension, administrative and titling data to process a survey plan from lodgement and registration, and is a good resource to show how LandXML can be adapted to represent even complex data.

---

[32] http://www.icsm.gov.au/index.html
[33] https://icsm.govspace.gov.au/files/2011/09/ePlan-Protocol-LandXML-Mapping-v2.1.pdf
[34] http://www.lpi.nsw.gov.au/
[35] http://www.lpi.nsw.gov.au/_data/assets/pdf_file/0019/138034/NSW_LandXML_formatv_7.5.2.pdf

A key driver to these developments is process improvement within organisations. Data acquistion and processing for the land registration process in NSW is typical of many other regions. Surveyors measure and log land dimensions and use drafting packages to convert these observations to plans. These plans are then produced as either paper or images documents, and lodged for registration. Once registered, the registration authorities need to convert these plan documents back into digital bearing and distances measurements to use for their internal processes. There is considerable duplication of effort in these conversion processes, digital to paper and back to digital. There have been many efficiencies identified from the lodgement of the orginal digital information directly, and LandXML is an ideal candidate for standardising this survey data so that it can be reused for many external and internal land management tasks.

There are other signficant advantages in using LandXML in the lodgement processes of surveys for title registration. These include:

Automated Validation of Survey Information XML data is well suited to automated validation Web services. This capability enables plan lodgement authorities (surveyors, land developers, etc.) to check the format and content of the plans they are registering against a predefined set of rules. These rules can cover both textual content of the plans and spatial information (closed polygons, lot area, lot attribution), greatly reducing the number of errors that may exist on plans. The NSW LPI has sucessfully implemented this type of service to validate plan content before accepting the plan for registration.

Automated Visualisation of Plans of Survey In much the same way as the validation services, a Web service has been developed by the NSW LPI to render LandXML survey data into plan documents and diagrams. To aid this plan generation, CgPoints defined in LandXML are used to define the extent of diagrams. In the case of the NSW LPI service, these elements can be existing points from the survey, defining extent of the survey, or additional elements created to define the diagram areas. These latter CgPoints have a status of *proposed* or *sideshot*. NSW LPI has produced a very useful document available online, the NSW Land XML Sample Plan. [36] There are three NSW schema files also availabe from their website, NSW Enumerated Types, NSW Reference Data and NSW ePlan Protocol Schema.

Use of Digital Survey for Internal Processes Maintaining a land registry is a complex issue. Processes MUST be robust, repeatable and correct at all times. Achieving this alongside processes modelling the real world in a computer system, and you have a wide mix of exact sciences with the approximation needed to represent the earth in a computer system. These processes are also traditionally labour intensive. Any technology that can automate the transfer and transformation of geospatial data from capture devices to storage reduces manual processing needs, providing more efficient and repeatable processes. LandXML is enabling this at NSW LPI.

---

[36] http://www.lpi.nsw.gov.au

Given the success of the LandXML solution at NSW LPI, it is likely that LandXML will replace TIFF as the file format for digital lodgement of plans in the plan lodgement process within NSW, Australia.

While it may not seem obvious, the use of LandXML for plan lodgement and validation is a clear implementation of Online GIS. The use of Web services to process these data, and the checking of textual and spatial content of the LandXML files, provides a powerful Online GIS system. Moreover, vendor software can be certified by LandXML.org as being standards compliant with LandXML, providing assurances for users integrating LandXML solutions with these vendor products that their developments are future proofed to some extent.

Finally, there are a significant number of sample LandXML files that are available online[37]. Not only does this website give readers some useful examples of LandXML, the links give possible adopters insight into the types of organisations that are using LandXML to support their business data needs. There are also a small number of user supplied LandXML transformation files,[38] which can be viewed in either VRML viewers, SVG viewers or Google Earth.

6.5.3   The Future of LandXML as a Supported Standard

There is considerable debate at the time of writing on the future of LandXML. This debate is centred around several aspects of the current LandXML implementation and its support into the future. In essence, the governing body of LandXML, LandXML.org, is not considered by some to be a formal standards organisation. This has led to a push from other standards organisations to provide a similar standard to support land related XML schemas, in particular, the OGC.

6.5.3.1   OGC Search for a Land Infrastructure XML

A Land Information Domain Working Group (DWG) was formed at OGC to find the best approach for incorporating the LandXML schema into the OGC standards base and to explore ways to incorporate land related information into OGC standards, to develop a transformation between the two standards by using UML modelling to reverse-engineer LandXML to find out the viability of OGC support for LandXML. The goal was to enable GML compliant systems to read and consume data supplied in LandXML. The initial step was to create a GML Application Schema to be known as LandGML that could model everything found in the LandXML schema and then to create a transformation from these LandXML instance documents into instance documents that conform to the LandGML Application schema, using the XLST XML transformation language.[39] This attempt to make LandXML compliant to OGC's GML was unsuccessful.

There are other projects to test translations between GML and LandXML. The

---

[37] http://www.landxml.org/webapps/landxsamples.aspx
[38] http://www.landxml.org/schema/UserSamples/UserTransformSamples.asp
[39] http://www.w3.org/TR/xslt20/

LandGML Interoperability Experiment (2014)[40] is a two-phase interoperability experiment intended to test a GML 3.0 application schema for encoding LandXML 1.0 documents (LandGML). This trial has attempted to validate the translation of LandXML 1.0 documents into LandGML documents and to transform LandGML documents into LandXML 1.0. The outcomes of this experient are unknown.

Prior to this project, there were a number of issues raised by OGC Standards Working Groups regarding the current 2.1 implementation of LandXML.[41] These can be summarised as

Few mandatory elements The LandXML Standard is based on flexibility as opposed to interoperability. Most properties of LandXML are *optional*.

Structure of the Schema Document The Schema Document structure is not organised by logical packages to control which types are used within multiple element definitions, and modifying a type could break other parts of the schema.

Choice Compositor Because any element is selectable from the "*choice list*" but the supply of a mandatory value is not enforced, XML definitions of elements can be made using the Choice Compositor, which does not contain the mandatory values. These invalid elements may still be considered valid.

Station Data Type Is Ambiguous LandXML data types are sometimes ambiguous, such as the Double used in the Station data type which may contain different types of observations (absolute, derived or reduced Station observations, etc.), leading to non-standardised uses of the data type.

PointType Is Weakly Typed In a number of packages, points can include additional data to represent other information about the point instance, including offset distance, absolution elevation, slope percentage and horizontal distance.

Additional Items There are serveral other issues with the 1.2 standard identified, including inconsistency in name optionality, inconsistencies is charater *case*, use of plural names and unique identifiers.

Moreover, there were a number of incompatibilities identified between LandXML and OGC standards based on OGC reports.[42] Table 6.2 lists these incompatibilities.

6.5.3.2 Benefits of LandXML.org Autonomy

At the time of writing, LandXML is still being supported by a consortium of users, and not a formal standards organisation. Moreover, it continues to be used worldwide as a neutral data exchange format for describing land development, civil engineering

---

[40] http://www.opengeospatial.org/projects/initiatives/landgmlie
[41] http://portal.opengeospatial.org/files/?artifact_id=56299
[42] http://portalopengeospatial.org/files/?artifact_id=56299

TABLE 6.2: OGC Standards baseline and LandXML 1.2. TC211 is the ISO Geographic/Geomatics standard

OGC Standard	LandXML 1.2 Disparity Description
Modular Specification	Modularisation is not well defined
Modular Specification	In LandXML, almost everything is optional
Conceptual Model	Lacks a conceptual model, *concepts/definitions* ...
Feature Based	Supports a different *feature* model from OGC and TC211
Geometry Model	Does not support ISO19107 (Harmonised Geometry)
GML	Uses W3C XML rather than GML for geometries
Case Conventions	Does not consistently support CamelCase - TC211

and survey activities. But the current version (1.2) is inconsistent with the existing OGC standards and thus cannot be represented as an extenstion, or sub-classing thereof.

It may also be true that LandXML has gained in popularity because it is a flexible standard in both its definition and in the way it could be applied to problems in these land infracture related clusters of the spatial industry. The enforcement of a more rigorous standard may reduce the flexibility for users of the standard, and may stifle its suitability for creating XML grammars for land infrastructure projects. LandXML is trying to satisfy the data description needs of a wide range of disciplines where infrastructure and regulations are wide and varied. Moreover, LandXML does not appear to be defining a set of rules and principals for a particular data domain, but is proving a flexible framework, providing business benefits to organisations whose business drivers are more about internal ingestion, validation and manipulation of data than in making their data fully interoperable to all the clusters involved in land management. It may be that, provided organisations that are using LandXML do formally document their implementation of LandXML in the types of recipe books and sample LandXML files as has been done by LPI in NSW, Australia, the thinking behind industry needs and usage is sustainable. Certainly the Australian experience, where jurisdictions implement their own extensions to the LandXML standard, has been quite successful.

6.5.3.3   Future Directions for LandXML

A merger of LandXML with an OGC standard is now looking unlikely. OGC is moving towards a new standard, *LandInfra*,[43] in preference to developing interoperability of LandXML with OGC standards. LandInfra will support a subset of LandXML functionality, but it is likely to implement GML as its base datatypes. LandInfra will also be supported by a UML conceptual model. This proposal was at the conceptual stage at the time of writing, with OGC requesting comments on this conceptual model. However, the formalisation of LandInfra as a LandGML standard is looking probable.

---

[43] http://www.opengeospatial.org/pressroom/pressreleases/2160

6.6   Summary and Outlook

There are now a significant number of XML implementations across many computer based disciplines, and Online GIS has been no exception. This chapter introduced four popular XML standards which are in common use for Online GIS, starting from the simple GPX standard use or GPS devices, and finishing with the more complex GML standard from OGC and the LandXML standard being used by many organisations for data translation necessary for Online GIS.

Each of these standards provides practitioner implementations of XML, based on the languages discussed in detail in Chapter 5, and each section covered both the schema details of each implementation, followed by practical examples of each to give the reader insight into how important markup languages have become to Online GIS.

6.7   Appendix: Picnic Schema

This application schema relies on the specification in **tgml** (§6.8) derived from GML 3.2.1.

```xml
<?xml version="1.0" encoding="UTF-8"?>
<schema xmlns="http://www.w3.org/2001/XMLSchema"
      xmlns:elephant="http://www.opengis.net/junk"
      xmlns:dingo="http://www.opengis.net/giraffe"
      targetNamespace="http://www.opengis.net/gorilla"
      xmlns:pn= "http://www.opengis.net/gorilla"
      elementFormDefault="qualified"
      version="2.2.0">

  <annotation>
   <appinfo>picnic.xsd v2.03 2001-02</appinfo>
   <documentation xml:lang="en">
    GML schema for picnic site
   </documentation>
  </annotation>

  <!-- <include schemaLocation="gml/2.0.0/feature.xsd"/>
      -->

  <!-- import constructs from the GML Feature and
      Geometry schemas -->
  <import namespace="http://www.opengis.net/junk"
```

```
        schemaLocation="gmlpp4expt.xsd"/>
<import namespace="http://www.opengis.net/giraffe"
        schemaLocation="tgml.xsd"/>

<!-- ==========================================
     global element declarations
     ========================================== -->

<!-- <element name="PicnicModel" type="pn:
    PicnicModelType"/> -->
<!-- substitutionGroup="gml:_FeatureCollection" /> -->

<!-- <element name="picnicMember" type="pn:
    PicnicMemberType" -->
<!-- substitutionGroup="gml:featureMember"/> -->
<!-- <element name="Road" type="pn:RoadType" -->
<!-- substitutionGroup="pn:_PicnicFeature"/> -->

<!-- a label for restricting membership in the
    PicnicModel collection -->
<!-- <element name="_PicnicFeature" type="gml:
    AbstractFeatureType" -->
<!-- abstract="true" -->
<!-- substitutionGroup="gml:_Feature"/> -->

<!-- ==========================================
     type definitions for picnic model
     ========================================== -->

<complexType name="PicnicModelType">
  <annotation>
    <documentation> This is the type for the
    high level picnic model </documentation>
  </annotation>

  <complexContent>
    <extension base="elephant:AbstractJUNKType">

      <sequence>

        <element ref="elephant:Point"/>


        <!-- <element name="Property" type= -->
```

```
      <element ref="dingo:boundedBy"/>
      <element ref="pn:picnicFeature" minOccurs="1"
         maxOccurs="unbounded"/>
    </sequence>
  </extension>
 </complexContent>
</complexType>
<element name="picnic" type="pn:PicnicModelType"/>

<element name="picnicFeature" type="pn:
   PicnicFeatureType"/>

<complexType name="PicnicFeatureType">
 <!-- <annotation> -->
 <!--  <documentation> This is the type for the -->
 <!-- features which make up the picnic model </
    documentation> -->
 <!-- </annotation> -->
 <sequence>
   <element name="Description" type="string"/>
   <choice>
     <sequence>
       <element name="Property" minOccurs="0">
         <complexType>
           <simpleContent>
             <extension base="string">
               <attribute name="name" type="string"/>
               <attribute name="type" type="string"/>
             </extension>
           </simpleContent>

         </complexType>
       </element>
       <element name="Geometry">
         <complexType>
           <choice>
             <element ref="elephant:Point"/>
             <element name="posList"
                   type="dingo:DirectPositionType"/>
             <element name="lpPolygon"
                   type="pn:lpPolygonType"/>
           </choice>
```

```
            <attribute name="srsName" type="string"/>
            <attribute name="name" type="string"/>
          </complexType>
        </element>
      </sequence>
      <element ref="pn:picnicFeature" minOccurs="0"
          maxOccurs="unbounded"/>
    </choice>

  </sequence>
  <attribute name="featureType" type="string"/>
  <attribute name="fid" type="positiveInteger"/>
  <attribute name="name" type="string"/>


</complexType>


<complexType name="lpPolygonType">
  <sequence>
    <element name="posList"
          type="dingo:DirectPositionType"/>
  </sequence>
  <attribute name="srsName" type="string"/>
  <attribute name="name" type="string"/>
</complexType>


</schema>
```

6.8 Appendix: Application Schema for the River Scene

This schema follows the specification of GML 3.2.1.

```
<schema targetNamespace="http://www.opengis.net/giraffe
    "
      xmlns:gml="http://www.opengis.net/giraffe"
      xmlns="http://www.w3.org/2001/XMLSchema"
      elementFormDefault="qualified" version="3.2.1.2">
<!-- Software Notice -->
```

```
<!-- This OGC work (including software, documents, or
    other related
    items) is being provided by the copyright holders
        under the
    following license. By obtaining, using and/or
        copying this work,
    you (the licensee) agree that you have read,
        understood, and will
    comply with the following terms and conditions: -->

<!-- Permission to use, copy, and modify this software
    and its
    documentation, with or without modification, for any
        purpose and
    without fee or royalty is hereby granted, provided
        that you
    include the following on ALL copies of the software
        and
    documentation or portions thereof, including
        modifications, that
    you make: -->

<!-- The full text of this NOTICE in a location viewable
    to users of
    the redistributed or derivative work. -->
<!-- Any pre-existing intellectual property disclaimers,
    notices, or
    terms and conditions. If none exist, a short notice
        of the
    following form (hypertext is preferred, text is
        permitted) should
    be used within the body of any redistributed or
        derivative code:
    "Copyright  [$date-of-document] Open Geospatial
        Consortium,
    Inc. All Rights Reserved. http://www.opengeospatial.
        org/ogc/legal
    (Hypertext is preferred, but a textual
        representation is
    permitted.) -->
<!-- Notice of any changes or modifications to the OGC
    files,
    including the date changes were made. (We recommend
        you provide
```

```
        URIs to the location from which the code is derived
          .) -->


<!-- THIS SOFTWARE AND DOCUMENTATION IS PROVIDED "AS IS
     ," AND
     COPYRIGHT HOLDERS MAKE NO REPRESENTATIONS OR
         WARRANTIES, EXPRESS
     OR IMPLIED, INCLUDING BUT NOT LIMITED TO, WARRANTIES
         OF
     MERCHANTABILITY OR FITNESS FOR ANY PARTICULAR
         PURPOSE OR THAT THE
     USE OF THE SOFTWARE OR DOCUMENTATION WILL NOT
         INFRINGE ANY THIRD
     PARTY PATENTS, COPYRIGHTS, TRADEMARKS OR OTHER
         RIGHTS. -->

<!-- COPYRIGHT HOLDERS WILL NOT BE LIABLE FOR ANY DIRECT
     , INDIRECT,
     SPECIAL OR CONSEQUENTIAL DAMAGES ARISING OUT OF ANY
         USE OF THE
     SOFTWARE OR DOCUMENTATION. -->

<!-- The name and trademarks of copyright holders may
     NOT be used in
     advertising or publicity pertaining to the software
         without
     specific, written prior permission. Title to
         copyright in this
     software and any associated documentation will at
         all times
     remain with copyright holders. -->


<!-- boundedBy comes from feature.xsd -->
        <element name="boundedBy" type="gml:
          BoundingShapeType">
 <!-- nillable="true"> -->
              <annotation>
                    <documentation>This property describes
                        the minimum bounding box or
                        rectangle that encloses the entire
                          feature.</documentation>
              </annotation>
        </element>
```

```
    <complexType name="BoundingShapeType">
        <sequence>
            <choice>
                <element ref="gml:Envelope"/>
                <!-- <element ref="gml:Null"/>
                    -->
            </choice>
        </sequence>
        <!-- <attribute name="nilReason" type="gml:
            NilReasonType"/> -->
    </complexType>

<!-- Envelope comes from geometryBasic0d1d.xsd -->

    <complexType name="EnvelopeType">
        <!-- <choice> -->
            <sequence>
                <element name="lowerCorner" type
                    ="gml:DirectPositionType"/>
                <element name="upperCorner" type
                    ="gml:DirectPositionType"/>
            </sequence>
            <!-- <element ref="gml:pos" minOccurs
                ="2" maxOccurs="2"> -->
            <!--   <annotation> -->
            <!--    <appinfo>deprecated</appinfo>
                --> 
            <!--   </annotation> -->
            <!-- </element> -->
        <!--   <element ref="gml:coordinates"/> -->
        <!-- </choice> -->
        <!-- <attributeGroup ref="gml:
            SRSReferenceGroup"/> -->
    </complexType>
    <element name="Envelope" type="gml:EnvelopeType">
        <annotation>
            <documentation>Envelope defines an
                extent using a pair of positions
                defining opposite corners in
                arbitrary dimensions. The first
                direct position is the "lower
                corner" (a coordinate position
                consisting of all the minimal
                ordinates for each dimension for
                all points within the envelope),
```

```
                    the second one the "upper corner"
                    (a coordinate position consisting
                    of all the maximal ordinates for
                    each dimension for all points
                    within the envelope).
The use of the properties "coordinates" and "pos" has
   been deprecated. The explicitly named properties "
   lowerCorner" and "upperCorner" shall be used instead
   .</documentation>
            </annotation>
      </element>
      <complexType name="DirectPositionType">
            <annotation>
                  <documentation>Direct position
                    instances hold the coordinates for
                     a position within some coordinate
                     reference system (CRS). Since
                     direct positions, as data types,
                     will often be included in larger
                     objects (such as geometry elements
                     ) that have references to CRS, the
                      srsName attribute will in general
                      be missing, if this particular
                     direct position is included in a
                     larger element with such a
                     reference to a CRS. In this case,
                     the CRS is implicitly assumed to
                     take on the value of the
                     containing object's CRS.
if no srsName attribute is given, the CRS shall be
   specified as part of the larger context this
   geometry element is part of, typically a geometric
   object like a point, curve, etc.</documentation>
            </annotation>
            <simpleContent>
                  <extension base="gml:doubleList">
                        <!-- <attributeGroup ref="gml:
                          SRSReferenceGroup"/> -->
                  </extension>
            </simpleContent>
      </complexType>

<!-- We get doubleList from basicTypes.xsd -->
      <simpleType name="doubleList">
            <annotation>
```

```
                    <documentation>A type for a list of
                        values of the respective simple
                        type.</documentation>
            </annotation>
            <list itemType="double"/>
        </simpleType>

</schema>
```

7

Information Networks

CONTENTS

Advances in technology and business over the last decade have resulted in the creation of a rich and diverse geospatially aware Internet. Standards now exist that facilitate the publication and exchange of these data online, from many different data sources. The online mashing up of data from disparate datasets was only just being realised at the time the first edition of this book was being written. A lot has happened since then, but fundamental considerations such as custodianship, legal and security issues, funding and stability of technology platforms are still relevant today. Here, we will look at the dynamics of the Information Network space.

7.1   Introduction

The preceding chapters have looked in detail at the mechanics of placing geographic information online. However, the real power of the Internet lies in sharing and distributing information. Geographic information is intrinsically distributed. Whether it be mines in Canada, roads in Britain or lakes in USA, most geographic applications concern particular themes within particular regions. To compile an overview of (say) the worldwide distribution of a single theme, or else to overlay a number of themes for a given region or country usually requires collating information from a number of different sources.

The result is that, in considering geographic information online, we need to go beyond the methods of delivering geographic information from a single website and look at how to coordinate geographic information that is spread across many different sites. In the following chapters, therefore, we move on to look at the issues and technology involved in so doing. The first step, which we address in this chapter, is to understand the issues involved in creating and coordinating an Information Network, integrating the tools and techniques of previous chapters into giant online systems.

So far, we have seen Web technologies for spatial operations, for validating data and for markup of text material and graphics in standardised and non-proprietary form. Now we look at the concept of an *Information Network* as a synergistic union of many distinct websites. Information Networks require a range of supporting ideas, including quality control and criteria for being part of the network, indexing, maintenance, security, privacy and scalability.

We shall examine the tools which are making IN's feasible later. In Chapter 10 we look at the information system of the new millennium: the distributed data warehouse, and how these can empower IN within organisations. We will also look at the Spatial Data Infrastructure Portals in Chapter 9 that operate as part of the IN, and in the final chapters, look at some of the popular Online GIS that are an intrinsic part of any online geospatial ecosystem (Chapters 11, 12 and 13).

Cloud computing (§13.4.1) has become an important technology in the last decade. But it is quite different from an Information Network. It is a way of slicing up data and storing it across many nodes. But it is effectively just a big, virtual disc. We are concerned here with structured and organisation-linked storage.

7.2   What Is an Information Network?

How do we organise information on a large scale? One approach is to start at the source and organise publishing sites into an information network, *a set of sites on the Internet that coordinate their activity, focusing on a particular theme or topic*. In particular they operate under some common framework, especially the indexing of the information that they supply.

Information Networks are organisations for coordinating the development of online information and should not be confused with the communication networks that connect computers together. Just as computer networks link together computers, so Information Networks link together information, people and activity on particular topics. Such Information Networks are often loosely organised and heterogeneous in nature. Chapter 10 considers the more tightly bound networks involved in distributed data warehouses.

The Internet creates the potential to develop worldwide information systems. In a real sense, the World Wide Web itself is a giant information network. However, it fails to satisfy the second part of the definition above, namely, the need for a focus on a particular theme or topic. As the World Wide Web spread, the 1990s saw a proliferation of cooperative projects to implement Information Networks in a number of fields. Astronomers set up worldwide networks to link star charts and to communicate about new sightings; biotechnologists set up networks, such as the European Molecular Biology Laboratory (EMBL), to provide a seamless umbrella for databases being developed at different sites.

A number of international projects have focused on putting global resource and environmental information online. For instance, the International Organisation for Plant Information (IOPI) began developing a checklist of the world's plant species (Burdet 1992). The Species 2000 project has similar objectives (IUBS 1998). At the same time, the Biodiversity Information Network (BIN21 2013) set up a network of sites that compiled papers and data on biodiversity on different continents. There are now many online Information Networks that focus on environment and resources (BIN21 2013).

Similar initiatives took place in primary industry as early as the late 1990s. In 1996 the International Union of Forestry Research Organisations established an international information network and in 1998 began work to develop a global forestry information system (IUFRO 1998).

Perhaps the most widely used Information Networks are those supporting popular search engines. Many search engines either farm out queries to a number of supporting databases and pool the results, or else they index source data from databases that gather primary data about sites within a restricted topic, network domain or region. There is one in particular that probably deserves a mention, and that is the Google search engine. This popular Information Network indexes data from a wide range of data types and categories, including images, documents, trends, news, maps, books and scholarly papers and more.

The same principles have been applied in many other forms of data. Specific to geospatial data, there are now many online catalogues containing metadata records for geospatial data resources (data and services) that use a combination of farming out queries and creating localised indexes of these data and links (see Chapter 9).

7.3   What Can Information Networks Do?

Perhaps the first distributed geographic information system on the Internet was the World Wide Web index of websites by countries. This service, established by CERN as the Web spread, simply sorted the websites registered with CERN by country. As the number of websites grew, this system quickly became unmanageable. In 1993, CERN began to outsource indices for particular countries, effectively turning the service into a distributed geographic information network. The system was ultimately abandoned when control of the Web was transferred to the World Wide Web Consortium (W3C). By that stage, a combination of commercial competition and sheer growth was creating an anarchic situation in which it was impossible for any single organisation in each country to maintain an official register of sites.

Another example of an early information network with a geographic basis was the Virtual Tourist (Plewe 1997), or VT, described in Chapter 2. The VT[1] consisted of a single world index that pointed to a hierarchy of national and regional indexes, which in turn pointed to sites providing primary data. Unlike the sites register, though, the VT's index supported geographic searching from its inception and it now sources and integrates textual, image and mapping data from many databases for its online portal.

There are enormous benefits to gain from organising geographic Information Networks. As the above examples show, a geographic Information Network makes it possible to put together information services that no single organisation could develop and maintain. Some of these advantages include the following:

- The whole is greater than the sum of its parts. Combining different datasets makes it possible to do new things with them that could not be done individually. One example is the geographic index, with each dataset covering a separate region.

- Another is the potential to create overlays of different kinds of data, making possible new kinds of analysis and interpretation.

- Data is updated and maintained at the source. The organisations that gather the primary data can also publish it. This makes it easier to keep information up to date. It also overcomes many of concerns about ownership and copyright that have plagued cooperative ventures in the past.

- Information Networks are scalable. That is, more and more organisations and nodes can be slotted in at different levels without the system breaking down.

The ultimate geographic information network would be a worldwide system that provided links between all kinds of geographic data at all scales (see Chapter 13), but such a system is still a long way off. Meanwhile there are enormous advantages in being able to draw together geographic information over wide areas. Gathering data

---

[1] VT is now owned by Expedia, and has evolved into an great example of a global Information Network which uses geospatial information.

has always been the most time-consuming and frustrating task for GIS managers. Information Networks have the potential to make widespread geographic data of many kinds available on demand.

7.4 The Organisation of Information Networks

The earliest networking projects were cooperating sites that simply provided a common interface to lists of online resources. Another common model is a virtual library, which consists of a central index to a set of accredited services. A more ambitious model is a distributed data warehouse (Chapter 10). This system consists of a series of databases at separate sites on the Internet, with a common search facility.

An information system that is distributed over several sites (nodes) requires close coordination between the sites involved. The coordinators need to agree on the following points:

- Logical structure of the online information;

- Separation of function between the sites involved;

- Attribute standards for submissions (see next section);

- Protocols for submission of entries, corrections and updates;

- Quality control criteria and procedures (§7.5);

- Protocols for online searching of the databases;

- Protocols for mirroring the datasets.

For instance, an international database project might consist of agreements on the above points by a set of participating sites (nodes). Contributors could submit their entries to any node, and each node would either mirror the others or else provide online links to them.

7.5 Issues Associated with Information Networks

One advantage of Information Networks is that they can directly address issues that are crucial in building a reliable information system. The sites in the network must operate under one of the common frameworks described above. To achieve this, they need to address a number of key issues, including

- Standardisation (§7.5.1);

- Quality assurance (§7.5.2);

- Publishing model (§7.5.3);

- Stability of information sources (§7.5.4);

- Custodianship of data (§7.5.5);

- Legal liability and other legal matters (§7.5.6);

- Funding (§7.5.7).

7.5.1   The Need for Standards and Metadata

Coordinating and exchanging scientific information are possible only if different datasets are compatible with one another. To be reusable, data must conform to standards. The need for widely recognised data standards and data formats is therefore growing rapidly. Given the increasing importance of communications, new standards need to be compatible with Internet protocols. Four main kinds of standards and conventions are used:

1. Information design standards and information models describe in conceptual terms the information needs of an enterprise. All data and information are collected, stored and disseminated in the framework.

2. Attribute standards define what information to collect. Some information (e.g. who, when, where and how) is essential for every dataset; other information (e.g., soil pH) may be desirable but not essential.

3. Quality control standards provide indicators of validity, accuracy, reliability or methodology for data fields and entries.

4. Export formats specify how information should be laid out for distribution.

The markup languages SGML and XML (see Chapter 5) provide extremely powerful and flexible standards for formatting information for processing of all kinds. They are also extremely good for interchanging database records.

Software libraries now exist that provide tools to manipulate and reformat files, while other lightweight formats such as JSON have appeared (JSON 2014) (§8.3).

7.5.1.1   Metadata

Metadata is data about data. It provides essential background information about a dataset, such as what it is, when and where it was compiled, who produced it, and how it is structured. Without its accompanying metadata, a dataset may be useless. Metadata has gained considerable prominence an indexing tool, especially since the advent of large-scale repositories on the Internet. Because of the vast range of information online, the W3C promotes the principle that all items should be self-documenting. That is, they should contain their own metadata. Whatever the material concerned, metadata always needs to cover the basic context from which the information stems.

Since the advent of Google, there has been tremendous progress in searching for, and finding, information on the Internet. But it can still be difficult to refine a large number of hits on a particular topic. Metadata helps to refine searches, but also has other functions.

Broadly speaking, metadata needs to address these basic questions:

- HOW was the information obtained and compiled?

- WHY was the information compiled?

- WHEN was the information compiled?

- WHERE does the information refer to?

- WHO collected or compiled it?

- WHO owns it and who, if not the owner, is the custodian (§7.5.5)?

- WHAT is the information?

- WORTH of the information, its validity, accuracy, precision (§7.5.2).

For instance, the Dublin Core (§8.2), designed originally for online library functions, specifies a suite of fields that should be used in identifying and indexing Web documents (Weibel et al. 1998). An important side effect of XML (Chapter 5) is to make metadata an integral part of the organisation and formation of documents and data. Moreover, the Resource Description Framework (RDF) (§8.5) provides a general approach to describing the nature of any item of information RDF (Lassila and Swick 1999).

7.5.2 Quality Assurance

Quality is a prime concern when compiling information. Incorrect data can lead to misleading conclusions. It can also have legal implications. The aim of quality assurance is to ensure that data is valid, complete and accurate. It must also conform to appropriate standards so that it can be merged with other data. Errors in any field of a data record are potentially serious. Important aspects of quality include the following:

- Validity, accuracy and precision of observations;

- Data currency;

- Accurate recording;

- Conformity to standards.

7.5.2.1 Tests on Quality Assurance

The most direct method of assuring quality is to trap errors at the source. That is, the workers recording the original data need to be rigorous about the validity and accuracy of their results. If electronic forms are used (e.g., over the Web), then two useful measures are available. The first is to eliminate the possibility of typographical errors and variants by providing selection lists, wherever possible, as we saw in Chapter 4. For instance, selecting (say) the name of a genus from a list eliminates misspelling (though selection errors are still possible). For free text, scripts can be used to test for missing entries and for obvious errors prior to submission. Online forms, for instance, can use (say) JavaScript routines to test whether a name entered is a valid taxonomic family or whether a location falls within the correct range, or even for alternative (not necessarily correct) spellings. There also exist some software products now on the market that can check for homophones, similar sounding names, when comparing data entry to known features, based on their spelling (e.g., Sitty and City).

Errors in methodology pose the most serious concern. For data from large institutions, the professional standing of the organisation is often deemed to guarantee reliability. However, for data obtained from private individuals, other criteria need to be applied. Publication of results is often taken as a de facto indicator that all data records are correct. However, this is not always true.

Everyone makes mistakes, so quality testing of data is essential. Although the methods we saw in Chapter 4 can help to guard against some kinds of errors, they do not guard against errors of fact. Perhaps the most effective way to test for errors is to build redundancy tests into data records. Redundancy makes it possible to check for consistency between related fields. For example, does the location given for a field site lie within the country indicated? Does a named species exist? If a database maintains suitable background information, then outlier tests can reveal suspect records that need to be rechecked. A record that shows (say) a plant growing at a site with lower rainfall than anywhere else needs to be checked. Either it contains important information, or it is wrong. Both sorts of checks can be automated and have been applied to many kinds of environmental data.

In principle, an appropriate indicator of quality could accompany every data field. For instance, is location given to the nearest minute of latitude? Or degree? And how was it derived? By reference to a map or satellite image? A GPS (Global Positioning System)? Gathered in situ, or interpolated much later from a site description? Perhaps the most important are indicators of what checks have been applied to the original records to ensure accuracy and validity.

7.5.2.2 Protocols for Quality Assurance

Large data repositories adopt a formal quality assurance protocol for receiving, incorporating and publishing data. Some of these protocols include testing conformity to required standards, examples of standards, publication of methodology for standards currently in use, results and validity checks such as those described above.

Protocols for Information Networks are as yet less developed. One possibility is for a site to become part of an Information Network, it should receive an appropriate quality accreditation. For example, in the manufacturing industry and software development, the ISO9000 quality mechanisms and their descendants are widespread. To achieve ISO9000 accreditation requires first having adequate mechanisms in place to ensure quality of output; secondly, a formal accreditation body will investigate (normally on site) the quality processes to ensure they meet international standards. Accreditation for conformance will often have to be renewed at regular time intervals.

7.5.3   The Publishing Model

It is important to realise that making information available online is really a form of publication. Traditionally the term publishing has been closely associated with books and other printed matter. However, in the modern electronic era, there are now many other formats besides print for circulating ideas. With the rise of multimedia, the distinction between print, video, audio, etc. has become blurred. Today a more apt definition might be to describe publishing as

> *The act of disseminating intellectual material to its intended audience.*

One of the features of the 21st century Web is the advent of self-publishing, through websites, blogs, forums and social media generally. Slightly different, but also without central control, are the online encyclopedias, such as Wikipedia. Here the correctness and quality control arise from validating and modifying by the readers themselves.

Although the medium and the material may differ vastly, essentially the same common process is always involved in such publication. For online publications this model makes it possible to automate many of the steps involved. The model encompasses all the stages that occur in traditional publishing, but in a somewhat more formalised form. We can summarise the steps as follows:

- Submission - The author submits material to the editor.

- Acquisition - The publisher acquires material. Here we take this to include permissions. Details of the submission are recorded and an acknowledgement is sent to the author.

- Quality assurance - The material is checked. Errors are referred back to the author for correction.

- Production - The material is prepared for publication. This stage includes copyediting, design, typesetting, printing and binding.

- Distribution - The publication is shipped to stores, etc. for sale. It is publicised so that people know that it is available.

Note that the above procedure is completely general. It applies to any kind of information, whether it be data, text, images, video or sound. Although the details of the above process vary enormously from case to case, essentially the model is always the same. In a traditional magazine or journal, for instance, authors submit articles to an editor, who records them and assesses them for quality (this may involve outside referees) and then passes them on to a production unit, which prepares the publication in its final form. An advertising or marketing unit prepares announcements when the material is ready for distribution.

This essence of the above process (as captured in Figure 7.1) applies to publication of any kind of material, whether in a traditional context or online. For instance, adding data to a database involves submission of data records by the custodian to the manager of the database, who runs quality assurance tests over it, arranges for it to be entered or marked up, and then added to the database.

The importance of the publication model is that it provides a systematic framework for automating many editorial and publishing functions. When an author submits data for publication, several tasks must be performed immediately. Typically, these tasks might include

- Assign a reference number to the submission.

- Date stamp the submission.

- Create a directory for the new material (a directory is needed because several files will always be involved).

- Write the loaded file into the directory.

- Create a registration file containing all the details of the submission.

- Add a summary and links to the relevant incoming queue and editorial control files for later processing.

- Send a receipt back to the author.

- Carry out preliminary checks of the information, such as checking that all fields are completed.

- Notify the editor.

Almost the entire submission procedure can be automated. When an author submits an item for publication today, in, say, a conference, an automatic process stores the files, records the submission, returns an acknowledgement to the author, notifies the editor and generates a website for other participants, such as the referees, to use. It usually carries out elementary quality checks, such as ensuring that all pertinent information has been provided, or testing the validity of embedded URLs. Although initially a number of such software systems sprang up, there has been some consolidation, with sysems such as EasyChair[2] covering many conferences.

Another important consequence of the general publication model is that the same automatic processes that are needed for online publication will be shared by many

---

[2] http://www.easychair.org/

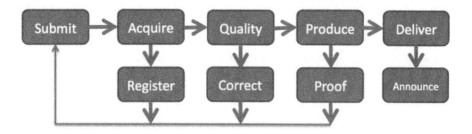

FIGURE 7.1: Summary of the steps involved in publishing material online.

information systems. This makes it possible to encapsulate those steps as publishing objects, complete on their own.

Finally, it should be noted that the above model really applies to publication of material on a single site. In the context of Information Networks, it applies only if the material can be centrally coordinated. In many contexts, a somewhat different approach of an accreditation model needs to be applied. We look at this in more detail in §7.6.1

7.5.4 Stability

The most frustrating problem, for users and managers alike, is that important sources of information frequently go stale. Existing mechanisms for reporting URL changes are simply not effective. However, the solution is not to concentrate information at a single centre. An important principle is that the site that maintains a piece of information should be the principal source.[3] Copies of (say) a dataset can become out of date very quickly, so it is more efficient for other sites to make links to the site that maintains a dataset, rather than take copies of it.

Publication of information online implies a measure of stability, that the information will be available for a long period. It will not suddenly disappear. Information Networks provide frameworks for reducing, if not eliminating, the problems involved.

Perhaps the most common reason for material disappearing is reorganisation of a website. The material is still there, but it is in a different location within the site. Another common problem is that websites often reflect the internal structure of the organisation that runs them. Each time the company or government department is restructured, so too the website needs to be changed. Publishers can avoid these problems by careful planning when the site is established, orienting the logical structure towards users rather than owners and using logical names for services instead of using machine names.

A more serious problem than address changes is the actual loss of material. Information goes offline for many reasons, but especially the closing of a site or service.

---

[3] This approach is referred to as a *federated model.*

Most often material disappears temporarily when a network failure occurs. Duplication is the best solution.

One approach is to run a mirror site, that is, a copy of all the material and services on another site. Some popular services have literally hundreds of mirror sites.

A mirror requires that information from the primary site be copied at regular intervals. This is done most economically through a system of updates, so that only new material is copied Another approach is a data archive. Archives became popular very early in the history of the Internet, well before the Web appeared. The most common examples were anonymous FTP sites that stored public domain software and shareware. Some government sites have established online archives for various kinds of data, such as weather records, satellite images or scientific data.

7.5.5   Data Custodianship

The word custodian has two meanings: a person who cares for something, and a person who controls something. The subtle difference between these two words is important in data management. Here we deal with the two issues in turn.

Data currency is often a major problem in GIS, as well as in many other kinds of databases. For instance, city councils need to keep the names of property owners up to date for purposes of taxation, voting registration and a host of other services and issues. Likewise, business directories and tourist databases need to keep contact details up to date or businesses will suffer. The task of maintaining the currency of data is hard enough for the primary agencies that compile the information. The problem is compounded when a dataset is passed on to other organisations. This issue is a great concern for GIS, which are often put together by combining data from many different primary sources. An important advantage of online GIS is the potential to access primary data sets directly from the source. This means that the data updates can be incorporated into a GIS directly at their source, rather than having to wait until a new version is received and uploaded. The working versions are also the primary sources.

Attempts to share data have often floundered on the issue of control. Datasets are often valuable to the organisations that develop them. As a result, agencies are understandably reluctant to hand over to other organisations datasets that they rely on for income. Possible remedies have been tried or recommended. These include changing the funding models, promoting and rewarding cooperation, tying contributions from researchers to grant funding and publication of results. Creating distributed databases online provides another potential solution to the problem. Publishing their data online has allowed many agencies to retain control and management of their data whilst also making it available externally.

The problem of recouping costs still remains an issue for agencies that make their data available online, especially if the data needs to be combined with other data layers, in automatic fashion. The problem is this. If an agency sells access to its data online, how does it charge other organisations that want to incorporate that data into other services? A good example would be (say) a geographic base layer that could serve as the base map for tourist data. At the time of writing, this question is still a major issue for online data.

Several approaches have been tried. Traditional models include syndication, in which an agency allows other organisations to use its data at some pro rata cost. Another is to pay royalties based on the rate of use of particular datasets.

The above model supposes that all of the players are aware of the way in which data is shared, and that systems are put in place to monitor this regular usage. However, the problem of micro payments becomes far more difficult in some of the automated scenarios that we shall consider in later chapters. For instance, as mentioned above, the use of common namespaces in XML raises the possibility of Information Networks in which different sites are not even aware of one another. So what happens, for instance, if an automatic agent grabs data from another site – a one-off transaction – in the course of addressing a query? In many cases the information access may require data processing. The same arguments extend to the processing time involved.

One possibility is to pay for data by some measure of the effort required to retrieve it. This could be the actual volume of data; it could be linked to CPU cycles to process a query or some other measure of query complexity. One problem is that most current standards originally appeared before the importance of electronic commerce became apparent. Hence they do not include models to accommodate issues such as those raised here. Despite this issue being relevant for over a decade, there remains an urgent need for a model that enables an agent to grab portions of geospatial datasets from wherever they are available, with minimal cost, or costing of just what it really uses. This could be very important for the growth of widespread Online GIS applications which need to access GIS data and services using mobile devices (see Chapters 11 and 12).

7.5.6 Legal Liability

Legal liability is always a cause for concern in publishing. Perhaps the most serious concern when publishing data online is the prospect of being held liable for any damage that may result from use of that data. In some cases, the fear of litigation has itself been enough to deter organisations from releasing their data. The biggest problem arises from false, inaccurate or incomplete data and from misleading interpretations. For instance, suppose that an error in the recorded values for latitude or longitude implied that an endangered species lay squarely in the middle of an area planned for commercial development. This information could be enough to halt the enterprise at a cost of millions of dollars. The developers might then seek to litigate against all parties they held responsible, including all the persons or organisations responsible for providing the false data.

Several steps are essential to reduce the risk of litigation. There is always an implicit assumption by the public at large that published information is correct. So publishers not only need to take every step possible to ensure that information is correct, but also that users are aware of limitations. Some of these steps might include the following:

- Carry out quality assurance on submitted data (see earlier sections).

- Along with each dataset provide a cover sheet that not only provides the relevant metadata, but also a clear statement that covers data quality, limitations and other caveats on use of the data. This is important because people tend to ignore limitations (e.g., precision) and assume that data can be used for any purpose.

- Along with the above caveats, provide a disclaimer that covers conditions of use and warning against responsibility.

The following simple example shows the sort of detail that may be included:

Users of the data contained herein do so at their own risk. Although the authors and editors have made every effort to verify the correctness of the information provided here, neither they nor the publishers accept responsibility for any errors or inaccuracies that may occur. Neither do they accept liability for any consequences that may result from any use that may be made of the information.

Such agreements now are commonplace: almost every Web activity has some set of disclaimers and implied agreements on signing up, whether money is involved or not. How often these agreements are actually read is a different matter altogether.

7.5.7   Funding

An important factor hindering many networking initiatives is how to pay for it. The economics of a single site are relatively straightforward. However, if there are many sites contributing to a particular service, then who pays for the activity? Attempts to set up Information Networks may fail because they entail a significant cost for the participants, without any obvious benefit. For this reason many site managers are reluctant to take part in such initiatives. This problem is particularly acute for non-commercial sites, which rely on limited insititutional funds for their existence. Even when direct funding has been available to fund a network, the problem is that the large number of players means that the funding available to each site in the network is likely to be small.

In most cases the only real payoff for contributing to sites is publicity. Being part of a network increases the number of pathways by which users can find a website (see details under accreditation below). This exposure is useful if it increases the number of users and commercial customers.

Another tricky issue arises with commercial networks in which users pay for the information that they access. How are payments made and how is the income distributed? In many attempts at creating online Information Networks, the agencies with data are often reluctant to contribute because they fear that the network will undermine sales of their data.

If the network is only loosely coordinated, then users might pay for individual services directly to the host site on a one-to-one basis. Problems arise where a service provided by a network makes use of elements from a number of different contributing sites.

For example, suppose that in a commercial geographic service, each map layer was maintained by a separate site and that a map building program retrieved data from each of the layers in the course of drawing a map. Presumably, the front end of the service would be maintained on a single site, which would also receive the payments from users. The contributing sites would of course expect payment for use of their data. The question then arises as to how to organise this. There are several possible price models. For instance, the provider might need to count the number of accesses to each layer and accumulate a micro-payment each time a particular data layer is used.

There are, however, several funding related factors contributing towards greater sharing of data. One is the implementation of policies around the world linked to funding models that are driving an Open Data policy across governments (Chapter 9). A big change to the commercial viability of data sharing online is of course the revenue that can be raised from online advertising.

Again, Google has been one of the pioneers in this regard and makes its extensive map facilities available for free to the general public, funded by the advertising revenue they generate. Another important development is services such as PayPal, which make internet financial transactions secure and private, and allow very small transactions to be made effortlessly, which is changing the way we shop. Online geospatial data plays a role in all of these.

7.6 Information Networks in Practice

Information Networks, unlike data warehouses, have a degree of self-organisation about them. On one hand, this enhances the volume and dynamic adaptability of the information source. On the other, it creates its own difficulties of internal coherence and accuracy.

7.6.1 The Accreditation Model

The publishing model presented earlier in this chapter provides a general approach to handling primary information. However, in many cases, Information Networks need to compile secondary information. That is, they simply link to existing information that is originally published on sites independently of the network organisation.

A particularly effective method of achieving status and visibility on the Web is the notion of endorsement, or accreditation (Green et al. 1998). In a sense, any hypertext link on the Web is a de facto endorsement. However, this is only partly true, because many sites maintain lists of relevant links without asserting anything about their quality. Even more so, search engines simply index sites that contain key words, without assessing quality in any way. Where accreditation differs is that the indexing site makes a specific statement about the quality of the indexed site (Green

1995). More to the point, it excludes sites that are not considered to be of sufficient quality. §8.4 takes up the self-description of resources.

Accreditation works as follows. An organisation that monitors information quality endorses a service provided by some content provider. In practice this means providing some form of badge or label that the endorsed site can place on its home page. Some organisations have made their reputation solely on the basis of providing an endorsement process. However, the process is a particularly effective way for organisations that already possess some form of authority to enhance and exploit their credibility in the Web environment (Green et al. 1998).

Accreditation is particularly suited to developing networks of geographic information. Almost any kind of information can be organised geographically. Businesses are always located somewhere and would benefit from the ability of users to search for them by city or region, as well as thematically. Likewise, it is relevant to be able to search governments, schools and most institutions geographically. This potential richness of information means that any geographic index needs to be able to link to, and index, information from a wide range of websites. The accreditation model provides a convenient approach to ensuring quality of the matter that the network indexes. The exact requirement for accreditation may vary; often it is simply how prominent the site is. However, for a network that stresses quality of service, some of the most common requirements are as follows:

- Relevance of the resource;

- Quality of the information (accuracy, validity);

- Absence of inappropriate material, such as pornography, seditious or criminal information.

- Conformance to any presentation requirements;

- Inclusion of essential metadata;

- Demonstrated stability of the resource, e.g., copies at mirror sites;

- Stability of addresses (using aliases);

- Notification of changes or closure;

- Freedom from concerns over legal liability, copyright, etc.

To test of the efficacy of accreditation, Green et al. monitored access rates to one of the services that we manage, the Guide to Australia (Green 1993), both before and after introducing accreditation early in 1999. The results showed that accreditation can increase the hit rate to a particular service by an order of magnitude. The reason for this is that it creates many new avenues by which users can navigate to the service concerned.

Accreditation can take several forms, for example:

- A reference to the item in an index;

- A reference to an item as though it were a publication on the local site, providing a badge of approval that the publisher can include on the site or in the item. Mozilla, makers of the Firefox Web browser, have introduced *Open Badges* (Mozilla 2014), which provide an easy way for accreditors to create and award badges, Web icons which contain embedded within them details of the accreditation process.

7.6.1.1 Why Use Accreditation?

There are many advantages in accreditation for the accrediting site and organisation: it

- reinforces recognition of the authority of the accrediting organisation;

- distributes the effort of developing material amongst other sites;

- encourages other sites to contribute material;

- encourages stability of contributing sites;

- helps to distribute the effort and cost of developing an information system;

- enables the accrediting site to impose standards and quality control on material published elsewhere;

- extends the number of links and references to the accrediting site;

- opens the potential for the accrediting site to set the agenda in the area concerned.

7.6.1.2 Advantages for the Accredited Site

Advantages for the contributor are similar to those for the accrediting organisation:

- Increasing the credibility of the publication;

- Extending the range of links and references to the site;

- Public recognition;

- Affiliation with an authoritative or high-profile organisation.

There are many low-level forms of endorsement on the Internet already. Some examples are

- The World Wide Web Virtual Library is simply a list of indexes that are held and managed elsewhere.

- Several sites have attempted to create a reputation for themselves simply by providing badges or "awards" to other sites. An example is the award of a badge to (say) *the 100 most popular websites.*

• Many sites that try to provide services on particular topics provide pointers to selected services elsewhere. This selection and indexing is a de facto form of endorsement.

By implication, accreditation implies a stamp of approval. Contributors who offer items for accreditation need to know what is expected of them. The list of requirements needs to ensure that the accredited item is suitable. On the other, hand it should not be so prohibitive that it discourages anyone from making contributions.

7.6.2   Examples of Geographic Information Networks

We have already alluded elsewhere to online geographic services that are effectively Information Networks of one kind or another. The Internet creates the potential to develop worldwide biodiversity information systems. As the World Wide Web spread, the 1990s saw a proliferation of cooperative projects to compile biodiversity information online.

A useful outcome of networking activity has been to put a lot of biodiversity information online, such as taxonomic nomenclature and species checklists. One of the first priorities was to develop consistent reference lists of the world's species. A number of international projects have focused on putting global biodiversity information online. For instance, the International Organisation for Plant Information (IOPI) began developing a checklist of the world's plant species (Burdet 1992). The Species 2000 project has similar objectives (IUBS 1998). At the same time, the Biodiversity Information Network (Green and Croft 1994) set up a network of sites that compiled papers and data on biodiversity on different continents. There are now many online Information Networks that focus on environment and resource (see Table 7.1).

The greatest challenges in collating biodiversity information have been human, especially legal and political issues, rather than technical problems. One outcome of the Convention on Biological Diversity was agreement on the concept of a Clearinghouse Mechanism ((UNEP) 1995). This scheme aimed to help countries develop their biodiversity information capacity. The longer-term goal was to enhance access to information through the notion of a system of clearing houses. These sites gather, organise and distribute biodiversity information. The greater challenge is to merge these clearing houses into a global network.

In 1994, the OECD set up a Megascience Forum to promote large science projects of major international significance (Hardy 1998). The Human Genome Project was one such enterprise. Another was the proposal for a Global Biodiversity Information Facility (GBIF). The aim of GBIF is to establish:

> "*A common access system, Internet-based, for accessing the world's known species through some 180 global species databases.*"

Primary industries have set up similar initiatives. For instance, in 1996 the International Union of Forestry Research Organisations established an international information network and in 1998 began work to develop a global forestry information system (IUFRO 1998).

TABLE 7.1: Some online biodiversity services and networks currently available

Organisation	URL
CIESIN	www.ciesin.org/
Clearing-House Mechanism of the Convention on Biological Diversity	www.cbd.int/chm
DIVERSITAS	www.icsu.org/DIVERSITAS/
Environment Australia	www.environment.gov.au/
European Environment Information and Observation Network (EIONET)	www.eionet.europa.eu
Global Biodiversity Information Facility (GBIF)	www.gbif.org/
International Legume Database & Information Service (ILDIS)	www.ildis.org
International Organization for Plant Information (IOPI)	plantnet.rbgsyd.nsw.gov.au/iopi/ iopihome.htm
International Union of Forestry Research Organisations (IUFRO)	www.iufro.org
Species 2000	www.species2000.org/ and www.sp2000.org
Tree of Life	www.tolweb.org
United Nations Environment Programme (UNEP)	www.unep.org/
USDA ITIS	www.plants.usda.gov/java/
World Conservation Monitoring Centre (WCMC)	www.unep-wcmc.org

7.7   Summary and Outlook

One might consider the Web to be one giant information network. With the sophistication of today's search software, information can be synthesised from across a gigantic domain. This chapter considered the principles underlying Information Networks.

- After a general introduction in §7.1, §7.2 asked what exactly an Information Network is.

- §7.3 considered what Information Networks can do.

- The organisation of Information Networks occupied §7.4 and various issues were discussed in §7.5.

- §7.6 considered Information Networks in practice.

The focus has been mostly on the practical and human issues involved in organising Information Networks. However, there have been many technical difficulties as well, such as coordinating updates, and developing and maintaining common indexes. We have glossed over most of these because the ad hoc approaches implemented in the past have been superseded by developments in Web technology.

These new developments are changing the focus of networking activity. For instance, instead of being effectively top down organisations, planned to deal with a particular issue, Information Networks can arise in bottom up fashion through sites adopting common approaches. For instance, the use of XML, with a common namespace, has the effect of immediately creating a de facto network of sites dealing with a common topic area. This effect has been foreseen by the W3C and standards, such as the Resource Description Framework (§8.5) have been developed to enhance the ability of metadata to contribute to the coordination and linking of online information resources.

The following chapters look at Information Network issues in more detail. Chapter 8 looks more closely at metadata standards for the Web, including new standards that have emerged. Chapter 9 looks in detail at the application of metadata standards by organisations that are implementing clearinghouse and portal components of the Information Networks that are making Online GIS a reality today.

8

Metadata on the Web

CONTENTS

The World Wide Web is now part of our everyday life, connecting systems and users to a wide range of services and information. In using any data over the Internet, it is important to have some information on the quality of the data you are using. This chapter discusses the general issue of metadata paradigms on the Web and a range of technologies: JSON for moving key-value pairs around, POWDER for providing information about content and the general powerful Resource Description Framework and its associated query language SPARQL.

8.1   Introduction

The World Wide Web has grown at an enormous rate. By the middle of the year 2000, there were around 50 millions sites worldwide, some 200 million forecast by the end of 2001. In fact the growth of the Web is strongly analogous to the growth in connectivity of a random graph. This phenomenon, first observed by Erdos and Renyi (1960), underlies many properties of complex interactive systems, as elsewhere demonstrated by Bossomaier and Green (2000).

In Figure 8.1, you can see the growth in the connectivity (the fraction of nodes connected together in one big component) as a function of the number of connections. As you can see, at a quite small number of possible connections, a connectivity, avalanche occurs. So it is with the Web. This unprecedented growth has brought various problems along with it:

- It's getting more and more difficult to find anything, but this seems to be a perennial problem; Google is currently dominant in the search engine space, but even it makes mistakes with subtle queries, or queries which get swamped by hits on some popular site, such as that of a celebrity;

- Some material may be offensive, yet whole scale censorship is undesirable;

- The authenticity of material may be questionable, an issue partly addressed with digital signatures (§8.7);

- It may be hard to decide on the quality or accuracy of Web pages;

- Personal data may be collected and used in ways not desired by the user.

Thus there is an urgent need for ways of describing websites. This in turn creates an urgent need for metadata.

HTML has had, since the early days, a META tag. This tag is still the only universally recognised source of metadata. But it has become hopelessly overloaded and several new directions have emerged:

- The Dublin Core workshops have generated a full set of bibliographic tags, which assign properties such as author and creation date to Web pages;

- PICS (Platform for Internet Content Selection) arose to label and rate content in specific ways, primarily for the use of parents and teachers. PICS was superseded by POWDER in 2009 (W3C 2014b) and (W3C 2014c) (§8.4);

FIGURE 8.1: Increasing connectivity shown during the formation of a random graph, as edges are added to join pairs of vertices. Notice the formation of groups (connected subgraphs) that grow and join as new edges (shown in bold lines) are added.

- XML grew rapidly as a more powerful and extensible alternative to HTML, which is potentially self-describing;

- the World Wide Web Consortium recommended in February 1999 the Resource Description Framework (RDF), which is a powerful structure for creating metadata for a wide variety of applications (§8.5).

Our main concern in this chapter is not specific metadata tag sets, such as Dublin Core, but the more general issue of metadata paradigms. Hence our focus will be on RDF (§8.5). The RDF working party had as part of its brief that it subsume PICS, so we shall briefly consider PICS and POWDER first. The XML issue is more complex. XML is like a protegé who outpaces the skills of his master. Although it began as a simplified form of SGML (Chapter 5), it is now taking on a much broader role, and taking on structural characteristics which go beyond the original DTD concept. XML frequently provides implicit metadata and is the language of RDF.

§8.4 covers the POWDER specification and §8.3 covers JSON, a compact format which can be used for moving metadata around. Finally, §8.7 briefly introduces XML signatures.

We introduced XML in Chapter 5, but in §5.4 we discussed the issue of namespaces. This is the mechanism of recording metadata tags for the many possible applications to which RDF might be applied. Having cleared the decks of preliminary material, we now look at the structure of RDF, its data model, its syntax and the schema mechanism through which data properties receive their definition.

TABLE 8.1: Dublin Core fixed attributes

Attribute	Value
Version	1.1
Registration Authority	Dublin Core Metadata Initiative
Language	English, according to IS0639
Obligation	Optional
Datatype	Character string
Maximum Occurrence	Unlimited

TABLE 8.2: Dublin Core variable attributes

Attribute	Value
Name	Creator
Identifier	Creator
Definition	The agent(s) responsible for the content of the resource
Comment	Resource examples which may include content

8.2 Dublin Core

The term Dublin Core (DC), perhaps surprisingly, refers not to Dublin in Ireland but to Dublin, Ohio, which is home of the OCLC, the Online Computer Library Center and the Dublin Core directorate. The DCMI (Dublin Core Metadata Initiative) began in 1995 and the first Dublin Core Workshop was held there, and subsequent workshops have been held around the world, including Warwick in the UK (source of the Warwick Framework for metadata design principles) and Canberra, Australia. Unlike the rest of this chapter, the Dublin Core is not about methods and techniques, but is a set of metadata requirements, derived from a library perspective. It is about semantics, rather than structure or syntax and aimed at facilitating resource discovery on the Internet. As such, DCMI has co-evolved with the structure and syntax mechanisms XML and RDF.

8.2.1 Specification of the Dublin Core Elements

The reader will soon become aware of the huge range of possible sets of metadata required for different applications, and in this chapter we are primarily concerned with how we describe metadata elements. But the relative simplicity of the DC specification means that a simple description mechanism is adequate. Each element has 10 attributes. The first six of these do not change for any of the elements in the current version. Table 8.1 lists them; they contain few surprises.

Table 8.2 gives the four variable attributes, again fairly obvious in intent, to define the Creator element.

TABLE 8.3: Dublin Core elements

Name	Identifier	Notes
Creator	Creator	
Subject and Keywords	Subject	
Description	Description	
Publisher	Publisher	
Contributor	Contributor	
Date	Date	Use of ISO8601 is recommended, i.e., YYYY-MM-DD
Resource Type	Resource	Nature or genre of the resource
Format	Format	Physical or digital manifestation
Resource Identifier	Identifier	URL, ISBN, etc.
Source	Source	
Language	Language	Use ISO0639 two-letter language codes with optional two-digit language code from ISO3166, e.g., "en-uk"
Relation	Relation	
Coverage	Coverage	May be temporal, spatial or administrative
Rights Management	Rights	Copyright, etc.

In most cases the name and identifier are the same, but occasionally the name is a somewhat more extended description than the (unique) identifier.

8.2.1.1 Dublin Core Elements

Having cleared the structural details of the definitions out of the way in the previous section, we can now look at the 15 DC elements themselves, given in Table 8.3 in simplified form.

Note that some of these descriptors are not complete in themselves, but require other standards or formal identification systems. In the case of URL and ISBN, these are in widespread use. But for other elements, such as Relation, there is no obvious standard to use. We see in the next section how we can incorporate these elements into an HTML document. There is no standard for this; however, an Internet Society memo exists (ISOC 2000).

8.2.1.2 HTML META Tag

An HTML document consists of two parts: the `head` and the `body`. The `head` does not contain information for display in the browser (although the `title` element often appears on the banner of the frame surrounding the Web page). In principle the browser can just retrieve the head of a page and determine whether it wants the full page. One immediate use here is to determine if the page has been updated since the browser last accessed it, although this usage has tended to be circumvented by other techniques. However, the head can contain information about what is in the page, the metadata, which can help determine if the user actually wants to download it.

When the Web first started, few people can have realised just how much and how rapidly it would grow. Thus the provisions for metadata were fairly limited, in fact, to just one empty tag[1] <META>. Each occurrence of the tag contains a property/value pair. Fortunately, multiple metadata tags can be included and it is sometimes used to carry the whole Dublin Core! The <meta> tag has several attributes:

name: refers to the name of the metadata property, such as the author of a document;

content: is the information specified in the name attribute, such as the author's name, i.e., the value of the property;

http-equiv: this gets a little bit more tricky and relates to the protocol for retrieving the page; we go into a little bit more detail below;

scheme: interpreting the value of the property defined in a name/content pair may require access to external information of some kind, referred to as a scheme;

lang: there are some other, global attributes, which can be included, which are not particularly important to our needs; one such is the language of the document.

Here is a straightforward example[2]:

```
<META name="author" content="Guiseppe Garibaldi"
              lang="en">
```

The attribute http-equiv is an alternative to the name attribute. It creates a special header in the form of the HTTP protocol used to fetch pages from Web servers, but the details would be a digression.

A frequent use of the meta tag is to provide keywords for the page, to facilitate the task of search engines, in the example below indicating the page contains map information in two different languages:

```
<META name="keywords" content="maps, spatial, OS, UK"
              lang="en">
<META name="keywords" content="cartes, spatial, France"
              lang="fr">
```

Here's a more complicated Dublin Core example:

```
<META name="DC.Creator" content="Terry Bossomaier">
<META name="DC.Creator" content="David Green">
<META name="DC.Creator" content="Brian A Hope">
<META name="DC.Title" content="Spatial Metadata and
                          Online GIS" lang="en">
<META name="DC.Publisher" content="Taylor and Francis">
<META name="DC.Language" content="en" scheme="ISO639">
```

---

[1] i.e., That is, no end tag.

[2] Note that this is an SGML (HTML DTD) construct, where the end tag is omitted. In XML this would not be permissible and the tag would end with / I>.

Note that we capitalise the element after the prefix and the elements may appear in any order. We have not included all the DC elements in this example, nor do we have to. The prefix DC, of course, refers to Dublin Core. But how does the Web page make this absolutely clear? It uses the <LINK> element.

```
<LINK REL="schema.DC" HREF="http://purl.org/DC/
                               elements/1.0">
```

Note that DC is just one of many possible prefixes which could follow the schema keyword in the REL attribute.

There is a little complication here with language which might be confusing. In the example above, we have referred to a scheme, in fact an ISO standard, which defines the language abbreviation. Now consider the following:

```
<META name="DC.Creator" content="Tomasi di Lampedusa">
<META name="DC.Title" content="Il Gattopardo" lang="it">
<META name="DC.Title" content="The Leopard" lang="en">
<META name="DC.Language" content="it" scheme="ISO639">
<META name="DC.Source" content="ISBN 88-07-80416-6">
<META name="DC.Type" content="novel">
```

This script describes an online version of a novel by Lampedusa (it is distinguished from Visconti's superb film of the book, which could in principle be embedded within a Web page, by the DC.Type value). The resource itself is in Italian, as indicated by the DC Language attribute value. But there are two titles, one in English, the other in Italian, indicated by lang attributes within each meta element. Finally, this online version was derived from an original printed novel, which we describe by its ISBN number.

Now a couple of issues should become apparent here which will become important later. First, this is a flat format, just an unstructured list of property-value pairs. So, as the metadata gets more extensive and complicated, it is a bit difficult for a human reader to digest. This may not be so much of a problem for a machine reading the data, but it can be a problem for a human author creating it.

Secondly, on a website or some other document collection there is likely to be a lot of repetition. Take, for example, the website of a computing department in a university. There are a number of generic properties for a university: it carries out teaching and research and gives degrees. A university will have several faculties, arts, science, health and so on. A search engine may wish to exploit this knowledge in its search, but obviously we don't want to repeat the higher level (university, faculty) information at the department level. Repetition risks errors; it increases download times and increases the workload for authors. So we need some sort of hierarchy mechanism, where we can inherit metadata from above. POWDER, which we consider shortly in §8.4, makes a start in this direction.

Dublin Core was in many ways a library initiative and is well known amongst librarians and archivists. But metadata is only of any use if people use it and if everybody agrees on what the terms mean. We have seen that Dublin Core has a simple

```
<foaf:Group>
    <foaf:name>Lilliput Tourists</foaf:name>
    <foaf:member>
        <foaf:Person>
            <foaf:name>Gulliver</foaf:name>
            <foaf:homepage rdf:resource=
                "http://swift.com/stars/gulliver.html"/>
            <foaf:mbox rdf:resource=
                "http://mailto:gulliver@swift.com"/>
        </foaf:Person>
    </foaf:member>
</foaf:Group>
```

FIGURE 8.2: A simple FOAF description of Gulliver as a member of the group of Lilliput tourists.

mechanism for standardising the description of its elements. But it's very general, as befits a minimalist description. As we want to provide more complex descriptions, we hit the problem of advertising. How do we share our metadata? POWDER provides a suitable mechanism (§8.4).

8.2.2   Friend of a Friend

Dublin Core dates back a long way, well before the Semantic Web. But the basic idea is sufficiently powerful that extended variants have appeared. One of the most prominent is **FOAF**, Friend of a Friend, which aims to make data about people and things machine readable on the Semantic Web.

FOAF is not a standard, from ISO or W3C. It relies heavily on W3C and all documents have to be correct RDF (§8.5). They can be expressed in XML or other RDF variants such as turtle (§8.5.9).

FOAF has two components: **foaf classes** and **foaf properties**. The first two rows of Table 8.4 show the elements of each. We can also group the FOAF terms in a different way: **core** terms, which have a lot in common with systems such as Dublin Core (§8.2) and **social web**, which relate to the way people interact over the Web, with items such as `accountname` and `skyleID`. Note that some `social web` terms are properties and some are classes.

Figure 8.2 shows a simple example of a FOAF `group`.

Not all elements are stable at the time of writing. For example, the `sha1` element refers to a textual representation of a hash of the contents, but is not yet sufficiently clearly defined (Brickley and Miller 2014).

FOAF is a `Linked Data System` (Berners-Lee 2014), a set of distributed resources, linked together but without necessarily any particular structure.

TABLE 8.4: FOAF elements, classes and properties

FOAF Terms	Elements
Classes	Agent; Document; Group; Image; LabelProperty; OnlineAccount; OnlineChatAccount; OnlineEcommerceAccount; OnlineGamingAccount; Organization; Person; PersonalProfileDocument; Project.
Properties	account; accountName; accountServiceHomepage; age; aimChatID; based_near; birthday; currentProject; depiction; depicts; dnaChecksum; familyName; family_name; firstName; focus; fundedBy; geekcode; gender; givenName; givenname; holdsAccount; homepage; icqChatID; img; interest; isPrimaryTopicOf; jabberID; knows; lastName; logo; made; maker; mbox; mbox_sha1sum; member; membershipClass; msnChatID; myersBriggs; name; nick; openid; page; pastProject; phone; plan; primaryTopic; publications; schoolHomepage; sha1; skypeID; status; surname; theme; thumbnail; tipjar; title; topic; topic_interest; weblog; workInfoHomepage; workplaceHomepage; yahooChatID.
Core	Agent; Person; name; title; img; depiction (depicts); familyName; givenName; knows; based_near; age; made (maker); primaryTopic (primaryTopicOf); Project; Organization; Group; member; Document; Image.
Social Web	nick; mbox; homepage; weblog; openid; jabberID; mbox; sha1; sum; interest; topic_interest; topic (page); workplaceHomepage; workInfoHomepage; schoolHomepage; publications; currentProject; pastProject; account; OnlineAccount; accountName; accountServiceHomepage; PersonalProfileDocument; tipjar; sha1; thumbnail; logo.

8.2.3 Profiles and Schemas

The HEAD element may contain a profile attribute, [3] which specifies a set of defini-tions or other material relating to the meta tag properties. For example, the profile might specify the Dublin Core. So, what goes into a profile? Apart from page-specific material, it could contain links to other profiles at a more general level. To make this work efficiently, the browser or search engine needs to be able to cache profile infor-mation to avoid continually downloading it for each page.

It should perhaps be clear already that the HTML meta tag is being asked to do a lot of work! Worse still, it has been the subject of abuse: keywords crammed into the document head of an HTML document, typically in the meta tag itself, are used to fool search engines into selecting a page, possibly independent of its actual content. Thus some search engines now ignore the meta tag completely.

8.3   JSON Protocol for Data Interchange

Although XML and its derivative formats dominate the metadata scene, there are other formats and protocols that are being used for successful data interchange in Online GIS systems. One of these is **JSON**, the *JavaScript Object Notation* (JSON 2014). JSON began as a component of JavaScript but is now a language entity in its own right. The language is similar to other "C" type languages and its content is easy for humans and applications to read and manipulate. The JSON language uses two common data structures to hold its data:

Name/Value pairs Attribute/value datasets as an object, record, struct, dictionary, hash table, keyed list or associative array.

Arrays Ordered list of values in an array data structure.

These simple data structures make JSON a simple language to learn, understand and implement, even for the programming novice.

One of its key strenghts for use in Online GIS, and why JSON is mentioned in this future directions chapter, is the way it can be stored in relational databases as a single record containing just the object itself. This is unlike the traditional way data is denormalised and stored across multiple records and attributes. It is a much simpler way to hold objects as single entities that can be retrieved by client appli-cations and processed. Some object relational databases are now treating JSON as a native datatype, optimising its storage and retrieval. This has significant advantages when simplifying the management of objects in Online GIS systems and why it is a protocol to consider when implementing Online GIS.

---

[3] This has been removed from HTML5.

8.3.1 JSON Protocol

JSON was created as a lightweight alternative to XML, and any additional overhead of XML is miniscule on the scale of memory, storage space and typical bandwidth available today. Unlike XML, it is not a W3C standard, but an **ECMA International** standard, ECMA-404.

ECMA began in 1961, long before the WWW as the **European Computer Manufacturers Association**. But its activities became increasingly global in scope and it changed its name in 1994 to *Ecma International – European association for standardising information and communication systems.* JSON also has a schema structure, akin to XML, which is the subject of an IETF draft (JSON 2014).

It is quite a simple standard based on unordered key-value pairs, somewhat like an ordinary associative list, called an object. Objects are enclosed in curly brackets and arrays are ordered lists of values enclosed in square brackets. Text is Unicode, but curly and square brackets are reserved, along with colon and comma outside of strings. It also has three reserved words: true, false and null.

Strings are delineated with double quotation marks and may include the following escape characters, similar to those used in UNIX or C:

\" Quotation marks

\\ Backslash (reverse solidus)

\/ Forward slash

\b Backspace

\f Form feed

\n Linefeed

\r Carriage return

\t Tab character

\u followed by 4 hexadecimal digits Hex symbol

Keys and values are separated by a colon. Key-value pairs are separated by commas. Values may be strings, objects, arrays, numbers, true, false, null. Thus objects may be nested to an arbitrary or implementation-dependent degree.

A strong indicator of the future of JSON has been the implementation of JSON by major database vendors. These vendors are adding JSON as native data types within their object-relational database products and the native use of JSON as a format for storing Big Data, Business Analytics and Online GIS is making this format an important one to consider for the light-weight interchange of spatial data objects over the WWW.

8.3.2   JSON Example Code

Here is a simple example for the JSON code used to describe a badge in the Mozilla Open Badge framework – the badge is earned for giving up bread.

```
{ "recipient": "earner.name@gmail.com",
  "evidence": "http://google.com",
  "issued_on": "2012-05-21",
  "badge": {"version": "1.0.0",
    "name": "Explorer",
    "image": "http://baked-bread.png",
    "Bronze Badge": "Has not eaten bread for five days",
    "criteria":"http://dietcentral.com/bread.html",
    "issuer": {"origin":"http://dietcentral.com/
                                      dietmaster.html",
      "name": "Thin MN",
      "org": "Gamification for Health",
      "contact": "thinman@dietcentral.com"
    }
  }
}
```

There are many other examples of JSON code now easily discoverable on the Web that show how JSON is being used for Online GIS data collection and rendering processes.

8.4   PICS and POWDER

PICS (Platform for Internet Content Selection) began as the need to protect children from unpleasant material, mainly pornography, on the Web. Although there are politicians who see censorship as necessary, many others feel that the Internet should have no restrictions provided nobody suffers unintentionally. Hence the idea behind PICS was that sites would either carry label data or their URLs would be listed in label bureaus, which would define their content. Browsers could then be configured not to accept data with particular labels. An underlying requirement was that PICS should be easy to use: parents and teachers would be able to use it effectively to block site access.

Voluntary censorship can be quite successful: managers of pornographic sites do not want to be closed down for the interests of the majority. Being able to restrict material to specific clientele is quite satisfactory. On the other hand, some sites with a strong marketing focus might be not so willing to self-regulate. Thus we need an additional mechanism of third-party rating. The academic world, for example, relies very strongly on peer group refereeing. Respectable journals contain papers which have been scrutinised by experts in the field before publication goes ahead, if at all. The PICS label bureau model looked after this.[4]

The W3C website `<http://www.w3c.org/pics>` has full lists of documentation, mailing lists and media commentaries on PICS. The website still exists but the standard has been subsumed by POWDER, as we noted at the beginning of the chapter.

There were three technical recommendations:

1. Service descriptions, which specify how to describe a rating services vocabulary; rating services and systems, which became a W3C recommendation on October 31, 1996.

2. A label format and distribution, which deals with the details of the labels themselves and how to distribute them to interested parties; label distribution and syntax also became a recommendation on October 31, 1995.

3. PICSRules, which is an interchange format for the filtering rule sets to facilitate installation or send them to servers; the rules specification became a W3C recommendation somewhat later, on December 29, 1997.

An additional recommendation was proposed on signed labels. The development of this recommendation is presumably entwined with the work on digital signatures themselves.[5]

POWDER is much more powerful and flexible than PICS, but one significant feature was dropped. PICS specifications could be embedded within HTML, using the `http-equiv meta` tag. POWDER requires a separate XML document, to which a link can be provided in the `HEAD LINK` tag. The PICS website still exists and PICS specifications should still work. As a W3C recommendation, some measure of future security has to exist. Nevertheless, the subsequent discussion now focuses on POWDER.

There are three things to consider in developing a POWDER specification:

1. What goes into the POWDER document itself? (§8.4.1)

2. Where does the POWDER document go and how does a website link to it? (§8.4.3)

3. How does the end user know that what is in the website actually matches the POWDER specification? (§8.4.2).

---

[4] There is a still unresolved issue here. If labels are kept on a site or within a page, then the network impact will be negligible. But if we have bureaus carrying labels for lots of Web pages or sites, then they potentially become network hotspots. Obviously, label bureaus need to be mirrored, but, as they grow in popularity, ensuring fast access may become very difficult.

[5] Part of the specification includes details of recording faithfully a check on the veracity of a document. The so-called MD5 message digest serves this purpose. It then has to be signed, and therein lies an ongoing tale of control, part patent, part government. Garfinkel (1995) gives a readable account, but see also Section 8.7 on XML signatures.

8.4.1   POWDER Description

As with many of the other standards we discuss in this book, the POWDER standard could fill an entire book. Thus we can at best give the key ideas and some illustrative examples. The rigour of the actual W3C recommendations is beyond the scope of the book. POWDER actually comes in two forms: a simpler, human readable version, and fuller version with formal semantics, POWDER-S, designed for enhanced artificial intelligence functionality. Our cursory overview will consider only POWDER.

A POWDER document needs two blocks of information:

1. The acquisition status: the issuer, the date issued and the authority.

2. The Description Resources (DR), which describe the resources (websites, pages, etc). There can be multiple such DRs within a POWDER document and they may refer to multiple websites.

The internet entities used in POWDER are actually **Internationalized Resource Indicators (IRI)** (IETF 2014a). These generalise the earlier **Uniform Resource Indicators (URI** (IETF 2014b) to be able to use non-ASCII character sets. The URLs with which we are familiar in accessing Web pages are in turn a subset of URIs. This distinction we shall gloss over, using URLs in most cases.

Figure 8.3 illustrates the basic structure. The first namespace is that of POWDER itself. The second is that created for the national mapping authority in Lilliput. The attribution block is then followed by two resources inside a list tag `ol`, exactly the same as used in HTML.

Inside the `attribution` tags we have the `issuedby` and `issued` tags. The issuer, in this case, is the Lilliput POWDER certification organisation. This is a URL which points to a profile (and some of the trust requirements can be met by pointing to an established or statutory authority). Figure 8.4 is defined in RDF (§8.5) and uses the Friend of a Friend namespace.

8.4.2   POWDER Description Resource

The meat of POWDER is in the one or more **Description Resources** (DR). Each DR has two components:

1. An IRI set, `iriset`, which specifies the resources (websites, pages) to which the DR applies; and

2. A set of information about the resource, `descriptorset`.

One driver behind the original PICS was to protect children from adult material. With sexting now a regular teenage pasttime, mobile phones and selfies have made sexually explicit material very widespread. Suppose Amanda wants to restrict part of her homepage, `amanda-x.com.lp`, to a somewhat risqué model shoot. The POWDER for her website now has two `iriset`s: the first deals with family friendly material, the second with the somewhat more delicate material. Each `iriset` starts

```
<?xml version="1.0"?>
<POWDER xmlns="http://www.w3.org/2007/05/POWDER#"
        xmlns:lp="http://national-mapping.org.lp">
    <attribution>
        <issuedby src="http://POWDER-cert.org.lp/
                            mapping.rdf#director"/>
        <issued>2014-11-13T05:07:00</issued>
    </attribution>
    <ol>
        <dr>
        </dr>
        <dr>
        </dr>
    </ol>
</POWDER>
```

FIGURE 8.3: The POWDER document skeleton.

```
<?xml version="1.0"?>
<rdf:RDF xmlns="http://www.w3.org/1999/
                        02/22-rdf-syntax-ns#"
        xmlns:foaf ="http://xmlns.com/foaf/0.1"/>
    <foaf:name xml:lang="en">
            Lilliput POWDER Authority
    </foaf:name>
</rdf:RDF>
```

FIGURE 8.4: Profile of the issuing authority for the example in Figure 8.3, where the namespace for RDF comes first, followed by the FOAF namespace.

with the address of her homepage, but the second specifies the sub-path with the model shoot.

```
<iriset>
    <includehosts>amanda-x.com.lp></includehosts>
</iriset>
<iriset>
    <includehosts>amanda-x.com.lp></includehosts>
    <includepathstartswith>model-shoot
                        </includepathstartswith>
</iriset>
```

Now she needs to specify that the model-shoot path is not for small children. To do this she uses the namespace, xrate, defined by the Lilliputian government for labelling salacious and other senstive material, which is found at classification.gov.lp. The rest of her website is family friendly.

```
<powder xmlns="classification.gov.lp">
...
<dr>
  <iriset>
   <includehosts>amanda-x.com.lp></includehosts>
   <includepathstartswith>model-shoot
                        </includepathstartswith>
  </iriset>
  <xrate:sex>explicit</xrate:sex>
<iriset>
<dr>
  <iriset>
   <includehosts>amanda-x.com.lp></includehosts>
  </iriset>
  <xrate:sex>family-friendly</xrate:sex>
<iriset>
</powder>
```

Now an agency, such as the government, can provide a set of guidelines and a namespace, but a Web browser accessing this information needs to know that the site actually conforms to what is claimed. The POWDER attribution gives us a release date, but websites are highly volatile. In the future, possibly quite soon, it might be viable to use an artificial agent in a Web crawler to check for compliance at frequent intervals. But for the time being, we have to rely on a combination of integrity of the website designers and administrators and perhaps random checking by the issuing authority.

We can now assemble all the pieces and the final POWDER document is shown in Figure 8.5.

Now let's consider one more example, making use of a particularly useful denotation, mobileOK, which, as the name suggests, means that a Web page or site

```
<?xml version="1.0"?>
<powder xmlns="http://www.w3.org/2007/05/powder#"
       xmlns:lp = "http://national-mapping.org.lp"
       xmlns:xrate="classification.gov.lp">

  <attribution>
    <issuedby src="http://powder-cert.org.lp/mapping.rdf#
        director"/>
    <issued>2014-11-13T05:07:00</issued>
  </attribution>
  <ol>
    <dr>
      <iriset>
        <includehosts>amanda-x.com.lp></includehosts>
        <includepathstartswith>model-shoot
                        </includepathstartswith>
      </iriset>
      <descriptorset>
        <xrate:sex>explicit</xrate:sex>
      </descriptorset>
    </dr>
    <dr>
      <iriset>
        <includehosts>amanda-x.com.lp></includehosts>
      </iriset>
      <descriptorset>
        <xrate:sex>family-friendly</xrate:sex>
      </descriptorset>
    </dr>
  </ol>
</powder>
```

FIGURE 8.5: POWDER document describing Amanda's website.

is suitable for mobile access. So in the Lilliput mapping site we have two sub-trees, one for high resolution imagery, and the second for mobile phone use, as shown in Figure 8.6

In this example we introduce a couple of new tags:

- `supportedby` is used to denote the program which is used to do the checking that the content of the website conforms to the POWDER specification.

- `displaytext` and `displayicon` are text and icons, respectively, which can be displayed to end users.

- `typeof` is the key element here. It refers to the mobileOK specification of the W3C.

8.4.3   Distribution and Validation of POWDER Description

Having written a POWDER description for our website, we now have to put it somewhere, and tell people where to find it. Where it is located is arbitrary. It could even go on the website if that does not conflict with what is said in the description. But, as we have already seen, it's desirable for the POWDER to have some level of independence from the website, to enhance its trustworthiness. As noted above, the ensuring of conformance is still not a fully developed protocol.

But even if the POWDER description is located away from the website itself, the website still needs to tell an enquiring browser where it is. Ideally the browser should get this information in the header, since it would not want to download pornographic images to find out what they were. Hence the most straightforward approach is to put the POWDER IRI in the HEAD element.

The HEAD part of an HTML document has the link element which serves this purpose precisely:

```
<link rel ="describedby"
      href="http://powder-docs.org.au/amanda-powder.xml"
      type="application/powder+xml"/>
```

The rel attribute explains that the file is resource description and the type attribute gives the mime type as POWDER. This approach has the disadvantage of requiring information in all pages and being applicable only to the HTML pages themselves, not to other formats (such as images, etc.).

A better alternative is to add the link in the HTTP response header, requiring a compliant server and administrator level access. This addresses the HTML only restriction and, even better, allows the POWDER to be accessed before any of the content is downloaded. The details are very similar to the link element:

```
Link: <powder-docs.org.au/amanda-powder.xml>;
      rel ="describedby";
      type="application/powder+xml";
```

```
<?xml version="1.0"?>
<powder xmlns="http://www.w3.org/2007/05/powder#" >
  <attribution>
    <issuedby src="http://powder-cert.org.lp/mapping.rdf#
        director"/>
    <issued>2014-11-13T05:07:00</issued>
    <supportedby src="http://lilliput-tools.lp/powder-checker
        "/>
  </attribution>
  <ol>
    <dr>
      <iriset>
        <includehosts>land-maps.gov.lp></includehosts>
        <includepathstartswith>high-res
                          </includepathstartswith>
      </iriset>
      <descriptorset>
        <displaytext>This site requires high resolution
         imaging devices and high bandwidth</displaytext>
      </descriptorset>
    </dr>
    <dr>
      <iriset>
        <includehosts>land-maps.gov.lp></includehosts>
        <includepathstartswith>phone-pages
                          </includepathstartswith>
      </iriset>
      <descriptorset>
        <typeof src="http://www.w3c.org/2008/06/mobileOK#
            conformant"/>
        <displaytext>The maps and images in this site
                   conform to mobileOK</displaytext>
        <displayicon src="http://www.lilliput-logos.lp/mobile-ok
            .png"/>
      </descriptorset>
    </dr>
  </ol>
</powder>
```

FIGURE 8.6: The Lilliput mapping website, which has two different areas, one for high resolution data and the other which is mobile friendly.

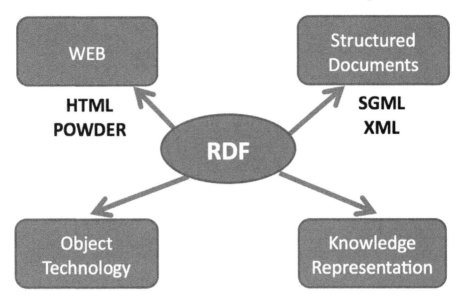

FIGURE 8.7: Concepts and technologies which play a part in the Resource Description Framework.

8.5 Resource Description Framework (RDF)

RDF has three distinct components. First, we have the RDF data model, which describes in a graphical notation the relationship between document components. Secondly, we have the serialisation of the model and a specific grammar (§8.5.7). The grammar tokens are just tokens at this stage. Their meaning and interrelationships are covered by the third component, the RDF schema. The data model and syntax became a recommendation of the World Wide Web Consortium (W3C 1999) in February 1999 and the schemas became a recommendation a month later.

RDF has utilised concepts and technologies from several different areas, as illustrated in Figure 8.7. In general, this is helpful but it can cause some confusion. Extensive use of object-oriented programming makes the structure easy to understand, but the term object also appears in the description of statements and the new term schema is used to describe an annotated class hierarchy. We will try to point out these sources of confusion as we go.

One of the key features to the success of object-oriented programming has been in software reuse. Exactly the same philosophy underlies the object-like model of RDF – reuse of metadata, particularly via inheritance mechanisms. So, for example, generic GIS data might be qualified by individual country-specific sub-classes.

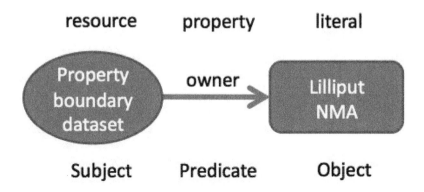

FIGURE 8.8: RDF ownership.

8.5.1 Labelled Digraph Model

A convenient way of representing the data model is via labelled diagrams. There are three components:

1. *The resource* modelled by an ellipse; and

2. *A property* modelled by an arc directed to

3. *The property value*, which may be literal, represented by a rectangle, or a further resource.

The combination of resource, property and value may be thought of as a statement in which the resource is the subject, the property the predicate and the value the object. Note that this is one of the areas where confusion of the use of object may occur.

Consider the statement "Lilliput NMA is the owner of the Property Boundary Dataset." This situation is represented graphically in Figure 8.8. Where the object concerned is a further resource, we can see that the second resource may be either labelled or unlabelled, as shown in Figure 8.9.

8.5.2 Container Model

It is perfectly reasonable to have repeated resources associated with a single subject. But sometimes we would like the set of resources to have an identity in its own right. We would then put them into a container class, known as *bag*. Thus the staff who designed and do a lot of the teaching in the Bachelor of Spatial Information Systems course (BSIS) might be referred to as an entity, with particular meetings, etc. Thus we would have on the website describing the course:

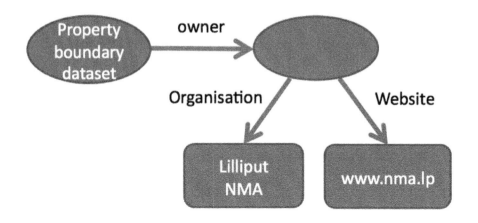

FIGURE 8.9: RDF joint ownership.

```
<rdf:RDF xmlns:sit="http://clio/sit-ns">
<rdf:description about="/bsis-staff>
    <csu:staff>
        <rdf:bag>
                <rdf:li resource=/bsis/Bill">
                <rdf:li resource="/bsis/Kate">
                <rdf:li resource="/bsis/Xihua">
        </rdf:bag>
    </csu:staff>
</rdf:description>
```

The syntax here is self-explanatory, the individual bag elements making use of the HTML list element. The `alt` container element expresses an alternative. For example, hard copy maps might be obtainable from several sources:

```
<rdf:RDF xmlns:lpi="http://lpi-ns.vir">
<rdf:description about="nsw-maps.vir/central-west">
   <nsw:mapHardCopy>
   <rdf:alt>
      <rdf:li resource=NSW Government Shopfronts">
      <rdf:li resource="Land and Property Information">
   </rdf:alt>
   </nsw:mapHardCopy>
</rdf:description>
```

The syntax here is again self-explanatory. However, the serialisaton syntax in XML does not make clear the notion of a type. Thus in Figure 8.11 the empty ellipse is a type. It does not have the same power of the type concept in object-oriented software technology.

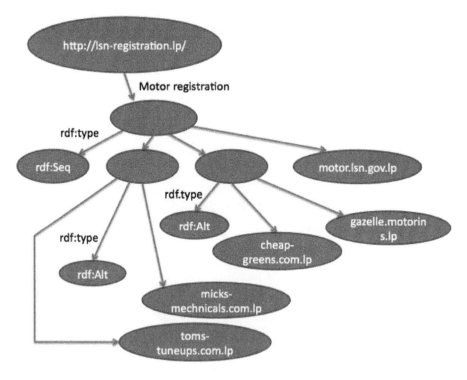

FIGURE 8.10: RDF sequence.

The final container model is a sequence for which the syntax and diagrams are exactly the same! For example, in NSW, the registering of a car involves three distinct operations, visiting three different places: testing the car at an approved testing station; purchasing third party insurance, the green slip; taking a test certificate and green slip to the registration office, or registering using these online. With digital signatures, not yet legally binding in NSW, but acceptable in Lilliput's Network State, we need only one visit to the garage. The rest happens over the Web. We select a garage, from which we receive a test certificate. An MD5 digest is signed by the garage using its private key. The signed digest is now entered into a second site, which is one of a number of insurance providers. A second MD5 signed digest is returned and the two[6] are now entered into the third and last government website where the certification is obtained. Figure 8.10 shows this sequence of events and the RDF serialisation is

```
<rdf:RDF xmlns:lrta="http://www.lrta.gov.ll/ns">
<rdf:description about="lrta-registration.ll/central-west">
  <lrta:motorRegistration>
```

---

[6] In principle, we do not actually need the original test certification, as this was a prerequisite for obtaining the insurance.

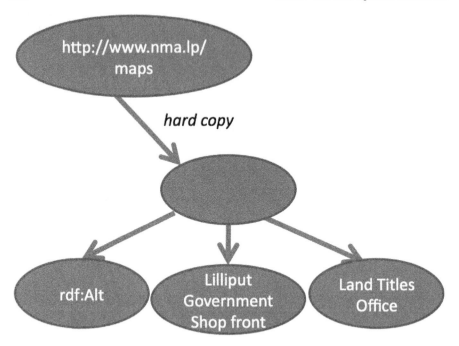

FIGURE 8.11: RDF container.

```
<rdf:seq>
  <rdf:li>
    <rdf:alt>
      <rdf:li resource="http://www.micks-mechanicals.com.ll
         ">
      <rdf:li resource="http://www.toms-tuneups.com.ll">
    <\rdf:alt>
  <\rdf:li>
  <rdf:li>
    <rdf:alt>"www.cheap-greens.com.ll">
      <rdf:li resource="http://gazelle-motorinsurance.com.ll
         ">
    <\rdf:alt>
  <\rdf:li>
  <rdf:li resource="http://www.motor.lrta.gov.ll">
  </rdf:seq>
</lrta:motorRegistration>
</rdf:description>
```

A processing agent (maybe belonging to your financial advisor) could process this description and carry out all the steps on your behalf. It would have your own criteria in choosing which of the alternatives, perhaps, say, always choosing the cheapest option, and could follow through this entire sequence autonomously. Your registration sticker is printed for you at the garage, almost the instant you enter your digital signature for transactions to proceed.

8.5.3 Formal RDF Model

The formal model as described by Lassila and Swick (Lassila and Swick 1999) has eleven features, but some of these concern statements about statements, a process referred to as reification. Ignoring reified statements, which we do not have the space to discuss, leaves us with:

1. There is a set called resources;
2. There is a set called literals;
3. There is subset of resources called properties;
4. There is a set called statements; each element (statement) is a triple pred,sub,obj, where pred is a property, sub is a resource and obj is either a resource or a literal;
5. There is an element of properties known as RDF:type;
6. Statements of the form RDF:type, sub, obj require that obj be a member of resources;
7. There are three elements of resources, which are not properties, known as RDF:bag, RDF:Seq and RDF:Alt;
8. There is a subset of properties called Ord, whose elements are referred to as RDF:_1, RDF:_2, RDF:_3, etc.

8.5.4 XML Syntax for the Data Model

RDF models have to be readable as text, for which we need a so-called serialisation syntax. This takes two forms: the first or basic syntax is the more straightforward, comprehensive and verbose; the second is the abbreviated form which handles a subset of statements in a more concise way. RDF interpreters should be able to handle arbitrary mixtures of the two.

So, let's look at a simple example.

```
<rdf:RDF>
    xmlns:rdf="http://www.w3.org/1999/02/22-rdf-syntax-ns"
    xmlns:cats="http://schemas.org/cats"
    <rdf:Description
          about="http://www.cats-of-the-world.org/
                  SuperCats/Garfield.html">
    <cats:Creator>Garfield</cats:Creator>
    </rdf:Description>
</rdf:RDF>
```

First, we specify the location of the RDF namespace. Next we specify a namespace *cats*. Now we define the metadata for a page about Garfield. The description element is the essential element. Garfield created his own page and is given as the creator. In fact we could have used Dublin Core to define the page creator too, but the beauty of the namespace framework is that we can use the term in a different way if we so desire.

8.5.5   Abbreviated Syntax

There are three strategies for abbreviation:

1. Instead of embedded description elements, we put all the attributes into a single description providing there is no name clash;
2. We make embedded descriptions resource attributes;
3. The type property becomes an element name directly.

8.5.6   Properties of Container Elements

Sometimes we want to make a statement about all the pages in a container, not about the container itself. Consider the following set of pages denoting maps in Lilliput.

```
<rdf:bag ID="maps">
    <rdf:li resource="http://landinfo.gov.ll/
                                redMountains">
    <rdf:li resource="http://landinfo.gov.ll/
                                forestLake">
</rdf:bag>
<rdf:Description aboutEach=#maps>
    <llm:mapMaker>Gulliver</llm:mapMaker>
</rdf:Description>
<rdf:Description about=#maps>
    <llm:custodian>
        Lilliputian Mapping Authority
    </llm:custodian>
</rdf:Description>
```

The effect here is to define Gulliver as the map maker for each map in the set using the `aboutEach` construction, while the custodian of the collection is the Lilliputian Mapping Authority.

8.5.7   RDF Serialisation

The primary model for RDF is the graph model, but for many purposes it is desirable to have an RDF description as a single sequence, such as the XML documents we have already seen. The XML serialisation is now just one option, and in 2014 two additional options reached W3C recommendation status: **Turtle** (§8.5.9) and **N-Triples** (§8.5.8). Both have precisely defined grammars, but, as elsewhere in this chapter, space will constrain us to picking out some key features with illustrative examples. Since N-Triples is the simpler, and is a subset of Turtle, we begin with it.

8.5.8 N-Triples

N-Triples became a W3C Recommendation in February 2014 (W3C 2014a). It is a line by line serialisation of an RDF graph, nothing more. Each line comprises a subject, predicate and object separated by spaces[7] terminated by a period:

- The subject is either an IRI (URL) or a blank node label (the empty boxes in Figure 8.10);

- The predicate is an IRI;

- The object can be an IRI, string (in quotation marks) or both. The string can have an optional language tag, such as @en. If the IRI is present too, it is joined to the string by ^^

```
"Garfield"@en^^<http://garfield.com>|
```

As each relationship in the graph is made into a triple, the same blank node may occur more than once. Hence the serialisation gives blank nodes arbitrary names, beginning with the reserved sequence, _:, such as _:fred. Thus we might have

```
_:garfield <http://xmlns.com/foaf/spec/#term_knows>
                                    _:everybody.
```

8.5.9 Turtle

Turtle extends the simple list of subject-predicate-object of N-Triples with additional features, which amongst other things, make complicated RDF documents easier to read. Before the list of triples a Turtle document can have a set of prefixes, which are just abbreviations for later use in the document. Each begins with the @ symbol, and the first line would usually be @base, the reference URI for the document as in the following example:

```
@base <http://lilliput-mapping.gov.lp/>
@prefix foaf: <http://xmlns.com/foaf/0.1/>
```

We can now abbreviate URLs (# introduces a comment):

---

[7] In the canonical form of N-Triples, there should be one and only one space.

TABLE 8.5: Turtle Lake code example in full and abbreviated form

```
@base <http://lilliput-mapping.gov.lp/>
$prefix lu: <http://land-use.lilliput-mapping.gov.lp/>
_:lake located "South Lilliput".
_:lake size "100 hectares".
_:lake lu:typeof" lu:lake.
_:lake lu:usage _:uses.
_:uses lu:transport lu:ferry.
_:uses lu:leisure-use _:watersport.
:_watersport lu:boating "Sailing".
:_watersport lu:boating "Canoeing".
# Now we rewrite these in abbreviated form
@base <http://lilliput-mapping.gov.lp/>
$prefix lu: <http://land-use.lilliput-mapping.gov.lp/>
[located "South Lilliput";
 size "100 hectares";
 lu:typeof" lu:lake]
 lu:usage
   [lu:transport lu:ferry;
    lu:leisure-use
      [lu:boating "Sailing";
       lu:boating "Canoeing"
      ]
   ]
]
```

```
@base <http://lilliput-mapping.gov.lp/>
@prefix foaf: <http://xmlns.com/foaf/0.1/>
<lakes> # same as
<http://lilliput-mapping.gov.lp/lakes>
<foaf:name> # same as
<http://xmlns.com/foaf/0.1/name>
```

Turtle allows more complex triples with multiple predicates and objects. A predicate may have multiple objects separated by commas. Multiple predicates, each of which could have multiple objects, are separated by semicolons.

The only other extensions we will include here are the grouping syntax for blank nodes and collections. A blank node may have a predicate list, enclosed in square brackets. A collection is a set of objects enclosed in normal parentheses '(' and ')'.

Figure 8.12 shows a more complicated example. Table 8.5 spelt out in full as a set of triples, and then in the abbreviated form.

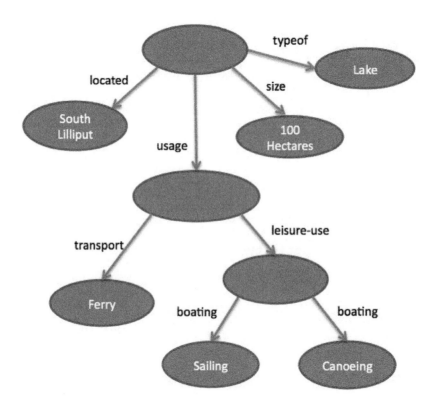

FIGURE 8.12: The Turtle Lake example.

8.6   Querying RDF: SPARQL

Given that the goal of the Semantic Web is for Web data to be machine readable and useable, then some form of query language is needed for RDF. That language is **SPARQL**, a recursive acronym for SPARQL Protocol and RDF Query Language (W3C 2014b).

Like many of the standards touched on in this book, SPARQL is complicated, comprising no fewer than 11 documents for SPARQL 1.1, the latest at the time of writing.

SPARQL queries look a lot like generic SQL. The output may have several forms, however: XML, JSON (§8.3) and spreadsheet-like formats such as CSV and TSV (comma and tab separated values), respectively. We will start with CSV, as the simplest to understand.

Using the lake example (Table 8.5), a basic query involves a SELECT keyword for the search item(s) and a WHERE keyword to specify the pattern as an RDF triple. We then have

```
@base <http://lilliput-mapping.gov.lp/>
$prefix lu: <http://land-use.lilliput-mapping.gov.lp/>
SELECT ?activity
WHERE
{
 ?x lu:boating ?activity
}
```

which returns

```
activity
sailing
canoeing
```

In XML the output looks much more complicated:

```
<?xml version="1.0"?>
<sparql xmlns="http://www.w3.org/2005/sparql-results#">
 <head>
  <variable name="activity"/>
 </head>
 <results>
   <result>
     <binding name="activity">
       <literal>Sailing></literal>
   </result>
   <result>
     <binding name="activity">
```

```
        <literal>Canoeing></literal>
    </result>
  </results>
</sparql>
```

Firstly, we have an XML header and then a SPARQL namespace wrapper `sparql`. Then we have two elements: the `head` element containing the variables (in this case `activity`) and the `results` element with the data retrieved.

The results in JSON are essentially the same, but with the more economical JSON syntax.

```
{
  "head": { "vars": [ "activity" ] } ,
  "results": {
    "bindings": [
      {
  "activity": { "type": "literal" , "value": "Sailing" }
      } ,
      {
  "activity": { "type": "literal" , "value": "Canoeing" }
      }
    ]
  }
}
```

Just like SQL, there are many structures for complex queries, involving multiple variables. The interested reader will find the W3C standard documents quite readable (W3C 2014b).

8.7 XML Signatures

XML signatures[8] are a specification for adding digital signatures to XML documents. For our present needs we need to consider three aspects:

1. Public/private key cryptography (§8.7.1), which provides the digital identification and validation;

2. Message digests (§8.7.2), which compress a document into a very small, unique digest file;

3. The XML signing schema (§8.7.3).

---

[8] http://www.w3.org/TR/xmldsig-core/

8.7.1 Public/Private Key Cryptography

The cornerstone of modern cryptography is the idea of public and private keys. The formative algorithm was introduced by Rivest, Shamir and Adleman in 1978 and is widely known as the RSA algorithm (Stallings 1995). This algorithm is propri-etary, but Phil Zimmerman[9] wrote a package to implement public/private keys, called PGP (Pretty Good Privacy)[10] (Garfinkel 1995). Although entangled in legal issues for some years, it became very widely used and ultimately a commercial product. A free version continues from the GNU project.[11]

A cryptography algorithm generates two keys. One is used to encrypt something, the other to decrypt it. If Amanda wants to send an encrypted message to Bertie, she encrypts it using Bertie's public key. Only Bertie can read it, using his private key. On the other hand, if she wants to send a message to Bertie, and maybe other people, she encrypts it with her private key. Anybody can now decrypt it with her public key. It is this latter option which underlies digital signatures.

To be secure, the private keys have to be unguessable, not just by a person, but by supercomputers using huge amounts of computer time. The algorithm which made all this possible was the RSA encryption algorithm. The idea is quite simple. The message M is represented as a number (say a plain text Ascii encoding) and is turned into the encrypted message using the formula

$$C = M^u mod N \tag{8.1}$$

where u and N are seriously large numbers, i.e., hundreds of digits long.

To retrieve the message we use the same equation, but with a different value of the exponent:

$$M = C^v mod N \tag{8.2}$$

How we find u, v and N is an exercise in number theory. Basically N is set equal to the product of two very large prime numbers, p, q, $N = pq$. The exponents are then found from number theory dating back to the great 18th century Swiss mathematician Leonhard Euler (Stallings 1995). The larger the prime numbers used, the more secure the code, more secure meaning the longer it would take a computer to find it by exhaustive searching.

8.7.2 Message Digests

The second building block we need is the idea of a message digest. For a digital signa-ture to be of any use, it must be possible to determine that the document has not been changed since the signature was applied. But for any reasonably sized document, encrypting the whole using the public-private key algorithm would take forever. The secret is to use a message digest.

---

[9] http://philzimmermann.com/EN/background/index.html
[10] http://www.pgpi.org/
[11] https://www.gnupg.org/

```
<?xml version="1.0" encoding="utf-8"?>
<Signature Id="director" xmlns="http://www.w3.org/2000/09/
    xmldsig#">
  <SignedInfo>
    <CanonicalizationMethod Algorithm="http://www.w3.org
        /2006/12/xml-c14n11"/>
    <SignatureMethod Algorithm="http://www.w3.org/2000/09/
        xmldsig#rsa-sha1"/>
    <Reference URI="http://maps.gov.lp/annualreport.xml#budget
        ">
      <DigestMethod Algorithm="http://www.w3.org/2000/09/
          xmldsig#sha1"/>
      <DigestValue>sflkwekrewhcae252esrw3csaw3rq3...</
          DigestValue>
    </Reference>
  </SignedInfo>
  <KeyInfo>
    <KeyName>DirectorKey</KeyName>
  </KeyInfo>
</Signature>
```

FIGURE 8.13: Example of an XML Signature.

The idea is simple. We pass the document through a *hash* function, which produces a relatively tiny *digest*. This digest, although much smaller, is very, very unlikely to be generated by any other document. Suppose the digest has 128 bits. Then the number of possible digests is $2^{128} \approx 10^{38}$, which is much larger than the World Wide Web or the total number of documents in human history. Hash functions are designed to make a collision, where two documents have the same hash value, almost impossible. So even a small change to a document would create a completely different digest (hash) value.

For a long time the message digest used was called MD5. But some security weakness in this algorithm has emerged, and the US Government, switched to the *Secure Hash Standard*, comprising algorithms such as SHA-1. [12] SHA-1 creates a digest of 160 bits, but others in the family create larger digests SHA-384, which is 384 bits, which is effective for documents of up to 2^{32} bytes, that's 100 million yottabytes, a yottabyte being a quadrillion (10^{24}) bytes, the largest data size with a name (so far).

8.7.3 Creating an XML signature.

Figure 8.13 contains a brief example of an XML signature. We repeat our usual mantra, that this is a very simplified view of what the full specification contains.

---

[12] http://csrc.nist.gov/publications/

The `SignedInfo` element contains all the information on how the document is digested and signed in a series of sub-elements: `CanonalisationMethod`; `SignatureMethod`; and `Reference` for the actual thing(s) being signed.

XML documents can contain a lot of slack, yet still validate against the appropriate schema. They may contain extra white space, the attributes may be listed in different orders and so on, without affecting the document's validity. The signature system shouldn't depend on these minor variations. Hence, before computing the digest, the XML is *canonalised* to bring it into a canonical form as specified in the W3C recommendation.[13] The `CanonalisationMethod` takes a canonalisation attribute containing a URI for the algorithm to be used.

The `SignatureMethod` contains the algorithm for signing, in this case the SHA-1 digest and the RSA cryptography algorithm. The `Reference` element contains the required `DigestMethod` and `DigestValue`, which as expected give the digest algorithm (SHA-1) and the digest itself, respectively. There can be more than one `Reference` element.

The `KeyInfo` is optional, if the document is only likely to be used by people who know where to find the key. There are a variety of choices for the sub-elements. This example uses the simplest, which is just a name for the key.

8.8   Summary and Outlook

This chapter focused on the technologies underlying Web metadata, without specific reference to Online GIS.

- §8.1 introduced the framework of Web metadata.

- §8.2 discussed one of the earlier, but still powerful, metadata frameworks, the Dublin Core. This is a keyword-value specification, predating XML descriptions.

- Of similar style to Dublin Core is a more general framework, also sitting outside the XML domain, the JavaScript Object Notation, discussed in §8.3.

- One key area of metadata application is in providing information about the accessibility and sensitivity of the content of Web pages. The earlier proposal PICS and the current standard POWDER are discussed in §8.4.

- The big standard for describing resources on the Web is RDF, the Resource Description Framework, discussed in §8.5, while §8.6 discusses SPARQL, used for querying RDF documents.

- Finally we considered in §8.7 XML signatures, important for commercial and other sensitive transactions on the Web.

---

[13] http://www.w3.org/TR/2007/CR-xml-c14n11-20070621/

9

SDI Metadata Portals and Online GIS

CONTENTS

Spatial Data Infrastructure (SDI) Metadata Portals are online marketplaces for the publication, search and discovery of geospatial datasets. These GeoPortals help users to find geographic datasets. They are built using interchangeable standards and play an important role in the socioeconomic development of countries through planning and management of critical infrastructure such as transport, land administration and water management. The standardisation of metadata is fundamental to the sharing of these datasets through online publication and discovery portals, and in providing users with meaningful information on how fit-for-purpose each spatial dataset is. In this chapter, we will discuss the state of geospatial metadata standards and review a small sample of global SDI implementations empowering Online GIS systems.

9.1 Introduction

The term Spatial Data Infrastructure (SDI) was used as early as 1990 in the US National Research Council Mapping Science Commmittee's report, Spatial Data Needs: The Future of the National Mapping Program.[1] It later formed part of the National Spatial Data Infrastructure (NSDI) initiative in the US and was soon followed by a number of other SDI initiativies in Australia, Canada, Germany and Europe. In principle, these SDI Metadata Portals are used to publish, search and discover spatial datasets online and will be discussed in more detail below in §9.5 and §9.6.

It is worth pointing out that Natural Resources Canada produced a very useful document in 2013 covering issues to be considered when planning, developing and implementing an SDI called Spatial Data Infrastructure (SDI) manual for the Americas'[2], and is a good introduction to SDI.

This chapter discusses SDI Metadata Portals and the standards used for these Online GIS marketplaces. Firstly, it revisits the ANZLIC Metadata framework covered in the first edition of this book. This early online metadata framework included

---

[1] http://www.nap.edu/openbook.php?record_id=9616
[2] http://geoscan.nrcan.gc.ca/cgi-bin/starfinder/0?path=geoscan.fl&id=fastlink&pass=&search=r=292897&format=FLFULL

a standard which used an SGML Document Type Definition (DTD) to define the ANZLIC metadata standard.

It then reviews the roles of prominent geospatial metadata standards from ISO, the Federal Geospatial Data Committee[3] (FGDC) and the OGC. These standards are in wide use across a number of metadata portals for geospatial dataset publication. Spatial Data Infrastructure is then discussed with reference to the geospatial metadata, and a small number of existing SDI Metadata Portals and Data Catalogues are introduced to give the reader an insight into the global nature, usage and importance that metadata standards are playing in Online GIS.

9.2   Geospatial Metadata Standards

There are three popular geospatial metadata standards from standards organisations that we will cover in this section. These standards are in use across a wide number of metadata portals for geospatial dataset publication.

However, before we look at each, we will review earlier metadata standards work in Australia as a way of introducing a simple XML based metadata standard.

9.2.1   Early Australasian Standard and DTDs

One of the early implementations of standardised geospatial metadata was the standards developed for Australia and New Zealand by the Australia New Zealand Land Information Council (ANZLIC).[4] This committee adopted a policy for the transfer of metadata in 1994 and formed a working group in April 1995 to develop a metadata framework. The process was a consultative one, leading to the shared ownership of the standards in this framework, which contributed to its strong uptake. This contrasts with the position in other countries where standards were imposed from above.

As discussed in Chapter 5, a Document Type Definition (DTD) describes each of the metadata record elements. As such, it is not the metadata records themselves, but the document that defines the structure of the documents that conform to the standard defined by the DTD.

ANZLIC released its initial Document Type Definition (DTD) in January 1998. Based on SGML, metadata records could be easily parsed and shredded into individual elements using the DTD. Using XML also gave organisations the ability to reject records that did not conform to the DTD. While this DTD has now been replaced by Catalogue Services for the Web (CSW) (see §9.4.3), we have retained some of the DTD content from version one of this book as a way of looking at the concepts of spatial metadata standards.

---

[3] https://www.fgdc.gov/
[4] A cross jurisdictional committee in Australasia.

9.2.1.1   ANZLIC DTD Outlook

The ANZLIC DTD was a strong implementation of a metadata standard at the time it was developed. The use of SGML made it was easy to understand and covered core data elements well, and as the DTD borrowed a range of elements from the HTML DTD for describing text structure, such as list elements and paragraphs, many users were already familiar with its definition. The use of thesauri was very effective, and the metadata processing tools leveraged the qualities of SGML and helped to check that the correct data elements are added at each stage. While the SGML concrete reference syntax specified a maximum of eight characters for an element name, this was overridden easily and the element names were largely aligned with US standards discussed in this section.

There were, however, limitations in the ANZLIC approach. With many of the software vendors based outside Australasia and each implementing local metadata standards, it was expensive to get the vendors to support the ANZLIC DTD. The element names are sometimes cryptic, much as in the computer programs of yesteryear in which short names were enforced by computer memory and other limitations. An alternative to specially developed checking tools would have been a detailed XML schema (Chapter 5) highlighting an important issue: much of the groundwork on spatial metadata has run alongside the development of new recommendations in metadata by the W3C. As a consequence, some of the tools are different or incompatible with emerging Web standards, which impeded the adoption of the ANZLIC standard.

A related issue was the way metadata documents may be clustered together. There was no specification, leaving the organisation to structure the website and Spatial Data Directory itself. A more comprehensive object-oriented model, such as implied within RDF (§8.5), did offer the advantages of

1. Easier searching through a hierarchy;

2. Reduced duplication of data where a group of datasets all belong to the same organisation.

Other standards developments improved interoperability, resulting in a move away from DTD for ANZLIC.

9.3   Metadata Standards: The Gory Details

The ISO/TC211 standard ISO19139 is now very comprehensive. We have already seen *gml*. The complete schema set now comprises eleven separate categories each with a number of schema documents (72 in total from the 2005 set): gco (3); gfc (1); gmi (6); gmd (17); gml (28); gmx (7); gplr (2); grg (2); gsr (2); gss (2); gts (2). Of these, the key metadata standard is **gmd,** which makes extensive use of **gco**, the Geographic COmmon (GCO) extensible markup language, and **gml**. Our study of a metadata standard will be thus based on *gmd*.

These schemas were not always crafted by hand but sometimes generated automatically from UML diagrams. This has an unfortunate consequence for readability, since the schema generators will often create deep type hierarchies of which elements in the tree may be largely empty. This is fine if the schemas are used by a metadata document generator tool, say wysiwyg, graphical or menu driven software, and they meet the Semantic Web needs for machine interpretable software (Berners-Lee 2014). But this is not much help for understanding the structure.

The UML diagrams corresponding to the ISO 19139 standard are themselves very complicated, and a detailed discussion of them, which requires UML expertise, would lengthen the text considerably. Thus we have adopted an intermediate approach, extracting elements of the schemas to create a minimalist version against which to parse the example in this section. In some cases, we have then simplified some elements to their prime content, which is usually `string` from the main XML schema specification.[5]

When we come to the `fileIdentifier`, discussed below, we see how general and precise the schemas are, perhaps at the expense of verbosity. A file identifier could be a variety of things, a simple string, a decimal or hexadecimal number and so on. So the `fileIdentifier` element encloses a `CharacterString` element. In fact the 2005 schema does not yet admit significant diversity in the file identifier!

Both the schemas and the XML files make extensive use of `import` and `include` tags(§5.4.2.2). Note that the attribute used here is different: `href` for including XML documents and `schemalocation` for including schema components. Note that we have to include the namespace in each separate file. This just goes in the root element for the file, be it an XML metadata file or a schema (which is XML, of course, anyway).

The schemas typically have a prefix for the type elements. The main ones, which we see here, are

MD_ denotes a metadata element

CI_ denotes citation information

LI_ denotes lineage (history)

DQ_ denotes data quality

The elements themselves derived from these types are a mixture of prefix names such as these and unprefixed camel case.

9.3.1   MD_Metadata Element

The metadata for our example comes from Jonathan Swift's imaginary country of Lilliput. The reduced schemas comprise *tgmd.xsd*, based on the `gmd` schema files; *tgco.xsd*, based on `gco`; and *tdq.xsd*, based on the data quality schema with `gmd`. Note that these schemas cross-reference each other in non-trivial ways, so allocating

---

[5] `http://www.w3.org/2001/XMLSchema`

an element of type completely to one schema is often not possible. For the purposes of this example we have consolidated all the schema fragments into a single namespace, denoted in the schema by `tgmd:` and in the XML files by `quoll:`.

Here is the top level metadata XML, which includes several subfiles for contact details, citation information, data quality, maintenance and distribution discussed in Sections 9.3.1.1, 9.3.1.2, 9.3.1.3, 9.3.1.4 and 9.3.1.5, respectively.

```
<?xml version="1.0" encoding="utf-8"?>
<quoll:MD_Metadata xmlns:quoll="http://www.opengis.net/giraffe
    "
                xmlns="http://www.w3.org/2001/XMLSchema"
                xmlns:xi="http://www.w3.org/2001/XInclude">
  <quoll:fileIdentifier>
   <quoll:CharacterString>
    lilliveg.xml
   </quoll:CharacterString>
  </quoll:fileIdentifier>
  <quoll:language>
   <quoll:CharacterString>
    English
   </quoll:CharacterString>
  </quoll:language>
  <xi:include href="md-contact.xml"/>
  <quoll:dateStamp>2010-02-01T11:00:00</quoll:dateStamp>
  <quoll:metadataStandardName>
   <quoll:CharacterString>
    LilliputMapMetadata
   </quoll:CharacterString>
  </quoll:metadataStandardName>
  <quoll:metadataStandardVersion>
   <quoll:CharacterString>
    1.2
   </quoll:CharacterString>
  </quoll:metadataStandardVersion>
  <quoll:dataSetURI>
   http://www.lilliput-mapping.gov.lp/53908.html
  </quoll:dataSetURI>
  <xi:include href="md-citeinfo.xml"/>
  <xi:include href="md-dataqual.xml"/>
  <xi:include href="md-maintenance.xml"/>
  <xi:include href="md-distinfo.xml"/>
</quoll:MD_Metadata>
```

The tags in the top level file are largely self-evident. Note that we have to specify the namespace for the *XInclude* schema. This is normally picked up by the processing engine (such as *xmllint*) (§5.4.2).

9.3.1.1   Contact Element

The contact element is straightforward:

```
<quoll:contact xmlns:quoll="http://www.opengis.net/giraffe">
  <quoll:individualName>
    Tobias Quunk
  </quoll:individualName>
  <quoll:organisationName>
    Lilliput Surveying and Land Information Group (LPLIG)
    Inquiries to Data Sales Staff.
  </quoll:organisationName>
  <quoll:positionName>
    Director Mapping Services
  </quoll:positionName>
  <quoll:contactInfo>
    <quoll:phone>
      Lilliput Landline Number
      +71 6332 4455569
      Australia Freecall
      1800 800 173
      Lilliput Fax Number
      +71 6332 4455567
    </quoll:phone>
    <quoll:address>
      <quoll:deliveryPoint>
        24 Short Street
      </quoll:deliveryPoint>
      <quoll:city>Lillyput</quoll:city>
      <quoll:administrativeArea>Scriblerus</quoll:
          administrativeArea>
      <quoll:postalCode>37387</quoll:postalCode>
      <quoll:country>Lindalino</quoll:country>
      <quoll:electronicMailAddress>
        onlinedatasales@lilliput.gov.li
        onlinemapsales@lilliput.gov.li
      </quoll:electronicMailAddress>
    </quoll:address>
  </quoll:contactInfo>
</quoll:contact>
```

It has the `CI_ResponsibleParty_PropertyType`, which has sub-elements of `individualName`, `organisationName` and `positionName`. Note that these are given in the schema as a `sequence` and thus have to be in this precise order. However, each has the attribute `minOccurs` set to zero, which means one or more may be omitted.

9.3.1.2   IdentificationInfo Element

The `identificationInfo` has very little surprising in it:

```
<quoll:identificationInfo xmlns:quoll="http://www.opengis.net/
    giraffe">
  <quoll:citation>
```

```
  <quoll:CI_Citation>
    <quoll:title>
      Vegetation: Pre-European Settlement (1788)
    </quoll:title>
  </quoll:CI_Citation>
 </quoll:citation>
 <quoll:abstract>
  Shows a mapping of population density in Lilliput
  as it would have been prior to 1700. Villages
  on the island of Lilliput with populations over
  500 are shown by location, along with the spatial
  extent of these villages. Attribute information
  includes village name, estimated population and
  village location (x,y) and area in square chains.
 </quoll:abstract>
 <quoll:descriptiveKeywords>
  POPULATION Census Mapping
  SETTLEMENT Mapping
  Lilliput
 </quoll:descriptiveKeywords>
 <quoll:language>Eng</quoll:language>
 <quoll:topicCategory>environment</quoll:topicCategory>
 <quoll:supplementalInformation>
  Source: Lilliput Online Metadata Portal.
  -------------------------------------------------
  Population density impacting urban sustainability
  -------------------------------------------------
  1:10,000
  -------------------------------------------------
  RESTRICTIONS ON USE
  None
  -------------------------------------------------
  20GB to 40GB depending on file format
  -------------------------------------------------
  PRICE and ACCESS
  RRP $99
 </quoll:supplementalInformation>
</quoll:identificationInfo>
```

This metadata describes a dataset of Lilliput population prior to the arrival of Lemuel Gulliver. Note that some elements have content which is only other elements, giving us a cascade of opening and closing tags.

The identificationInfo contains in sequence the citation, abstract, descriptiveKeywords and language and topicCategory. Plain text suffices for the abstract, keywords and language. The citation contains sub-elements of CI_Citation, which is itself a wrapper for the title. The topicCategory element is of type MD_TopicCategoryCode_Type, which is an enumerated type, rather than a free string. So there are other options to *environment*, such as *boundaries, climatology, meteorology, atmosphere, economy, elevation*, but the options are limited and have to be given **exactly** in the form given in the schema. Finally, we have supplementary information, which here is just an arbitrary block of text.

9.3.1.3 dataQualityInfo Element

Finally we come to the data quality, an important chunk of metadata for any dataset:

```
<quoll:dataQualityInfo xmlns:quoll="http://www.opengis.net/
    giraffe">
 <quoll:DQ_DataQuality>
  <quoll:scope>
   <quoll:DQ_Scope>
    <quoll:level>
      <quoll:MD_ScopeCode
         codeList="http://lilliputmap.gov.lp/profileinfo/
             scopecodes"
         codeListValue="lpcodes">
           Lilliput code lists
      </quoll:MD_ScopeCode>
    </quoll:level>
   </quoll:DQ_Scope>
  </quoll:scope>
  <quoll:report>
   <quoll:nameOfMeasure>
    ????
   </quoll:nameOfMeasure>
   <quoll:measureDescription>
    ????
   </quoll:measureDescription>
   <quoll:evaluationMethodDescription>
    ????
   </quoll:evaluationMethodDescription>
   <quoll:dateTime>2015-03-10T09:00:00</quoll:dateTime>
   <quoll:result>Passed
   </quoll:result>
  </quoll:report>
  <quoll:lineage>
   <quoll:LI_Lineage>
    <quoll:statement>
      Captured from mapping material used to produce
      Lilliput Large Scale Mapping Series 1699.
    </quoll:statement>
   </quoll:LI_Lineage>
  </quoll:lineage>

 </quoll:DQ_DataQuality>
</quoll:dataQualityInfo>
```

There is a cascade of tags here, each of which may have additional elements and attributes. The ones we have selected here are mostly easy to understand, but again, note from the schema, that they are a `sequence` and must be in the precise order specified. Some tags, such as `dateTime`, have a restricted format, the ISO8601 date format.[6] Here we start to see some of the difficulties of handling legacy data: a lot of the characteristics are just not known, were never recorded or have been lost.

Dates were given in ISO8601 format for the earlier ANZLIC DTD. Note that there was a move within the online community against using text for months at the time as purely numeric formats were preferred, viz. 2015-04-01.

9.3.1.4 Maintenance

Currency and update information is crucial to the usefulness of a dataset. This information is provided in the `metadataMaintenance` element:

```
<quoll:metadataMaintenance xmlns:quoll="http://www.opengis.net
    /giraffe">
  <quoll:maintenanceAndUpdateFrequency>
   <quoll:MD_MaintenanceFrequencyCode
      codeList="http://lilliputmap.gov.lp/maintenance.html"
      codeListValue="maintenanceSchedules">
     Maintenance data
   </quoll:MD_MaintenanceFrequencyCode>
  </quoll:maintenanceAndUpdateFrequency>
  <quoll:dateOfNextUpdate>2015-07-01T12:00:00</quoll:
     dateOfNextUpdate>
</quoll:metadataMaintenance>
```

Here we have specified the maintenance update schedule (or actually the URLs where this may be found) and the date of the last update.

9.3.1.5 Distribution Information

The manner and means of distribution are diversifying daily. Here we have two relatively standard options:

```
<quoll:distributionInfo xmlns:quoll="http://www.opengis.net/
    giraffe">
  <quoll:distributionFormat>
   ARCInfo, Vector Data
  </quoll:distributionFormat>
  <quoll:distributor>
   Lilliput land information
  </quoll:distributor>
  <quoll:transferOptions>
   Online, DVD purchased from kiosk
  </quoll:transferOptions>
</quoll:distributionInfo>
```

---

[6] http://www.iso.org/iso/home/standards/iso8601.htm

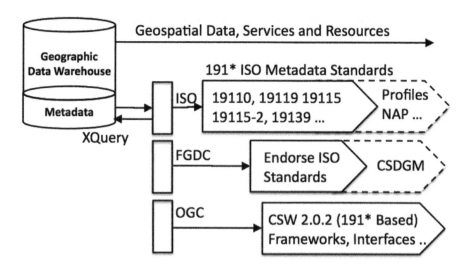

FIGURE 9.1: Metadata standards hierarchy.

The data is in available as Vector data in Esri Shape file format. It is available in the following `transferOptions`: online or for purchase from a mapping kiosk on a DVD.

9.4 Prominent Metadata Standards

Underpinning the success of Online GIS were the geospatial metadata standards developed by organisations including the International Organisation for Standardisation (ISO), Federal Geographic Data Committee (FGDC), the Open GIS Consortium (OGC), the International Society for Photogrammetry and Remote Sensing (IS-PRS), the European Spatial Data Research Network (EuroSDR) and the Committee on Earth Observation Satellites (CEOS). The standards developed by these organisations facilitate the effective discovery and distribution of geospatial information.

In Figure 9.1, the data services and resources are shown at the top, coming from a Geographic Data Warehouse. Below this arrow are the three standards organisations, with ISO providing a number of metadata standards through a family of standards known as 19100. FGDC endorses these standards and promotes their use within the USA. FGDC also has its own metadata standard, CSDGM, although it believes that, over time, metadata in this format will be phased out for the ISO standards. Finally, OGC provides a number of standards for metadata publication, based on the 19100 standards (amongst others). OGC also provides a number of frameworks and interfaces specific to metadata publication and query, one of these being for cataloging of metadata records.

In the following subsections (§9.4.1, §9.4.2 and §9.4.3), we discuss three main standards-based organisations that are developing and maintaining global metadata standards, and the relationship between these organisations. These organisations are the ISO, FGDC and OGC. The number of standards now in use across the geospatial industry is significant, too many to cover in any one publication, let alone this book. While many of these are based on XML, many of the industry-authored standards will not be discussed here.[7]

9.4.1   ISO and International Geospatial Metadata Standards

The International Organisation for Standardisation (ISO) (ISO 2014) is an international standards organisation headquartered in Geneva, with 165 member countries. Working as an independent and non-governmental organisation, it has published almost 20,000 standards,[8] making it the world's largest developer of international standards. Among these many standards, ISO has published a number of metadata-specific standards to describe geospatial information. These metadata standards now play a significant role around the globe, enabling publication and discovery of geospatial datasets through online catalogues and spatial marketplaces. The standards have also provided a standardised way of providing important metadata as part of the normal geospatial data supply chain and as part of Web services based data supplies and online geospatial data rendering.

These metadata standards for the publication of geographic datasets are contained in ISO's 191** series of standards, and include

ISO 19115:2003 ISO 19115 was an early standard developed to describe geographic information. The standard provided organisations with a standard way to document vector data types found in GIS systems, and a documentation standard to describe geospatial Web services, catalogues[9] and data models. This initial standard formed the basis of the other ISO metadata standards that followed.

ISO 19115-2:2009 ISO 19115-2, ratified in 2009, extended 19115 to include raster and grid data such as digital terrains. The standard also contains elements to describe GPS data and data generated by survey instruments, monitoring stations and other devices. Because this standard also contained all elements from the earlier ISO 19115 standard, organisations used this newer standard, as it was able to describe in one standard all of their geospatial data needs.

ISO 19110:2005 The 19110 standard from ISO was termed a methodology standard for geographic data. This standard can be used in conjunction with ISO 19115 to document geospatial feature types such as roads, rivers and boundaries, or it can be

---

[7] Strictly speaking, FGDC is focused on improving spatial infrastructure in the US; however, it has played a leadership role in development and adoption of metadata standards globally for decades.

[8] http://www.iso.org/iso/home/about.htm

[9] Collections for descriptions of datasets where each description is a metadata record.

used independently of 19115 to document data models. The standard uses an entity and attribute model to describe geographic objects in a similar way to FGDC's CSDGM.

ISO 19119:2005 This standard supports the detailed description of digital geospatial services including geospatial data portals, Web mapping applications, data models and online data processing services. As with 19110, this standard can be used in conjuction with 19115 to describe services belonging to a geographic dataset described by 19115, or can be used independently of 19115 to document a geospatial service.

ISO 19139:2007 This metadata standard is the `implementation` standard for 19115. Defined by an XML schema, developers use this XML document when developing applications that process 19115 metadata records.

ISO 19115-1:2014 ISO 19115-1 is a metadata standard that has only been recently ratified (2014). This standard describes both geographic data and services. It provides information about "the identification, the extent, the quality, the spatial and temporal aspects, the content, the spatial reference, the portrayal, distribution, and other properties of digital geographic data and services."[10]

The standard defines a number of mandatory and optional elements, entities and sections, and can be applied to other non-digital geographic data such as maps and charts. There are a number of changes to base elements in 19115-2 and support for additional geospatial data types which makes it applicable to many types of Online GIS metadata applications in online catalogues and spatial marketplaces. 19115-2 metadata records can also contain elements and attributes from other ISO metadata standards.

9.4.2 FGDC and Content Standards for Digital Geospatial Metadata

FGDC is a US Federal government agency responsible for the enforcement of metadata standards in the US when organisations are capturing geographic datasets. One of its primary objectives is to endorse a number of ISO standards and to promote their use across government. It firmly endorses ISO metadata standards to ensure the usefulness of these data to other organisations.

While this *enforcement* could be considered voluntary, using these standards in accordance with the FGDC Policy on Recognition of Non-Federally Authored Geographic Information Standards and Specifications, the Office of Management and Budget's Circular A-119 established clear policies linked to funding on Federal use and development of such standards.

In addition to the endorsed ISO standards, FGDC has implemented a standard known as the Content Standards for Digital Geospatial Metadata (CSDGM). This

---

[10] www.iso.org/iso/home/store/cataloguetc/cataloguedetail.html?csnumber=53798.

TABLE 9.1: Categories of data for the CSDGM

Category	Number of Elements
Identification Information	44
Data Quality Information	33
Spatial Data Organisation Information	14
Spatial Reference Information	25
Entity and Attribute Information	32
Distribution Information	43
Metadata Reference Information	13
Citation Information	15
Time Period Information	8
Contact Information	18

metadata standard documents GIS vector, raster and point data, and is the official federal metadata standard in the US. It has been applied to all geospatial data created by federal agencies since January 1995 and both state and local governments are also encouraged to use this standard through agreements to fund capture.

The standard covers four categories of information about spatial data. These categories are the *availability* of data at a given location, the *fit-for-purpose* of data, data *access* information for obtaining the data and information needed to process and use the data, termed *transfer* information.

One of the strenghts of the CSDGM is that it enables the creation of extended elements. These extensions can be defined by individuals and are how organisations made CSDGM relevant to their needs. There are also endorsed extensions to CS-DGM Version 2 (FGDC-STD-001:1998) for remote sensing metadata. These support the documentation of raw data from airborne digital sensors, including platform and sensor metadata.

The CSDGM also enables the creation of profiles, which enable the definition of terms and data elements not found in the base standard. These profiles enable the application of CSDGM to a broader range of spatial and non-spatial metadata without the necessity to have these elements in the base profile.

The CSDGM has 11 categories of data containing several hundred entries. There are a total of 324 elements in the standard.

These elements are made up of individual elements and compound elements, where compound elements are groups of data elements. Table 9.1 shows the information categories and the number of elements in each, which totaled 245.

Not all entries are mandatory and there is often considerable overlap across categories. There are also two user defined items. Creating such metadata is a skilled job. It requires not only a full understanding of the data itself, but also a detailed understanding of the content standard. It involves considerable personnel costs in creating and updating the information.

There are a number of CSDGM resources that at the time of writing could be found at the FGDC website,[11] including guidelines for extending the CSDGM standard and for creating metadata profiles using CSDGM.

9.4.3   OGC and Catalogue Services for the Web

OGC figures prominently throughout this book, as it plays a key role in development of interoperability standards for geospatial data. Its role in online metadata standards is no different. If an organisation is looking to publish geospatial services online, it will need to consider a standard format for the data's metadata records. One such standard to support online publication and discovery is the OGC catalogue services standard.

"The OGC catalog service implementation standard specifies the interfaces, bindings, and a framework for defining application profiles required to publish and access digital catalogs of metadata for geospatial data, services, and related resource information" (Nebert and Whiteside 2005; Nebert et al. 2007).

One commonly used profile of the OGC catalogue services standard is the Catalogue Services for the Web (CSW). Formally known as the OpenGIS Cataloge Services Specification 2.0.2 - ISO Metadata Application Profile, Catalogue Services for the Web (CSW) is a *profile* of the OGC Catalogue Services and the part of the standard, which we will discuss here.

CSW is essentially a standards-based approach to exposing a *catalogue*. It operates using HTTP requests (Chapter 3) over the Internet and supports requests in one of three encodings:

1.   GET requests that contains URL parameters;

2.   POST requests with form-encoding payload;

3.   POST requests with XML playload, like SOAP.

Responses to these requests are XML format fitting the `csw:Record` element. Moreover, the csw:Record structure maps directly to the 15 Dublin Core elements (§8.2), making Dublin Core the basis of OGC's CSW standard. This mapping to the simple Dublin Core elements makes the `csw:Record` a subset of most metadata standards.

There are several CSW operations, similar to those that can be used to query and retrieve information of other OGC standards, such as Web Mapping Service (WMS) and Web Feature Service (WFS). These CSW operations include functions to retrieve properties of the catalogue Web service (GetCapabilities), and others to return record IDs (GetRecords) and to get the records based on their IDs (GetRecordsById). There is also a Harvest function that can be used to *pull* metadata from a CSW service, and a Transaction function that can be used to *push* metadata to CSW service. The available operations with the transaction operation are insert, update and delete. Transactions also require access controls to manage the service calls.

There are a large number of global GIS companies that have implemented

---

[11] http://www.fgdc.gov/metadata

CSW 2.0.2 support in their Online GIS products. These companies include large GIS companies such as ESRI and Integraph, as well as a number of smaller GIS companies and companies offering metadata portal products.[12] The compliance of these vendors to such standards demonstrates the importance of standardisation of metadata if we are to enable integration of metadata from different organisations and portals.

It is worth pointing out that the OGC catalogue service standards do not define the metadata that is used in a custodian's catalogue. Its does, however, define a standard that these custodian metadata elements should be translated to in order that the CSW compliant services are able to interpret the metadata records in a programatic way.

9.5   Spatial Data Infrastructure and Metadata Portals

As discussed earlier in this chapter, SDIs originated in the early 1990s, attracting significant global interest following the *U.S. President Executive Order 12906* in 1994. This presidential order recognised the importance of geographic information to

> promote economic development, improve our stewardship of natural resources, and protect the environment.[13]

More importantly for Online GIS, this order established a coordinated NSDI in the US, providing funding and resources for the greater coordination and sharing of digital geographic information. It also highlighted the importance of standards to the success of an NSDI, and in defining an SDI as the

> technology, policies, standards, and human resources necessary to acquire, process, store, distribute, and improve utilisation of geospatial data

and it strengthened the need to develop suitable standards for geographic information sharing as core to its NSDI objectives.

Some of the early literature on Spatial Data Infrastructure, however, appeared to focus on the physical systems used to transfer geospatial information, an almost *hardware and copper* view of systems, and the research efforts were towards the large vector datasets for topographic and cadastral data,[14] as the effective distribution and sharing of these provided the most benefits to organisations. A number of researchers, including Williamson and Grant (1999) and (Williamson et al. 2003) were also working on ways to remove the barriers to sharing of spatial information, and establishing SDIs was acknowledged as an important part of this vision.

---

[12] http://www.opengeospatial.org/resource/products/byspec/?specid=236
[13] http://www.archives.gov/federal-register/executive-orders/pdf/
     12906.pdf
[14] Cadastral data are data showing land ownership and occupation.

As SDIs evolved further, the main objectives of an SDI were clearly to *facilitate the sharing of data that describes the position, distribution and attributes of objects in space* (Manning and Brown 2003a). These objectives were to allow the efficient collection, management, access and integration of spatial data (Manning and Brown 2003b) by organisers, contributors and users of SDIs. SDI also provided a basis for achieving Chapter 40 objectives of Agenda 21 (2002) from the United Nations Conference on Environment and Development (UNCED) in Bathurst, NSW Australia, enabling ready access to spatial information to support better decision making (Williamson and Grant 1999). As such, the profile of SDIs was increased and their value for publication, discovery and procurement of spatial data assets in a way that contributed to better decision making by government and across the wider spatial industry continued to gain momentum.

At that time, there were also a number of portal applications under development. These portals became the vehicles for implementing SDI metadata portals and provided several capabilities for SDI implementations, including

- Environments for data custodians to publish searchable metadata records describing their products;

- Services to render data samples of the metadata records;

- An online query capability for users to discover available datasets that included contact and licencing information;

- Binding capabilities, where these data resources are offered as map objects through Web services such as WMS, WFS, WTS.

There were many factors that contributed to SDI successes; however, the development of open standards, an increased profile of metadata and the availability of portal technologies have resulted in the establishment of a number of spatial marketplaces globally. A small number of these online spatial marketplaces that support Online GIS are covered in §9.6.

9.6   Examples of Online SDI Metadata Portals

Governments from all over the globe are opening up access to their data resources. They are doing this through the implementation of SDI metadata portals. These include governments at national, state and local levels and a major driver for this is the GOV 2.0 movement which has highlighted the many benefits of sharing this information. We will look at some of these implementations here.

There are also a number of organisations promoting the development of SDI metadata portals. Global Spatial Data Infrastructure Association (GSDI) is one of the prominent organisations which works to promote SDI at the global level. With a membership consisting of organisations and individuals, GSDI provides a focal point for the sharing and collaboration of SDI projects at all levels of society, local, national and international through a number of conferences, webinars and publications. A number of SDI metadata links can be found on the GSDI website.[15]

Some of the early SDI metadata portals are now branded as data.gov websites. The US was the initial country to move to this branding, followed by Australia and New Zealand, and data.gov websites are also appearing in the United Kingdom. As discussed, this section contains a small number of SDI metadata portals and data.gov portals from around the globe. Each of these portals will be accessible using a standard Web browser, and they will all have metadata search capabilities, data format and projection information with online mapping visualisation tools to show users what each spatial dataset looks like.

9.6.1   SDI Metadata Implementations across North America

North America has played a major international role in developing mapping data for all sectors of the community. As major technological leaders in satellites and aeronautics, home of the world's largest GIS and imaging companies, the world's computer giants, and many other leading government and private sector companies, this is hardly surprising.

As discussed in §9.4.2, the FGDC is the Federal body in the US charged with metadata definition and implementation. It received presidential endorsement as Circular A-16 in October 1990 and was rechartered in 2002. The FGDC has twelve major sub-committees and six working groups, endorsing over 200 standards relating to geographic data. The next important administrative step for the USA came in April 1994, when the National Geospatial Data Clearing House, NGDC, was created by Executive Order 12906 to oversee the development of an NSDI. The challenge for these organisations was significant. In May 1994 the NSDI began a standards project for a standard grid reference system, as there were problems not only on the formats of spatial data but in the very coordinate systems in which it was defined. The Universal Grid Reference Systems was subsequently released on May 24, 1999, providing a common coordinate system for all USA geospatial data.

A number of SDI metadata portals from North America are referenced here. These include the US Federal *Data.Gov* and geospatial data platforms, the Canadian SDI and sample state and municipality portals.

9.6.1.1   United States *Data.Gov*

The NSDI clearing house was originally branded the Geospatial One-stop Shop (GOS). When GOS was retired, it was integrated into *Data.Gov*,[16] which is a centralised clearinghouse for the federal government's Open Data.

President Obama's Executive Order and the Administration's Open Data Policy in 2013 provided the governance around development of a new centralised clearinghouse for government data, with all newly generated government data required to be

---

[15] http://www.gsdi.org/SDILinks
[16] http://www.data.gov

made available in open, machine-readable formats while continuing to ensure privacy and security. When creating these data, the federal organisations are also required to publically list any data assets and contacts for these data that are approved for general publication, and to provide feedback mechanisms to promote citizen engagement. *Data.Gov*[17] provides all of these capabilities, supporting the US government's Open Data Policy.

In addition to publishing federal government data, *Data.Gov* also provides a hosting service for state, local and tribal governments that uses nightly harvesting for these organisation's open data catalogues. At the time of writing, *Data.Gov* was harvesting from states, counties and city portals.[18]

Data.Gov contains open data from hundreds of federal, state and local government authorities and other organisations. There were around 125,000 datasets listed on *Data.Gov* in early 2015.

Data.Gov supports a number of open standards. There are almost 50 different data formats available for download and viewing, including text-based formats through to image formats and standards-based Web services. We will not cover these here, but there are datasets available from a wide range of contributors at the *Data.Gov* portal.

The metadata standard formats available as downloads at *Data.Gov* include CSV, JSON (§8.3) and RDF (§8.5). However, it also promotes the use of a reduced metadata standard for organisations to use when publishing. One in particular is the Project Open Data Metadata Schema, currently at version 1.1.[19] This schema consists of a number of different fields, catalog fields, datasets fields and dataset distribution fields, and is currently distributed as JSON (§8.3). The metadata schema is exensible, like many other of the schemas discussed in this book, and the schema can be extended using the grammar from well-known standards such as Dublin Core (§8.2), FGDC, ISO-19115 and XML schema, as this will improve the sharing and understanding of *Data.Gov* metadata records. Interestingly, the code base used for *Data.Gov* is all open source code. The portal is based on two open source projects:

CKAN: The Comprehensive Knowledge Archive Network (CKAN) is an open-source data portal software suite, which is commonly used for Web-based data portals. It consists of a database repository for metadata and tooling that makes it easy to publish, discover, store and share datasets with users. Because it is open source, it can be extended to suit individual organisation needs. The CKAN open source project contains components of *Data.Gov*.

WordPress: This project contains the source code for the data catalogue. It is freely available online[20] and has already been used for data catalogues in other SDI metadata portals.

These source codes are all managed in GitHub, arguably the world's largest

---

[17] www.data.gov

[18] http://www.data.gov/opendata/get-local-government-data-gov/

[19] https://project-open-data.cio.gov/v1.1/schema/

[20] http://catalog.data.gov

source code repository. *Data.Gov* code is easily accessible online at GitHub[21] and there are examples of the tooling used to manage metadata records in *Data.Gov*, which are freely available from the *Data.Gov* website.[22] The current version of the *Data.Gov* schema is called Project Open Data Metadata Schema V1.1 and all agencies were expected to be at this schema by February 2015.

Of particular interest to Online GIS, are two CSW service endpoints published by *Data.Gov*.[23] [24] Implemented using *pycsw* (where py is the standard abbreviation for Python, and CSW is self-explanatory) and OGC CSW 2.0.2, these service endpoints work over HTTP using GET, POST and SOAP requests (§3.1.1). Using these standards and OGC certified language, *Data.Gov* has created a truly standards-based SDI metadata catalogue.

9.6.1.2 *GeoPlatform.gov* in the US

Like *Data.Gov*, the *GeoPlatform.gov* portal is a comprehensive catalogue of data covering the US. It provides federal, state, local and tribal governments with an online portal for publication, as well as private sector, academic and general citizens. Unlike *Data.Gov*, the *GeoPlatform.gov* only contains geosptial data, services and applications.

GeoPlatform.gov (Figure 9.2) has been developed by member organisations of FGDC and it provides access to authoritative and trusted geospatial datasets. It also provides a shared infrastructure that can host both datasets and metadata records, and applications that can be used for visualisation and mashup.

As with *Data.Gov*, the portal has been implemented using an open source Web content management system. This SDI portal has been implemented using Drupal,[25] a PHP application distributed under the GNU General Public License.

There is a wide range of datasets discoverable and available from *GeoPlatform.gov*. At the time of writing, these datasets numbered around 74,000 geospatial datasets from a large number of contributors. There are also a number of APIs that are linked to the portal which can be used for visualisation of the datasets.

GeoPlatform.gov uses the CSW endpoints published by *Data.Gov* to enable its integration, retrieving dataset information and records using the GetCapabilities, DescribeRecord, GetDomain, GetRecords and GetRecordsById CSW operations. More details on the operations and some sample GET, POST HTTP requests can be seen at the GeoPlatform website.[26] There are also a number of useful tutorials and training manuals with examples of further examples of common Online GIS metadata records.

The geospatial data that can be discovered at GeoPlatform also comes in a wide range of data formats. These formats include text files such as HTML, XML, GML and CSV; geospatial datasets such as ArcGIS Shape files and file geodatabases

---

[21] http://gitub.com/GSA/data.gov
[22] http://Labs.Data.Gov
[23] http://catalog.data.gov/csw
[24] https://catalog.data.gov/csw-all
[25] https://www.drupal.org/
[26] https://www.geoplatform.gov/csw-resources

FIGURE 9.2: *GeoPlatform.gov* in the US, which is seamlessly integrated with *Data.Gov* through CSW.

(FGDB) and can be accessed from Web services such as WMS, WFS, REST and SOAP, just to name a few. We will not go any further into these data formats here; however, there are a couple of interesting features of *GeoPlatform.gov*.

One is the integration between *GeoPlatform.gov* and *Data.Gov*. Despite being different groups, these teams have worked closely together to integrate datasets and metadata between their portals, resulting in a seamless integration between the sites. GeoPlatform has now adopted the CKAN open-source product used by *Data.Gov* and it has integrated the *Data.Gov* master catalogue as a service within *GeoPlatform.gov* to expose the *Data.Gov* geospatial datasets through *GeoPlatform.gov* online searches. This cooperation has effectively *chained* the two portals together in a real sharing of Open Data for searching and discovery.

Another great feature of *GeoPlatform.gov* is its implementation of a marketplace. This function allows users to search the portal using textual or spatial searches to find out what datasets are being planned for acquisition by other organisations. This abiltity may lead to considerable savings of taxpayer funds where this type of capa-

FIGURE 9.3: New York State Spatial Data Infrastructure Clearing House.

bility leads to better sharing of data resources across the spatial industry and reductions in multiple purchases by government of the same data.

9.6.1.3   New York State Metadata

The SDI metadata portal in New York State in the US was another success story of the early Online GIS portals. Complexity and the high cost of establishing metadata to populate online portals opened up a movement towards a simplified standard MetaLite.[27] At the time, MetaLite was essentially the mandatory items of the CSDGM (§9.4.2).

More recently, New York State has moved to an online GIS clearinghouse, shown in Figure 9.3,[28] to publish some of the state's open data geospatial data and services. These data and services are provided by eight participating organisations and are offered as ArcGIS Server URL, WMS URL and REST Endpoint service formats. Some of the data is hosted on NYSGIS for download; however, many of the datasets can also be accessed from the contributing agency websites or Web services.

One contributor to the NYSGIS Clearinghose is the Adirondack Park Agency (APA). The APA manages the Adirondacks, a huge park in New York State including the Adirondack mountains and host to the Lake Placid Winter Olympics. APA has published metadata for 14 datasets through the NYSGIS, and these data are currently hosted in the ArcGIS Online, a cloud-based solution provided by ESRI Inc.[29] These cloud hosted datasets can be viewed and accessed through standards-based REST endpoints.

---

[27] Metalite had just 29 main elements, specifying data by name and place and providing contact details.

[28] http://gis.ny.gov/gisdata/ (c) New York State Office of Information Technology Services. All rights reserved.

[29] http://www.esri.com

APA also lists a number of datasets on its own website,[30] providing alternate access methods. Supported data formats include services that can be viewed in standards-based Map APIs and Esri Clients. REST tile services, REST Feature Services and several of the census datasets, covering places, towns, cities and schools can be downloaded as KMZ files (KMZ files are zip files containing compressed KML files that are automatically unzipped at the client, reducing download times). Metadata statements are listed for some of the datasets; however, APA has published these using the *FGDC Content Standards for Digital Geospatial Metadata* metadata standard. This metadata standard is verbose compared to many of the open metadata standards now available, although it is a good example of an organisation holding its metadata in whatever format it would like. The next step for APA may be to translate these records into CSW supported standards so that it can be discovered programatically by potential clients.

9.6.1.4 Natural Resources Canada

One of the successful national initiatives in North America is the Canadian Geospatial Data Infrastructure (CGDI).[31] It is an Open Data technology infrastructure operating through a consortium of governments, private sector, non-government organisations (NGOs) and academia which is supported by and initiative called GeoConnections (Figure 9.4).[32] This infrastructure provides data, services and technologies for on-demand access to online geospatial datasets, using publically available standards and specifications.

There are a large number of resources available from the Earth Sciences tab of the CGDI website. These include policy and strategy documents, architecture documents for CGDI as well as research papers, overviews and reports. One video resource that may be of interest as in elementary introduction can be found at

```
http://www.nrcan.gc.ca/earth-sciences/geomatics/
   canadas-spatial-data-infrastructure/8906
```

Publically available standards used in the early CGDI (1.0) standards included ISO-19107, ISO-19108, ISO-19115 and ISO-19136 and the Web services standard ISO-19128. With the evolution of CGDI 2.0 between 2005 and 2010, metadata in CGDI conforms to ISO19115:2003 - Geographic Information North American profile and some datasets use Digital Geospatial Metadata (DSM) when less detail is required to describe these datasets.

The CGDI also provides data and service using a number of domain standards from organisations including GML, KML, GeoRSS, CityGML, Land Cover and Land Admin.[33] These syntax and encoding standards allow the translation of data between standards so that they can be transferred between systems.

---

[30] http://www.apa.ny.gov/gis/index.html

[31] http://www.nrcan.gc.ca

[32] http://www.geoconnections.org

[33] http://www.isotc211.org/WorkshopQuebec/Workshop_presentations/
Habbane-SIA-workshop-Quebec-2009.pdf

FIGURE 9.4: Canadian Government's GeoConnections Webpage, supporting the on-going evolution of the CGDI.

Service-based standards available in CGDI include Web Mapping Service (WMS) and Web Feature Services (WFS), the Web Services Processing Service (WPS), Web Coverage Service, Web Map Tile Service (WMTS) and Catalogue Services for the Web (CSW). Many of these standards have already been discussed. CGDI does provide a number of Web services based on these standards that can be used to publish, query and render geospatial data.

One of the interesting offerings from CGDI is the implementation of a GeoNews service for emergency management professionals. This GeoNews service used the RDF encoding of geography in the CPNewsML XML schema to create GeoRSS news feeds that could be viewed in a number of Online GIS map APIs. The resulting news articles were geo-referenced in WSG84 and could be visualised on websites such as Google Earth, which gave critical spatial context to each of the stories, which was critical for EMPs. The new feeds were also available as live data streams over the Internet, so they could be accessed wherever users could get Internet connectivity. The CGDI continues to evolve and has had significant environmental, social and economic benefits for Canadians.

Return to Open
Data home page
Terms of Use

Open Data catalogue

A B C D E F G H I J K L M N O P Q R S T U V W X Y Z

Name & Information about Data	CSV	XLS	DWG	KML	SHP	ECW	Other Formats	✳ Google Maps	✳Bing Map
0 - 9									
3-1-1 case location details	✓ CSV	✓ XLS							
3-1-1 case volume	✓ CSV	✓ XLS							
3-1-1 contact centre metrics	✓ CSV	✓ XLS							
3-1-1 interaction volume	✓ CSV	✓ XLS							
A									
Accessible parking			✓ DWG	✓ KML	✓ SHP				
Address labels for map display			✓ DWG	✓ KML	✓ SHP				
Alleyways			✓ DWG	✓ KML	✓ SHP			G	◌
Animal Inventory - Deceased Animals	✓ CSV	✓ XLS					✓XML ✓JSON		

FIGURE 9.5: City of Vancouver Open Data Catalogue listing datasets and supported file formats.

9.6.1.5 Vancouver SDI Metadata Portal

The City of Vancouver in Canada launched their Open Data website in September 2009, following their support for the Open3 motion in May earlier that year. This motion intent was to make data open and accessible to citizens, business and other jurisdictions, while respecting citizen privacy. To support this decision, the City of Vancouver defined Open Data as *non-personally indentifiable data that is made freely available to everyone in one or more open and accessible formats.*

The Vancouver Open Data Catalogue, which can be accessed directly at (Figure 9.5),[34] has a rich suite of datasets published. These datasets can be accessed through a number of common file formats including textual formats CSV, XLS, JSON, GeoJSON, XML and geospatial formats of DWG, KML, SHP (Esri Shape) and ECW for imagery data. The catalogue has a easy to read graphical table that shows the supported datasets available for each of its datasets in its Open Data Catalogue. It also has links to Google and Bing (Figure 9.6) map viewers, where those viewers are supported by the datasets. Here, a dataset from the Vancouver City Data Catalogue has been linked to a view in Bing Maps, followed by a search for directions to the town of Hope in British Columbia from Block 61 in the City of Vancouver. The result was a rendered map showing the user directions to Hope; all done in a standard browser, in less than a minute, from the other side of the world.

---

[34] http://vancouver.ca/your-government/open-data-catalogue.aspx

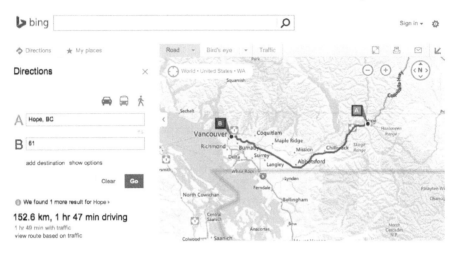

FIGURE 9.6: City of Vancouver Open Data rendered in the Bing Maps API, complete with routing information between two locations.

To achieve this, both map viewers are passed KML dataset information (§6.3), which is automatically opened within the browser, and the dataset is automatically zoomed to. From there, the users have full access to tools such as zoom and pan, and can run other searches and queries against the dataset.

The website is powered by metadata records for each of the datasets. In September 2013, the City adopted the Canadian Open Government Licence (OGL).[35] It gives users consistent terms for accessing and using government data, making it easier to combine, integrate and share data from multiple providers across Canada. This is a major step towards encouraging wider use of open data and delivering more value to the public. Vancouver joins the Government of Canada, Provinces of BC, Alberta, and Ontario, and the cities of Nanaimo, Edmonton, and Toronto in adopting the OGL.

9.6.1.6 Toronto City Portal with Open Data

The City of Toronto in Canada is another great example of what can be offered by a municipality through an online portal.[36] The success of this online portal has been through formalisation of a number of Open Data policies and a commitment to implementing Open Government across its municipality.

Their portal includes a data catalogue in the same way other Online GIS Portals do, with search, view and download capabilities. The Toronto catalogue includes a wide range of different data types, from textual data in spreadsheets and csv files to

---

[35] http://open.canada.ca/en/open-government-licence-canada
[36] http://www.toronto.ca/open

Esri Shape files and Integraphs DGN[37] files containing vector data, JSON and XML datasets. Erdas Imagery[38] data is also available from the catalogue.

While the city of Toronto only covers a small area in comparison to the larger federal and jurisdictional organisations discussed here, it also has a strong Online GIS content available from its website. Toronto began releasing datasets online in November 2009 and offers a number of machine readable formats and file formats that are discoverable through these catalogue searches and viewer tools (Figure 9.7). These datasets include data about Toronto city services, environment management, financial and health data, etc., and the data can be accessed as either digital data, which requires the signing of a digital data agreement, or as standard PDF-based map products.

As an example, you can access the Toronto Fire Station location at WGS84 geographic data in DGN, Excel spreadsheet and Esri Shapefile directly online. These data can then be used to mashup against other geographic data and Online GIS systems. There are also data available for permit and license holders across a wide range of data types. Data on registered cats and dogs is available. From the primary breed datasets it was listed that there are 1,737 minature poodles, 1 less than the number of Parson (Jack) Russell terrier dogs registered in the Toronto City Municipality. These data, for obvious reasons, do not contain geo-location information, as this would identify individuals who owned specific cat and dog breeds; however, the data is available online.

The website does not appear to offer access to their data through any online Web services, such as WMS, WFS, SOAP or REST endpoints, but it does provide easy ways for users to search, discover and download available datasets for any public use. As a result, there are a number of mobile applications that have been built by third parties including transport apps for bus, bike and route information, and others that provide Toronto festivals and events information to smartphone and tablet devices. So, users can freely download GIS datasets from the Toronto City website and use these data to develop apps, which make an active markeplace for Online GIS users.

9.6.2 European SDI Metadata Implementations

The European countries have had highly sophisticated mapping technologies and national organisations for a long time and the European Union has been working towards an integration capability across all the member countries. European advancements with metadata have been an important part of creating this capability, with a number of catalogue instances now in place.

The early European initiatives have their origins in SDI work, including some cross-border initiatives, JRC EU Portal, EUROSTAT, Inspire and the German SDI Initiative, where each of these initiatives is based on application profiles from ISO, including ISO 19115 metadata for geospatial data and ISO 19119 for geospatial services.

---

[37] http://www.integraph.com
[38] http://www.hexagongeospatial.com

FIGURE 9.7: Imagery and GIS data available from the City of Toronto.

The European standards were from a broader IT initiative, INFO2000,[39] and come in two different levels, CEN[40] and ENV, representing different levels of development. CEN, *Comite European de Normalisation*, are essentially the mature standards. But recognising the incredible speed of technological progress, the need for fast-track standards led to the ENV, *Euro-Norm Voluntaire*, specifications.

In parallel to these standard mechanisms, there are various organisations linking together the national mapping agencies (NMAs) of the member countries. At the top of these was CERCO *(Comite European des Responsables de Ia Cartograpltie Officielle)*,[41] a group of NMAs represented by the head of each and including 31 member organisations. In 1993, CERCO spun off the daughter organisation MEGRIN (Multipurpose European Ground Related Information Network),[42] but in January 2001, the two organisations merged to form Eurographics. MEGRIN was a non-profit organisation with the status in French law of a GIE *(Group d'lnteret Economique)* with a number of key roles, two of which were

1. To assist the trade of spatial data across national boundaries and to ensure adequate metadata (such as the GDDD);

---

[39] http://cordis.europa.eu/news/rcn/10289_en.html
[40] https://www.cen.eu/Pages/default.aspx
[41] http://www.eurographics.org/cerco
[42] http://wwweurographics.org/megrin

FIGURE 9.8: Europe's Inspire GeoPortal.

2. To harmonise data across national boundaries (such as SABE).

Europe had considerable success with the SDI metadata portal INSPIRE (§9.6.2.1).

9.6.2.1 INSPIRE SDI Metadata Portal

The Infrastructure for Spatial Information in the European Community (INSPIRE) GeoPortal, shown in Figure 9.8, was one of the early regional SDI metadata portals. It was established in 2007 as a result of Directive 2007/2/EC of the European Parliament and of the Council of 14th March 2007, and as a result, INSPIRE is an example of a legistative approach to SDI. Like many other portals around the globe, INSPIRE provides a search and view capability for spatial datasets and services, to enable the sharing of spatial information across language and geographic boundaries.

The directive was originally to ensure that implementing rules (IR) were adopted for metadata, data specifications, network services, data and service sharing, monitoring and reporting, and these still hold true today. However, there are other major EU projects taking place at the moment in Europe which support Spatial Data Infrastructure. For this reason, we will not look at the existing INSPIRE portal, but review these projects at a high level:

GIS4EU: Is creating harmonised European datasets and online services for administration boundaries, hydrography, transport and elevation, without the need to create a centralised database and replicate data. The project will use standards that reflect those currently in INSPIRE and will deliver simplified data access and sharing.

HUMBOLT: Is providing capability for EU members to prepare and publish their geospatial data through INSPIRE and Global Monitoring for Environment and Security (GMES).

EURADIN: (European Addresses Infrastructure) Has as its focal point, the harmonisation of addressing across the EU. Its current activities are working on the establishment of a metadata profile for addresses and the development of tooling for automated metadata extraction.

NatureSDI+: Has as its focal point the harmonisation of national datasets across the EU countries so that they are easily interchangeable between the GIS systems of each country. Being primarily a data modelling exercise, the project will rely heavily on the use of standards such as GML, and on quality metadata as part of the national datasets.

eSDI-NET+: This projects is working on cross-border communication and information sharing throughout the EU to promote SDI best practices.

What should become apparent to the reader is that SDI is much more than just metadata portals and clearinghouses (markeplaces); however, each of the European projects listed is only possible due to the standards-based approach to sharing of geospatial data and because of the creation of quality metadata statements to describe each dataset (source, owner, scales, capture processes, currency, date, licencing, etc.). All of these projects support an Open Data framework.

9.6.2.2 United Kingdom *Data.Gov.uk*

A national approach to the Open Data movement in the United Kingdom[43] gained prominence in 2013, following the Shakespeare Review of Public Sector Information.[44] This report made several recommendations for the creation of a National Data Stragegy (NDS), led in principle by the public sector to maximise economic value of public assests while stimulating private sector growth and innovation.

In response, the government undertook to move the NDS forward through their Information Economy Strategy and stated that *The vast majority of government data will be published as open data and available through Dataportals such as Data.Gov.uk.*[45] Moreover, the government's response recommedations acknowledges that metadata is a critical component of any published data so that users can be assured that it is fit for purpose if used. Open standards and metadata will play a signficant role in making the datasets discoverable and useable.

One of the early actions was a declaration to define and implement a National

---

[43] http://data.gov.uk
[44] https://www.gov.uk/government/uploads/system/uploads/attachment_data/file/198752/13-744-shakespeare-review-of-public-sector-information.pdf
[45] https://www.gov.uk/government/uploads/system/uploads/attachment_data/file/252172/Government_Response_to_Shakespeare_Review_of_Public_Sector_Information.pdf

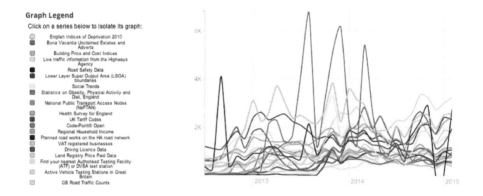

FIGURE 9.9: *Data.Gov.uk* site analytics.

Information Infrastructure (NII). This action has taken the form of a prototype document that has been published on the *Data.Gov.uk beta* website for public comment, demonstrating a nationwide consultative process across public sector, private sector, citizens and academia. The *Data.Gov.uk* portal also contains links to data resources (published and unpublished), licencing, data requests and some of the available Web apps that already use these datasets. It also contains information on the currently supported file formats for these data, including OGCs, WMS, XLS, RDF, DOC, CVS and PDF, to list a few.

It is worth noting that *Data.Gov.uk beta* contains a Graphical Map Searching tool that supports textual and spatial searching. While, at the time of writing, the functionality was limited, the site is simple in its design and contains an attractive map cache with quality cartographic data.

There are several statements in the NII documentation and blogs regarding metadata standards that will be used to publish data. The Open Standards Board and the UK Government Open Standards Principles will determine these in consultation with user groups, in particular, local authorities, and, at the time of writing, there is a clear direction to use standards currently endorsed and in use by SDI portals such as INSPIRE. NII itself will not develop standards for publication and discovery, but metadata content is already visible in the *Additional Information* tables and spatial searching is supported for a rectangular area of interest (AoI) or minimum bounding rectangle (MBR). It is hoped that open standards bodies such as ISO, OGC and FGDC will have a role to play in the NII standards, to increase *Data.Gov.uk* relevance to Online GIS.

One of the nice features of the portal is the recording and storage and presentation of site analytics (Figure 9.9). There are several tabs containing statistical data ranging from site visited to browser and operating systems used by visitors to the site. It also contains graphic representations of publisher usage statistics and data usage statistics, that can also be downloaded as CSV files.

From the online media, there appears to be a significant number of synergies and agreement between the recommendations of the Shakespeare Review and UK central government policy and directions, and it will be interesting to see how successful the UK can be in delivering Spatial Data Infrastructure that benefits all sectors in this next wave of their move towards successful Open Data infrastructure.

9.6.2.3   *London.gov.uk* **Clearinghouse**

London Datastore Mark II (LDSII),[46] a data clearinghouse for London in the UK, went live in 2014 (Figure 9.10). This *functional upgrade* to London Datastore included improved spatial query and searching, an ability to preview datasets and a number of API capabilities. More significantly, the LDSII continues to offer hosting capabilities for other local authorities to publish their data at the site. To eleviate potential concerns, the LDSII hosting is performed in a UK datacentre and includes secure access and update, backup and restore capability.

Data can be uploaded through a simple graphic interface to LDSII and the site offers metadata tooling for the creation of metadata for uploaded datasets. There exist a number of census-based reporting tools that provide data in a graphical format that is easy to read and understand.

At the time of writing, LDSII contained almost 600 datasets. These data included health, housing, eduction, arts and culture, and a number of other categories of data available for download and integration with develop applications. Supported file formats include CSV, XLS, PDF, HTML and Esri Shape file, although only 21 geospatial datasets were identified that provided boundary, zone and regions data. The website did, however, support a *find data by location* capability and does provide simple filtering of datasets to aid searching.

While it is still early days for the LDSII and it is clearly evolving, it is a good example of what governments can achieve when partnering with NGOs and private sector to improve services and create innovation. The focus of LDSII towards their Open Data initiatives is offering real and tangible benefits to organisations through the online sharing of information. These form of agreements and portals will become more crucial as the UK government decentralises power through arrangements such as the Manchester Devolution,[47] placing more responsibility at the municipality or city level for health, transport, housing, planning and policy.

9.6.3   **Australasian Metadata Portals**

A number of Australasian spatial infrastructure portals will be discussed in the next three subsections.

---

[46] http://www.london.gov.uk
[47] https://www.gov.uk/government/uploads/system/uploads/attachment\
   _data/file/369858/Greater_Manchester_Agreement_i.pdf

FIGURE 9.10: London Spatial Data Infrastructure Portal.

9.6.3.1   New Zealand *Data.Govt.nz*

The New Zealand SDI portal[48] is an *online directory of publicly available, non-personal New Zealand government held datasets.*[49] It does not store the datasets, but hosts metadata records describing datasets that are available from other sites. Users can search for published datasets, based on keyword searches, and are then redirected through the dataset Title links to the agency website where the data is hosted.

Released in 2009, the portal contains almost 3000 online government datasets across a number of categories including health, education, agriculture, transport and tourism, to name a few. The datasets can be queried and in many cases viewed through Online GIS sites such as the Land Information New Zealand (LINZ) Data Services.

To better explain the federated model implemented at *Data.Govt.nz*, a query of their website through a keyword search is available through the `Search New Zealand Government Data and Data Requests` key-in field at the top of the home page. Using the keywords `airport_poly`, this search returns a small number of results. At the time of writing, just two records were returned, NZ Mainland Airport Polyons (Topo, 1:50K) and NZ Chatham Is. Airport Polygons (Topo, 1:50). Selecting the second link, NZ Chatham Is. Airport Polygons,[50] loads up

---

[48] https://www.data.govt.nz/
[49] https://www.data.govt.nz/
[50] https://data.govt.nz/dataset/show/1995

FIGURE 9.11: The New Zealand Chatham Is. Airport Polygon Data Catalogue Record, showing a brief data description, available formats and the link to the LINZ website.

the *Data.Govt.nz* metadata record to the catalogue tab shown in Figure 9.11. This record contains some descriptive information about the datasets, including a link to its Data Dictionary, some format and agency information. The tab will also contain a "Dataset URL", in this case, `https://data/linz.govt.nz/layer/` `63-nz-chatham-is-airport-polygons-topo-150k/` which, when selected, will redirect your browser to the Land Information New Zealand website, passing the dataset information through as part of the URL so that the LINZ website opens on that data record.

The data record for New Zealands Chatham Is. provides the user with an overview of the data (description, license details, general information). The page also contains both metadata records for this dataset, which are ISO 19115/19139 and Dublin Core, and some other information. The LINZ Data Service also provides a viewer to visualise the coverage area of the datasets. Finally, download format options for this dataset include GIS, CAD, KML and CSV.

A number of other Open Data sources for New Zealand are also listed on the *Data.Govt.nz* portal. These include DigtialNZ, Open Data Catalogue, LRIS Portal and *Geodata.govt.nz*, highlighting the strong Spatial Data Infrastruture that exists in New Zealand.

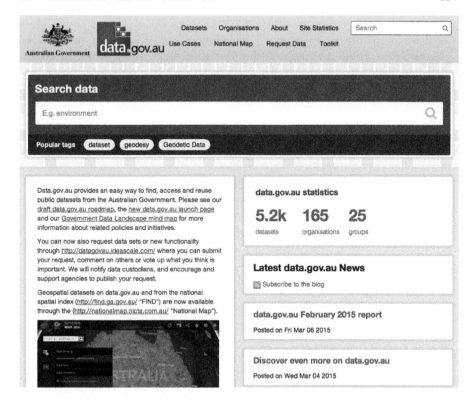

FIGURE 9.12: The Australian *Data.Gov.au* geospatial portal.

9.6.3.2 Australian *Data.Gov.au*

The *Data.Gov.au* website is an Australian portal[51] that was commissioned following the Australian Government's Declaration of Open Government.[52] Its primary purpose is to promote democracy through better access to government data and to provide a collaborative environment where innovative use of these data by the non-government organisations (NGOs) will contribute to stronger economic growth through three key principals: informing, engaging and participating.

Development of the portal was the central recommendation of the Government 2.0 Taskforce Report in 2010,[53] which was in response to an earlier Government 2.0 Taskforce Report.[54] The principal factor behind these was to promote an Open Data policy for Australian governments, similar to other *Data.Gov* initiatives.

At the time of writing, *Data.Gov.au* contained almost 5800 records of publicly

---

[51] http://data.gov.au

[52] http://data.gov.au

[53] http://www.finance.gov.au/blog/2010/07/16/declaration-open-government/

[54] http://www.finance.gov.au/publications/gov20taskforcereport/

available government datasets. One hundred and sixty-five government organisations are listed as contributors to the portal, including federal departments, state and local govdernements. These data are available in a wide range of common data formats, including EsriShape files, XML, GPX, CVS, JSON, GEOJSON, KML/KMZ, RDF, XLS, Doc, API, HTML, XLSX, PDF and MDB. Some of the datasets can also be accessed online through a number of online Web services including WMS and WFS, OGC standards.

Many of the organisations listing data on *Data.Gov.au* provide standards-based metadata records. One example is the Brisbane City Council Cemeteries datasets that has links to ISO19115/ISO19139 metadata statements. As these have previously been covered, we will not go back over the standards here.

Data.Gov.au provides two access models for its listed datasets. The first is a *federated access* model where the portal simply harvests metadata records from contributing portals and hosts only these metadata records for searching purposes. The second model, a *centralised access* model, is where contributing organisations also upload the dataset along with its metadata record, enabling both the searching and supply of these data directly from a central location. Let's take a closer look at two of the geospatial data records from *Data.Gov.au*, to demonstrate the two models.

The dataset of Rock Lobster Sanctuaries from South Australia,[55] discoverable at *Data.Gov.au*, is an example of federated access. This type of access is achieved by including an http URL as part of the metadata statement, which is used to redirect the user to the website run by the custodian of the dataset. A user's search (textual or spatial) returns this metadata record containing descriptive information on what the dataset contains, its extent and information about the publisher. The metadata record also contains URL links that redirect users to the custodian portal where standards and dataset information is held. This dataset is available in several formats, including GPX. The logical reason to provide these data in the GPX format is so that they can be uploaded into navigational devices to show boats where they are in relation to the sanctuary boundaries. It is worth pointing out that there are not polygon elements in the GPX data, as these are not supported in the standard, and boundary polygons are simply created from sets of linear features.

Examples of *central access* models include a number of geodetic datasets. Geodetic data is data used to provide survey network information that underpins the fixing of all location relative to the earth. There are over 1500 Geodetic datasets available as zip downloads directly from *Data.Gov.au*. These data are hosted in the portal itself and not accessed from the contributor's website.

All datasets published through *Data.Gov.au* are licenced under Creative Commons Attribution 3.0 Australian License and there are well-defined security and privacy policies set out to govern the use of material from the portal. *Data.Gov.au* also uses cloud-based infrastructure to store its content and at the time of writing, the portal was hosted in an Australian Amazon Cloud.

The Spatial Data Infrastructure evolution in Australia is very similar to the development of Open Data policies in both the US and the UK. There is a clear sharing

---

[55] https://data.sa.gov.au/dataset/rock-lobster-sanctuaries

of knowledge and resources across the organisations responsible for developing and maintain these portals. They use common standards, are using the same open source development tools and languages, even to the point that the portals share a common look and feel. This strong collaboration across the Data.Gov movement is certainly one of its strengths and demonstrated the power that open standards can bring to the sharing of datasets, including geospatial datasets.

9.6.3.3 New South Wales and Open Data Policy

New South Wales (NSW) is the most populated state in Australia. The Department of Finance, Services and Innovation (DFSI) in NSW has a number of initiatives supporting the move towards Open Data. Some of these initiatives will be covered here.

Like other jurisdictions around the world, DFSI host an Open Data portal[56] in support of its Open Government policy. This website is also powered by open source software (CKAN), and hosts an increasing number of datasets and services. In addition to the portal, NSW Spatial Data Catalogue[57] is hosted by Land and Property Information, a division of DFSI. As previous sections have already covered a number of similar initiatives around the globe, we will not cover these further here, but will focus on two of the trending initiatives being used effectively by DFSI.

One of the goals of *data.nsw.gov.au* is to stimulate innovation. To faciliate this, DFSI has *a program of events to encourage the use of NSW government data by creating innovative Web and mobile applications*. The program is called apps4nsw[58] and it provides a leading example of governments engaging their citizens to find new ways of using government data. The apps4nsw is a competition, where prize money is offered for the best apps. In 2015, it was run as a form of GovHack challenge and also as part of a major computer expo, CeBIT.[59]

Underpinning these types of initiatives are key government strategies supporting Open Data. In NSW, the state government has endorsed a number of key goals in its strategic planning, specific to Online GIS, including a NSW ICT Strategic Plan. It has been working towards a centralisation of govenment data warehouses[60] designed and implemented by DFSI specifically to host government data. It provides secure and reliable cloud services to *achieve more value from NSW, from limited government funds*.

Another area where DFSI is having success is in using social media to promote interest in NSW Spatial Data infrastructure. The apps4nsw competition, for example, uses YouTube[61] to promote the competition, with individual organisations creating short videos to advertise the different challenges in the competition. These videos were, in many cases, presented by government data users and people with needs that could benefit from the develoment of apps. There are also a number of tweets and

---

[56] http://data.nsw.gov.au
[57] https://sdi.nsw.gov.au
[58] http://data.nsw.gov.au/apps4nsw
[59] http://www.cebit.com.au
[60] http://www.govdc.nsw.gov.au
[61] https://www.youtube.com/watch?v=VpRXOBjTpTg

blogs on the Web being effectively used by NSW government to promoted Open
Data events and new ideas across the region.

With these types of spatial data initiatives around the world, the successful participation of government, industry and citzens will be key to the success. To achieve
this, there is a lot to do to provide collaborative environments and to make it easier for
users to conform to standards-based approaches. To that end, DFSI is formalising a
number of policies, procedures and practices that will make it easier for users to contribute in a consistent way. These initiatives are a good example of what government
can achieve and NSW Government is demonstrating this through its actions.

9.6.3.4   FIND: Australian National Spatial Data Catalogue

An early example of a truly distributed information system, discussed in the first edi-
tion of this book, was the Australia-New Zealand Land Information Council (AN-
ZLIC) Australian Spatial Data Directory (ASDD),[62] which was launched in 1998.
This SDI metadata portal was replaced by a new data catalogue called FIND[63] in
August 2014.

The FIND website (Figure 9.13) is the Australian Government's spatial data catalogue. This catalogue is a joint venture between Australia's National Mapping organisation Geoscience Australia (GA) and *Data.Gov.au* (§9.6.3.2) and used in conjuction with *data.gov.au*, this catalogue provides access to a network of government
data publishers across Australia, such as the NSW Spatial Data Catalogue (§9.6.3.3).

The home page of the clearinghouse portal is somewhat unique, appearing almost Google like, with a single key-in field for entry of search keywords and only
minimal text on the page to introduce the portal. There is also an advanced search
page on FIND that includes filters to refine the user search. These filters include text-
based searches using title, keyword and custodian searches; spatial searches using
a Bounding Box on a simple map API and a temporal search using a time window
(From - To).

The metadata records on FIND are harvested from a number of particpating
spatial data catalogue nodes, including federal and state government organisations.
Search results are returned as metadata records for datasets meeting the search criteria and these records include links to participating nodes.

FIND is implemented using GeoNetwork, an open source platform used internationally, and the metadata records also include GetCapatilities requests that can
be used independently to retrieve information about the service from CSW service
endpoints at the participant's catalogues. The standards underlying FIND were developed by the Open Geospatial Consortium and FIND is a catalogue service that
conforms to the HTTP protocol binding of the OpenGIS Catalogue Service ISO
Metadata Application Profile specification (version 2.0.2).[64]

A sample GetCapabilities response from the FIND Metadata Portal is

Geoscience Australia Product Catalogue A catalogue service for Geo-

---

[62] http://asdd.ga.gov.au/asdd/
[63] http://find.ga.gov.au
[64] http://www.opengeospatial.org/standards/cat

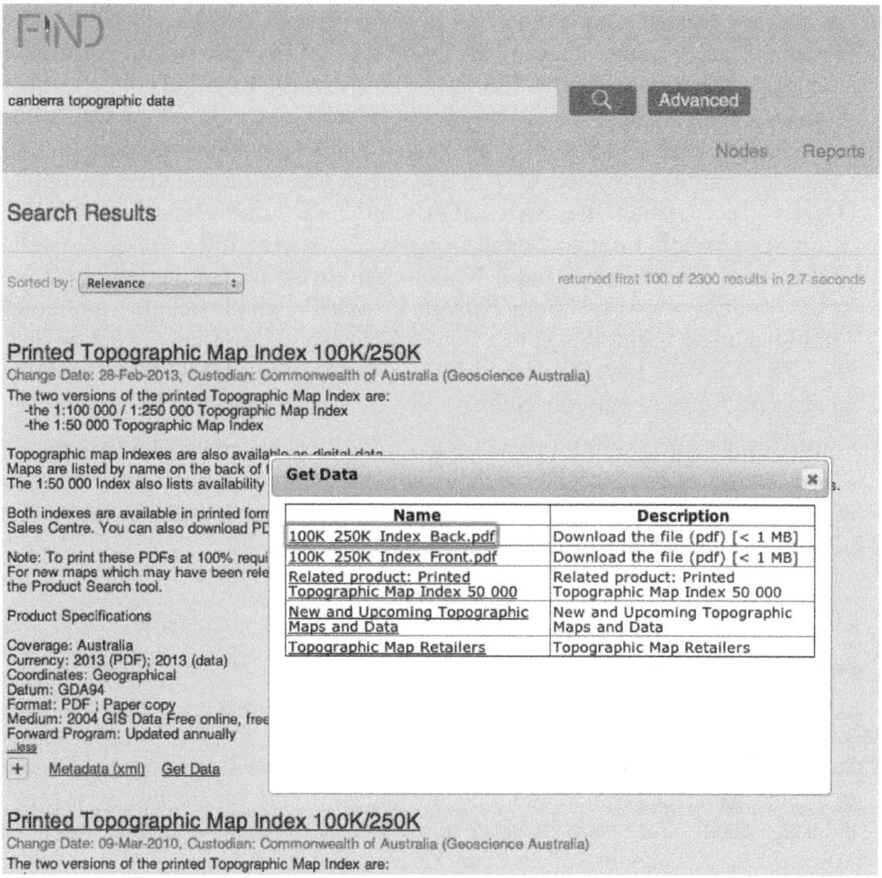

FIGURE 9.13: Australia's Online Spatial Data Catalogue, FIND.

science Australia products Earth Sciences GA Publication Record geology geophysics seismics petroleum exploration geochemistry gravity mineral deposits theme CSW 2.0.2 GeoNetwork opensource webmaster Administrator GPO Box 378 Canberra ACT 2601 au webmaster@ga.gov.au gov ServiceIdentification ServiceProvider OperationsMetadata FilterCapabilities XML XML SOAP csw:Record gmd:MD × Metadata application/xml http://www.w3.org/TR/xmlschema-1/ csw:Record gmd:MDMetadata XML XML SOAP hits results validate application/xml http://www.opengis.net/cat/csw/2.0.2 http://www.isotc211.org/2005/gmd csw:Record gmd:MDMetadata FILTER CQLTEXT XML Operation Format OrganisationName Type ServiceType DistanceValue ResourceLanguage RevisionDate OperatesOn GeographicDescriptionCode AnyText Modified PublicationDate ResourceIdentifier ParentIdentifier Identifier CouplingType Top-

icCategory OperatesOnIdentifier ServiceTypeVersion TempExtentend Subject
CreationDate OperatesOnName Title DistanceUOM Denominator AlternateTitle
Language TempExtentbegin HasSecurityConstraints KeywordType Abstract
AccessConstraints ResponsiblePartyRole OnlineResourceMimeType OnlineRe-
sourceType Lineage SpecificationDate ConditionApplyingToAccessAndUse
SpecificationDateType MetadataPointOfContact Classification OtherConstraints
Degree SpecificationTitle XML SOAP http://www.opengis.net/cat/csw/2.0.2
http://www.isotc211.org/2005/gmd   application/xml   hits   results   validate   brief   summary   full   XML   http://www.opengis.net/cat/csw/2.0.2
2.0.2   http://www.isotc211.org/2005/gmd   SOAP   gml:Envelope   gml:Point
gml:LineString   gml:Polygon   EqualTo   Like   LessThan   GreaterThan
LessThanEqualTo LessThanOrEqualTo GreaterThanEqualTo GreaterThanOrE-
qualTo NotEqualTo Between NullCheck

While the GetCapabilities response looks like a lot of unrelated words, because
it is standards based and is CSW compliant, the response is easily parsed and usable
in a computer system that is CSW compliant to find out what data is available from
the Web service and how to access these data.

9.7   Metadata Standards and the Future of SDI and Online GIS

GeoPortals are now an important part of any Spatial Data Infrastructure. The ability to find spatial data resources online, including the provider's contacts, ownership
and usage details, has opened up a previously specialised field to anyone with a Web
browser. These catalogues of data can be accessed using manual searches entered
online, or can be accessed programmatically by other computer systems and Web
services to harvest the data directly. As demonstrated in Section 9.6, there have been
a number of implementations of online spatial marketplaces across all levels of government.

From a practitioner perspective, however, it is unclear to online users just how
successful these portals have been. The earlier gusto of these implementations appeared to have reached a plateau by around 2010, before being reignited by additional standards and an increased understanding of spatial data in the community.
Because of the ubiquitous nature of spatial information, society depends heavily on
these data without knowing it. Many of the day to day things we do require location information, from driving to work to buying groceries at the local store. Spatial
information underpins so many things we do. And while spatial information has no
intrinsic value, the benefits of quality spatial information are well documented. It will
be important for organisations to publish statistics on the use of these portals in the
future and to promote their products, not just from a community benefits perspective,
but to realise the true value of spatial information.

If SDI metadata portals and data catalogues are the online marketplaces for spatial information, then markup languages, metadata and Web standards have been the

enablers. These technologies, which have been the focus of this book, underpin the open standards and open GIS that have made Online GIS possible. In the final chapters of this book, we will look at some of the current trends and future directions of Online GIS.

9.8 Summary and Outlook

This chapter is a key chapter, as it pulls together two streams of metadata and programming languages discussed throughout this book through describing current examples of Online GIS implementations from around the world.

- The first half of this chapter, (§9.1, §9.2 and §9.4), pulled together details on the metadata standards in use with Online GIS systems, to give a solid grounding of the current state of metadata standards and their importance in Online GIS.

- The second half of this chapter provided the reader with leading examples of Online GIS implementations, including SDI metadata portals (§9.5) and examples of how organisations are using metadata to power their search and discovery engines, and for their online publication of geospatial assets through services (§9.6).

- The future of Online Spatial Data Infrastructures is then discussed in §9.7.

9.9 Appendix: Simplified Schemas

9.9.1 TGMD.XSD

```
<?xml version="1.0" encoding="UTF-8"?>
<xs:schema targetNamespace="http://www.opengis.net/giraffe"
        xmlns:tgmd="http://www.opengis.net/giraffe"
        xmlns:xs="http://www.w3.org/2001/XMLSchema"
        elementFormDefault="qualified" version="3.2.1.2">
  <!-- ======================================= Annotation ======================================= -->
  <xs:annotation>
      <xs:documentation>This file was generated from ISO TC/211 UML class diagrams == 01-26-2005 12:40:04 ====== </xs:documentation>
  </xs:annotation>
  <xs:include schemaLocation="tgco.xsd"/>
  <xs:include schemaLocation="tdq.xsd"/>
  <xs:complexType name="MD_Metadata_Type">
    <xs:annotation>
      <xs:documentation>Information about the metadata</xs:documentation>
    </xs:annotation>
    <xs:complexContent>
      <xs:extension base="tgmd:AbstractObject_Type">
        <xs:sequence>
          <xs:element name="fileIdentifier" type="tgmd:CharacterString_PropertyType" minOccurs="0"/>
          <xs:element name="language" type="tgmd:CharacterString_PropertyType" minOccurs="0"/>
          <xs:element name="contact" type="tgmd:CI_ResponsibleParty_Type" maxOccurs="unbounded"/>
          <xs:element name="dateStamp" type="xs:dateTime"/>
          <xs:element name="metadataStandardName" type="tgmd:CharacterString_PropertyType" minOccurs="0"/>
          <xs:element name="metadataStandardVersion" type="tgmd:CharacterString_PropertyType" minOccurs="0"/>
          <xs:element name="dataSetURI" type="xs:anyURI" minOccurs="0"/>
          <xs:element name="identificationInfo" type="tgmd:MD_DataIdentification_Type" maxOccurs="unbounded"/>
```

```
          <xs:element name="dataQualityInfo"
                   type="tgmd:DQ_DataQuality_PropertyType"
                   minOccurs="0" maxOccurs="unbounded"/>
          <xs:element name="metadataMaintenance"
                   type="tgmd:MD_MaintenanceInformation_Type"
                   minOccurs="0"/>

          <xs:element name="distributionInfo"
                   type="tgmd:MD_Distribution_Type"
                   minOccurs="0"/>
       </xs:sequence>
     </xs:extension>
   </xs:complexContent>
</xs:complexType>
<!-- ............................................................. -->
<xs:element name="MD_Metadata" type="tgmd:MD_Metadata_Type"/>
<!-- ............................................................. -->
<!-- <xs:complexType name="MD_Identification_PropertyType"> -->
<!-- <xs:sequence minOccurs="0"> -->
<!-- <xs:element name="MD_" type="xs:string"/> -->
<!-- </xs:sequence> -->
<!-- <xs:attributeGroup ref="tgmd:ObjectReference"/> -->
<!-- <xs:attribute ref="tgmd:nilReason"/> -->
<!-- </xs:complexType> -->


<!-- From gmd/identification.xsd -->

<!-- From gmd/citation -->
<xs:element name="CI_Citation" type="tgmd:CI_Citation_Type"/>
<xs:complexType name="CI_Citation_Type">
  <xs:annotation>
    <xs:documentation>Standardized resource reference</xs:documentation>
  </xs:annotation>
  <xs:complexContent>
    <xs:extension base="tgmd:AbstractObject_Type">
      <xs:sequence>
        <xs:element name="title" type="xs:string"/>
      </xs:sequence>
    </xs:extension>
  </xs:complexContent>
</xs:complexType>

<!-- ............................................................. -->
<xs:complexType name="CI_Citation_PropertyType">
  <xs:sequence minOccurs="0">
    <xs:element ref="tgmd:CI_Citation"/>
  </xs:sequence>
  <xs:attributeGroup ref="tgmd:ObjectReference"/>
  <xs:attribute ref="tgmd:nilReason"/>
</xs:complexType>

<!-- To save space and to make the examples more readable, we have replaced -->
<!-- CharacterString_PropertyType with xs:string. -->

<xs:complexType name="CI_ResponsibleParty_Type">
  <xs:annotation>
    <xs:documentation>Identification of, and means of communication with, person(s) and organisations associated with the dataset</xs:documentation>
  </xs:annotation>
  <xs:complexContent>
    <xs:extension base="tgmd:AbstractObject_Type">
      <xs:sequence>
        <xs:element name="individualName" type="xs:string" minOccurs="0"/>
        <xs:element name="organisationName" type="xs:string" minOccurs="0"/>
        <xs:element name="positionName" type="xs:string"
                   minOccurs="0"/>
        <xs:element name="contactInfo" type="tgmd:CI_Contact_Type" minOccurs="0"/>

      </xs:sequence>
    </xs:extension>
  </xs:complexContent>
</xs:complexType>


<!-- ............................................................. -->

<!-- ======================================== Reference of a resource from gcoBase======================== -->
<!-- ================ Reference of a resource from gcoBase=============== -->
<!-- The following attributeGroup 'extends' the GML gml:AssociationAttributeGroup -->
<xs:attributeGroup name="ObjectReference">
  <!-- Took out simpleAttrs since can't find this in xlink -->
  <!-- in fact it's in XMLSchema!! It doesn't do very much either -->
  <!-- <xs:attributeGroup ref="xlink:simpleAttrs"/> -->
  <xs:attribute name="uuidref" type="xs:string"/>
</xs:attributeGroup>

<xs:complexType name="MD_DataIdentification_Type">
  <xs:complexContent>
    <xs:extension base="tgmd:AbstractMD_Identification_Type">
```

```xml
      <xs:sequence>
        <xs:element name="language" type="xs:string" maxOccurs="unbounded"/>

        <xs:element name="topicCategory"
                    type="tgmd:MD_TopicCategoryCode_Type"
                    minOccurs="0" maxOccurs="unbounded"/>
        <xs:element
            name="supplementalInformation"
            type="xs:string"
            minOccurs="0"/>
      </xs:sequence>
    </xs:extension>
  </xs:complexContent>
</xs:complexType>
<!-- ......................................................... -->
<xs:complexType name="AbstractMD_Identification_Type" abstract="true">
  <xs:annotation>
    <xs:documentation>Basic information about data</xs:documentation>
  </xs:annotation>
  <xs:complexContent>
    <xs:extension base="tgmd:AbstractObject_Type">
      <xs:sequence>
        <xs:element name="citation" type="tgmd:CI_Citation_PropertyType"/>
        <xs:element name="abstract" type="xs:string"/>


        <xs:element name="descriptiveKeywords" type="xs:string" minOccurs="0" maxOccurs="unbounded"/>
      </xs:sequence>
    </xs:extension>
  </xs:complexContent>
</xs:complexType>
<!-- .....changed keywords to strings........................................................ -->
<xs:element name="AbstractMD_Identification" type="tgmd:AbstractMD_Identification_Type" abstract="true"/>

<xs:simpleType name="MD_TopicCategoryCode_Type">
  <xs:annotation>
    <xs:documentation>High-level geospatial data thematic classification to assist in the grouping and search of available geospatial datasets</xs:documentation>
  </xs:annotation>
  <xs:restriction base="xs:string">
    <xs:enumeration value="farming"/>
    <xs:enumeration value="biota"/>
    <xs:enumeration value="boundaries"/>
    <xs:enumeration value="climatologyMeteorologyAtmosphere"/>
    <xs:enumeration value="economy"/>
    <xs:enumeration value="elevation"/>
    <xs:enumeration value="environment"/>
    <xs:enumeration value="geoscientificInformation"/>
    <xs:enumeration value="health"/>
    <xs:enumeration value="imageryBaseMapsEarthCover"/>
    <xs:enumeration value="intelligenceMilitary"/>
    <xs:enumeration value="inlandWaters"/>
    <xs:enumeration value="location"/>
    <xs:enumeration value="oceans"/>
    <xs:enumeration value="planningCadastre"/>
    <xs:enumeration value="society"/>
    <xs:enumeration value="structure"/>
    <xs:enumeration value="transportation"/>
    <xs:enumeration value="utilitiesCommunication"/>
  </xs:restriction>
</xs:simpleType>

<xs:complexType name="MD_MaintenanceInformation_Type">
  <xs:annotation>
    <xs:documentation>Information about the scope and frequency of updating</xs:documentation>
  </xs:annotation>
  <xs:complexContent>
    <xs:extension base="tgmd:AbstractObject_Type">
      <xs:sequence>
        <xs:element
            name="maintenanceAndUpdateFrequency"
            type="tgmd:MD_MaintenanceFrequencyCode_PropertyType"/>
        <xs:element name="dateOfNextUpdate" type="xs:dateTime" minOccurs="0"/>
      </xs:sequence>
    </xs:extension>
  </xs:complexContent>

</xs:complexType>
<xs:element name="MD_MaintenanceFrequencyCode" type="tgmd:CodeListValue_Type"/>
<!-- ......................................................... -->
<xs:complexType name="MD_MaintenanceFrequencyCode_PropertyType">
  <xs:sequence minOccurs="0">
    <xs:element ref="tgmd:MD_MaintenanceFrequencyCode"/>
  </xs:sequence>
  <xs:attribute ref="tgmd:nilReason"/>

</xs:complexType>
<!-- ================================================================== -->
<!-- The type hierarchy has been simplified here for clarity -->
```

```xml
<xs:complexType name="MD_Distribution_Type">
  <xs:annotation>
    <xs:documentation>Information about the distributor of and options for obtaining the dataset</xs:documentation>
  </xs:annotation>
  <xs:complexContent>
    <xs:extension base="tgmd:AbstractObject_Type">
      <xs:sequence>
        <xs:element name="distributionFormat" type="xs:string" minOccurs="0" maxOccurs="unbounded"/>
        <xs:element name="distributor" type="xs:string" minOccurs="0" maxOccurs="unbounded"/>
        <xs:element name="transferOptions" type="xs:string" minOccurs="0" maxOccurs="unbounded"/>
      </xs:sequence>
    </xs:extension>
  </xs:complexContent>
</xs:complexType>

<xs:complexType name="CL_Contact_Type">
  <xs:annotation>
    <xs:documentation>Information required enabling contact with the responsible person and/or organisation</xs:documentation>
  </xs:annotation>
  <xs:complexContent>
    <xs:extension base="tgmd:AbstractObject_Type">
      <xs:sequence>
        <xs:element name="phone" type="xs:string" minOccurs="0"/>
        <xs:element name="address" type="tgmd:CL_Address_Type" minOccurs="0"/>
      </xs:sequence>
    </xs:extension>
  </xs:complexContent>
</xs:complexType>
<!-- ........................................................ -->
<xs:complexType name="CL_Address_Type">
  <xs:annotation>
    <xs:documentation>Location of the responsible individual or organisation</xs:documentation>
  </xs:annotation>
  <xs:complexContent>
    <xs:extension base="tgmd:AbstractObject_Type">
      <xs:sequence>
        <xs:element name="deliveryPoint" type="xs:string"
                    minOccurs="0" maxOccurs="unbounded"/>
        <xs:element name="city" type="xs:string" minOccurs="0"/>
        <xs:element name="administrativeArea" type="xs:string" minOccurs="0"/>
        <xs:element name="postalCode" type="xs:string" minOccurs="0"/>
        <xs:element name="country" type="xs:string" minOccurs="0"/>
        <xs:element name="electronicMailAddress"
                    type="xs:string" minOccurs="0"
                    maxOccurs="unbounded"/>
      </xs:sequence>
    </xs:extension>
  </xs:complexContent>
</xs:complexType>
</xs:schema>
```

9.9.2   TGCO.XSD

```xml
<?xml version="1.0" encoding="UTF-8"?>
<xs:schema xmlns:xs="http://www.w3.org/2001/XMLSchema"
           xmlns:xlink="http://www.w3.org/1999/xlink"
           xmlns:gml="http://www.opengis.net/gml/3.2"
           xmlns:tgco="http://www.opengis.net/giraffe"
targetNamespace="http://www.opengis.net/giraffe"
elementFormDefault="qualified" version="2012-07-13">
       <!-- ============================= Annotation ============================= -->
<!-- targetNamespace="http://lpmeta.org.lp" -->

       <xs:annotation>
       <xs:documentation>Geographic COmmon (GCO) extensible markup language is a component of the XML Schema Implementation of Geographic
Information Metadata documented in ISO/TS 19139:2007. GCO includes all the definitions of http://www.isotc211.org/2005/gco namespace. The root document of this
           namespace is the file gco.xsd. This basicTypes.xsd schema implements concepts from the "basic types" package of ISO/TS 19103.</xs:documentation>
       </xs:annotation>

       <xs:complexType name="AbstractObject_Type" abstract="true">
               <xs:sequence/>
               <xs:attributeGroup ref="tgco:ObjectIdentification"/>
       </xs:complexType>
       <!-- ================= Element ================= -->
       <xs:element name="AbstractObject" type="tgco:AbstractObject_Type" abstract="true"/>

       <xs:attributeGroup name="ObjectIdentification">
               <xs:attribute name="id" type="xs:ID"/>
               <xs:attribute name="uuid" type="xs:string"/>
       </xs:attributeGroup>

       <xs:element name="CharacterString" type="xs:string"/>
       <!-- ........................................................ -->
```

```xml
<xs:complexType name="CharacterString_PropertyType">
    <xs:sequence minOccurs="0">
        <xs:element ref="tgco:CharacterString"/>
    </xs:sequence>
    <xs:attribute ref="tgco:nilReason"/>
</xs:complexType>


<!-- From gcoBase for nilReason, but gml/basicTypes.xsd for the nilReasonType -->

    <xs:attribute name="nilReason" type="tgco:NilReasonType"/>

    <xs:simpleType name="NilReasonType">
        <xs:annotation>
            <xs:documentation>tgco:NilReasonType defines a content model that allows recording of an explanation for a void value or other
exception.
```
tgco:NilReasonType is a union of the following enumerated values:
— inapplicable there is no value
— missing the correct value is not readily available to the sender of this data. Furthermore, a correct value may not exist
— template the value will be available later
— unknown the correct value is not known to, and not computable by, the sender of this data. However, a correct value probably exists
— withheld the value is not divulged
— other:text other brief explanation, where text is a string of two or more characters with no included spaces
and
— anyURI which should refer to a resource which describes the reason for the exception
A particular community may choose to assign more detailed semantics to the standard values provided. Alternatively, the URI method enables a specific or more
complete explanation for the absence of a value to be provided elsewhere and indicated by — reference in an instance document.
tgco:NilReasonType is used as a member of a union in a number of simple content types where it is necessary to permit a value from the NilReasonType union as an
alternative to the primary type.

```xml
            </xs:documentation>
        </xs:annotation>
        <xs:union memberTypes="tgco:NilReasonEnumeration xs:anyURI"/>
    </xs:simpleType>
    <xs:simpleType name="NilReasonEnumeration">
        <xs:union>
            <xs:simpleType>
                <xs:restriction base="xs:string">
                    <xs:enumeration value="inapplicable"/>
                    <xs:enumeration value="missing"/>
                    <xs:enumeration value="template"/>
                    <xs:enumeration value="unknown"/>
                    <xs:enumeration value="withheld"/>
                </xs:restriction>
            </xs:simpleType>
            <xs:simpleType>
                <xs:restriction base="xs:string">
                    <xs:pattern value="other:\w{2,}"/>
                </xs:restriction>
            </xs:simpleType>
        </xs:union>
    </xs:simpleType>




</xs:schema>
```

9.9.3   TDQ.XSD

```xml
<?xml version="1.0" encoding="UTF-8"?>
<xs:schema xmlns:xs="http://www.w3.org/2001/XMLSchema"
        xmlns:xlink="http://www.w3.org/1999/xlink"
        xmlns:tgmd="http://www.opengis.net/giraffe"
        targetnamespace="http://www.opengis.net/giraffe"
        elementFormDefault="qualified" version="2012-07-13">
    <!-- ================================= Annotation ================================= -->
    <xs:annotation>
        <xs:documentation>
        These are elements extracted from the dataquality schemas.
        Minimum necessary to parse the md-full.xml example in the book.
        </xs:documentation>

    </xs:annotation>

    <!-- ..................................................... -->
    <xs:element name="DQ_DataQuality" type="tgmd:DQ_DataQuality_Type"/>
    <!-- ..................................................... -->
    <xs:complexType name="DQ_DataQuality_PropertyType">
        <xs:sequence minOccurs="0">
            <xs:element ref="tgmd:DQ_DataQuality"/>
        </xs:sequence>
        <xs:attributeGroup ref="tgmd:ObjectReference"/>
        <xs:attribute ref="tgmd:nilReason"/>
    </xs:complexType>
```

```xml
<!-- ============================================================== -->
<xs:complexType name="DQ_DataQuality_Type">
  <xs:complexContent>
    <xs:extension base="tgmd:AbstractObject_Type">
      <xs:sequence>
        <xs:element name="scope"
                    type="tgmd:DQ_Scope_PropertyType"/>
        <xs:element name="report"
                    type="tgmd:AbstractDQ_Element_Type"
                    minOccurs="0" maxOccurs="unbounded"/>
<!-- PropertyType" -->
        <xs:element name="lineage" type="tgmd:LI_Lineage_PropertyType" minOccurs="0"/>
      </xs:sequence>
    </xs:extension>
  </xs:complexContent>
</xs:complexType>
<!-- ..................................................... -->
<!-- <xs:element name="AbstractDQ_Element" -->
<!-- type="tgmd:AbstractDQ_Element_Type" abstract="true"/> -->

<!-- <xs:complexType name="DQ_Element_PropertyType"> -->
<!-- <xs:sequence minOccurs="0"> -->
<!-- <xs:element ref="tgmd:AbstractDQ_Element"/> -->
<!-- </xs:sequence> -->
<!-- <xs:attributeGroup ref="tgmd:ObjectReference"/> -->
<!-- <xs:attribute ref="tgmd:nilReason"/> -->
<!-- </xs:complexType> -->


<!-- ..................................................... -->
<!-- <xs:element name="DQ_DataQuality" -->
<!-- type="tgmd:DQ_DataQuality_Type"/> -->


<!-- ============================================================== -->
<xs:complexType name="DQ_Scope_Type">
  <xs:complexContent>
    <xs:extension base="tgmd:AbstractObject_Type">
      <xs:sequence>
        <xs:element name="level" type="tgmd:MD_ScopeCode_PropertyType"/>
        <xs:element name="levelDescription" type="xs:string"
                    minOccurs="0" maxOccurs="unbounded"/>
      </xs:sequence>
    </xs:extension>
  </xs:complexContent>
</xs:complexType>
<!-- ============================================================== -->
<!-- To simplify the type hierarchy this AbstractDQ_Element is not abstract -->
<xs:complexType name="AbstractDQ_Element_Type"> <!-- abstract="true"> -->
  <xs:complexContent>
    <xs:extension base="tgmd:AbstractObject_Type">
      <xs:sequence>
        <xs:element name="nameOfMeasure" type="xs:string" minOccurs="0" maxOccurs="unbounded"/>
        <xs:element name="measureIdentification" type="xs:string" minOccurs="0"/>
        <xs:element name="measureDescription" type="xs:string" minOccurs="0"/>
        <xs:element name="evaluationMethodType" type="xs:string" minOccurs="0"/>
        <xs:element name="evaluationMethodDescription" type="xs:string" minOccurs="0"/>

        <xs:element name="dateTime" type="xs:dateTime" minOccurs="0" maxOccurs="unbounded"/>
        <xs:element name="result" type="xs:string" maxOccurs="2"/>
      </xs:sequence>
    </xs:extension>
  </xs:complexContent>
</xs:complexType>


<!-- ..................................................... -->
<xs:element name="DQ_Scope" type="tgmd:DQ_Scope_Type"/>
<!-- ..................................................... -->
<xs:complexType name="DQ_Scope_PropertyType">
  <xs:sequence minOccurs="0">
    <xs:element ref="tgmd:DQ_Scope"/>
  </xs:sequence>
  <xs:attributeGroup ref="tgmd:ObjectReference"/>
  <xs:attribute ref="tgmd:nilReason"/>
</xs:complexType>

<xs:element name="MD_ScopeCode" type="tgmd:CodeListValue_Type" substitutionGroup="tgmd:CharacterString"/>
<!-- ..................................................... -->
<xs:complexType name="MD_ScopeCode_PropertyType">
  <xs:sequence minOccurs="0">
    <xs:element ref="tgmd:MD_ScopeCode"/>
  </xs:sequence>
  <xs:attribute ref="tgmd:nilReason"/>
</xs:complexType>
<!-- ============================================================== -->

<!-- ============================================================== -->
<!-- ================ The CodeList prototype ================ -->
<!-- It is used to refer to a specific codeListValue in a register -->
<!-- =============== Type =============== -->
```

```
<xs:complexType name="CodeListValue_Type">
  <xs:simpleContent>
    <xs:extension base="xs:string">
      <xs:attribute name="codeList" type="xs:anyURI" use="required"/>
      <xs:attribute name="codeListValue" type="xs:anyURI" use="required"/>
      <xs:attribute name="codeSpace" type="xs:anyURI"/>
    </xs:extension>
  </xs:simpleContent>
</xs:complexType>
<!-- - Lineage stuff - -->

<xs:complexType name="LLLineage_Type">
  <xs:complexContent>
    <xs:extension base="tgmd:AbstractObject_Type">
      <xs:sequence>
        <xs:element name="statement" type="xs:string" minOccurs="0"/>
      </xs:sequence>
    </xs:extension>
  </xs:complexContent>
</xs:complexType>
<!-- ......................................................... - -->
<xs:element name="LLLineage" type="tgmd:LLLineage_Type"/>
<!-- ......................................................... - -->
<xs:complexType name="LLLineage_PropertyType">
  <xs:sequence minOccurs="0">
    <xs:element ref="tgmd:LLLineage"/>
  </xs:sequence>
  <xs:attributeGroup ref="tgmd:ObjectReference"/>
  <xs:attribute ref="tgmd:nilReason"/>
</xs:complexType>


</xs:schema>
```

10

Data Warehouses

CONTENTS

Data warehousing has become a popular approach for organisations to gather their disparate data into one place. Geographic data is ideally suited to this type of aggregation. The ability of organisations to discover new relationships between the aggregated data has driven greater investment in data warehousing and so has the need for online publication of data. This chapter introduces some common data warehouse terms, architectures and approaches with reference to the management of geographical data in these infrastructures, in particular, spatial and temporal data management.

10.1   What Is a Data Warehouse?

William Inmon (1995) introduced the term *data warehousing* to describe a database system that was designed and built specifically to support the decision making process of an organisation. However, data warehousing goes well beyond the construction of a database. More importantly, data warehousing is a process, *not* a product. The process includes assembling and managing data from various sources for the purpose of gaining a *single detailed view* of part or all of an organisation.

A Data Warehouse is an organised collection of databases and processes for information retrieval, interpretation and display. These databases may be relational databases (RDBMS), object relational databases (ORDBMS) or file based databases. Data marts (§10.1.1), a delivery component of data warehousing, provides cached versions of geographical data as vector and raster data and have been the growth area of data warehousing since the first edition of the book.

Inmon (1995) defined a Data Warehouse as a managed database in which the data is

Subject Oriented There is a shift from application-oriented data (i.e. data designed to support processing) to decision-support data (i.e. data designed to aid in the decision making process). For example, sales data for a given application contains specific sales, product and customer information. In contrast, sales data for decision support may contain historical records of sales over specific time intervals.

Integrated Data from various sources is combined to produce a global, subject-oriented view of the topic of interest.

Time Variant Operations data is valid at the moment of capture; however, within seconds that data may no longer be valid. However, Data Warehouses can be developed to hold temporally aware versions of these data indefinitely. These *time slices* can then be provided through Online GIS to show *changes over time*.

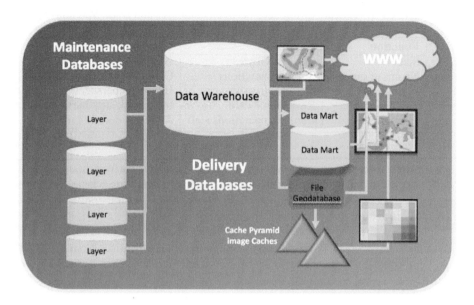

FIGURE 10.1: The integrated view of a geographic Data Warehouse, where layers from various sources are combined to produce a global subject-oriented view of a geographic region in the warehouse, and additional delivery environment consisting of data marts, file geo-databases and/or map caches of image tiles for fast delivery.

Non-Volatile New data is always appended to the database rather than replacing existing data. The database continually absorbs new data, integrating it with the previous data (Inmon 1995).

How does a Data Warehouse differ from a database? Perhaps the most important criterion is that it usually contains several distinct databases. The warehouse is an umbrella that links together many different data resources, as shown in Figure 10.1. Now a single database may contain many different tables, but they are tightly integrated within a single software shell. In contrast, a Data Warehouse is usually created by combining different databases, that already exist. These databases may use different software. They may be developed and maintained by completely separate working groups or organisations at different geographical locations.

Another important criterion is inherent in the subject-oriented nature of Data Warehouses. Whereas a database normally supports only simple (SQL) queries, a Data Warehouse usually provides additional tools to assist in interpreting and displaying data. This is a consequence of the motivation behind the warehouse. It is only worth the trouble to link different databases, if there is useful information to be gained in so doing. In the case of geographical data, there can be many benefits to business analytics generation and interpretation, and to data quality improvements. So it is always a high priority in a Data Warehouse to be able to generate the kinds of

TABLE 10.1: Scale differences between databases and Data Warehouses

Function	Database	Data Warehouse
Size	Mbytes–Gbytes	Gbytes–Tbytes
Nature of queries	Small transactions	Complex queries
Operations	Current snapshot	Historical perspective
Type of data	Raw source data	Integrated, summarised
Main users	Clerical, operations staff	Analysts, decision makers

information for which it was created. For example, many Data Warehouses are created as commercial marketing tools. such as when a corporation wishes to improve its marketing strategy by analysing sales figures. Thus it is imperative to provide tools to extract and analyse the required data.

The final difference between a database and a Data Warehouse is the scale of the system (Table 10.1). The whole notion of data warehousing arises from the rapidly growing volumes of raw data that are now available in many spheres of professional activity, the so-called Big Data trend.

10.1.1   Key Definitions

As with any new field of endeavour, data warehousing has developed its own taxonomy. Below is a collection of terms used in the remainder of this chapter:

Data Mart   A Data Mart or local Data Warehouse is a database that has the same characteristics as a Data Warehouse, but is usually smaller and is focused on the data for one workgroup within an enterprise or data used for one specific business need, such as online publication.

Data Transformation   The modification of data as it is moved into and out of the data warehouse. These modifications can include data cleansing, normalisation, transforming data types, encoding values, adding temporal attribution and summarising data into time periods, changing data projections or precision.

Data Mining   The non-trivial process of extracting previously unknown information or patterns from large datasets. A typical application of data mining is to determine what attributes (and values for these attributes) best describe a geographic region. Data mining falls into the more general category of intelligent systems, considered briefly in §10.9, and data mining in more detail in §10.10.

Knowledge Discovery in Databases (KDD)   KDD (Fayyad et al. 1996b) is a term quite often interchangeably used with data mining. However, KDD concentrates on the discovery of useable knowledge (quite often in the form of rules), where data mining also includes the detection of trends and models.

10.2   Geographic Data Warehouses

The idea of a Data Warehouse will be familiar to anyone who works with geographic information. Many geographic information systems are essentially Data Warehouses. This is because a GIS reduces different data sources to the common themes of geographic location and spatial attributes. It is common for different data layers to be derived initially from separate databases. For instance, cadastral data, relating to land ownership, could be rendered from a database of municipal information and be overlaid on data rendered from databases of environmental features, health or census records. This mashing up of disparate datasets has significant benefits for individuals, groups and organisations.

Conversely, although they may not be developed with GIS specifically in mind, many Data Warehouses do have a geographic dimension to them. This is certainly true of most government Data Warehouses where it is estimated that 85% of government information has a spatial context (ACIL-Tasman 2008).

Governments also play a critical role in the acquisition and management of large foundation spatial datasets that derive many economic benefits for jurisdictions. These large spatial datasets are ideally suited for aggregation into Geographic Data Warehouses.

The World Data Center for Paleoclimatology (Figure 10.2), for instance, acts as a repository for many kinds of environmental datasets that relate to climate change.[1] Datasets such as tree ring records, pollen profiles and ice cores all refer to particular geographic locations.

Many organisations have set up geographic Data Warehouses over the last decade. For example, the Canadian Government established a Data Warehouse Infrastructure Project in 2000 as a project under its National Forest Information System (NFIS), where[2]

> *The objective of the National Forest Information System is to provide a national monitoring, integrating and reporting system for Canada's forest information and changes over time. The current focus of the NFIS Data Warehouse is to provide information to meet Kyoto reporting requirements for forest carbon stocks and criteria and indicators. Over time, NFIS will improve access to and use of accurate and timely spatial and non-spatial information on Canada's forest resources. Some of the information required about the forest is not directly observable (critical wildlife habitat, water quality, risks of disturbance, etc.). The NFIS will integrate the information in the database above with a national modeling framework.*

What has evolved since then is a comprehensive series of public and private portals that are used to manage almost 500 million hectares of Canada's forest located across the entire country. The NFSI website contains links to a number of forestry

---

[1] http://www.ngdc.noaa.gov/paleo/data.html
[2] http://nfis.cfs.nrcan.gc.ca/warehouse

FIGURE 10.2: NOAA website content in 2014 showing a number of paleoclimatology datasets now accessible online.

related datasets including Land Cover, Invasives, Biodiversity and Conservation Areas with limited online access to some of these datasets. This restricted access model for Online GIS enables general users to render some data through Online GIS services while still giving forestry working groups access to sensitive environmental information over the Web. Their current website goes on to describe[3]

Responsibility for these forests falls under a great many jurisdictions and management agencies including governments, industry and other organizations. Forestry information is gathered in different ways, for different uses and is stored in different locations. As a result, accessing and integrating this information is extremely complex. NFIS is addressing these significant and wide-ranging differences by adopting international standards and by building a distributed network of servers and applications which allow access to forestry information held by independent agencies. NFIS provides Web tools, ranging from simple portrayal to sophisticated analyses, to users from anywhere in the world. It means users can discover, integrate, and display this current, authoritative and accurate information on Canada's forests and on sustainable forest management.

---

[3] http://ca.nfis.org/about_eng.html

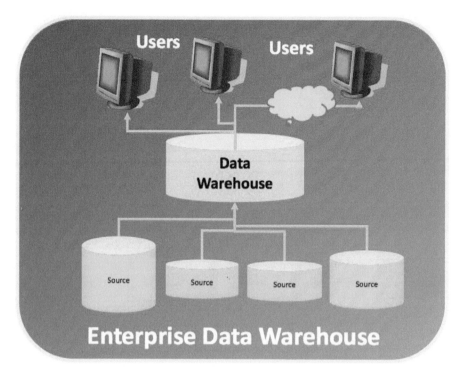

FIGURE 10.3: Enterprise Data Warehouse model.

10.3   Data Warehouse Architectures

Data warehousing systems are most effective when data can be combined from several operational systems or maintenance environments. The nature of the underlying systems where the data is being maintained, the level of integration required and the organisation structure should drive the selection of a Data Warehouse architecture or topology. Here we present three Data Warehouse topologies: Enterprise wide Data Warehouse, Independent Data Mart and Dependent Data Mart.

The **Enterprise Wide Data Warehouse** (Figure 10.3) integrates all information contained in an organisation's maintenance databases into a single global data repository. While the integration process may add additional metadata such as temporality to the database records, the primary role of an Enterprise wide Data Warehouse is to bring disparate databases together into a single database environment.

The **Independent Data Mart** topology implements a number of smaller Data Warehouses or data marts (Figure 10.4). This architecture integrates several databases to produce an independent data mart. This architecture works well where organisations have a number of departments with different data needs. A downside

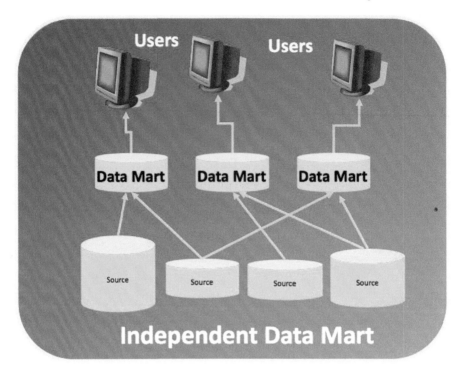

FIGURE 10.4: Independent Data Mart model, which builds data marts from other data repositories.

of this model is the data replication and management overheads needed to keep the independent data marts synchronised with the source databases.

The final topology discussed here is the **Dependent Data Mart.** Under this architecture, all the underlying databases feed into an enterprise Data Warehouse (Figure 10.5). A subset of the Data Warehouse is then replicated into a downstream data mart. This type of architecture is appropriate where an organisation has a business system that requires a particular aspect of the organisation's entire operations. This aspect may be the delivery processes used for Online GIS where the data needs to be optimised and not all data in the Data Warehouse is to be exposed externally. These databases may hold a *current view* only of data, and historical data stored in the Enterprise Wide Data Warehouse are not held in the Data Mart.

The architectures for Data Warehouses described here are not exhaustive by any means. As you could imagine, there are many hybrid systems. The overall architecture of a Data Warehouse is very much dependent on the needs of the organisation.

With the exception of Figure 10.3, all of the Data Warehouse architectures include data mart components. These data marts play an important role in the optimisation of Data Warehouse content for Online GIS delivery. In fact, there are a number

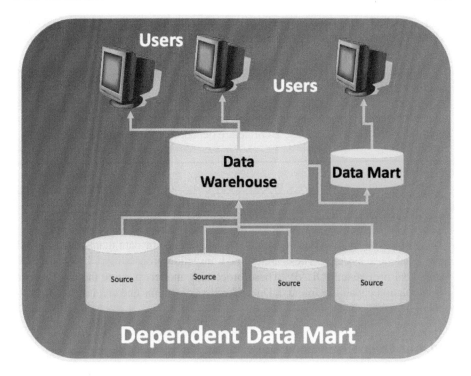

FIGURE 10.5: Dependent Data Mart model, where role-based data repositories are built up as subsets of the Data Warehouse information, often as non-temporal views of an organisation's data optimised for online publication.

of reasons why an organisation may choose to restrict access to their data warehouse. These include

- Data Warehouses are often large and difficult to scale.

- The spatial queries are often complex, especially where the queries need to have a temporal qualifier or date range.

- The primary role of a Data Warehouse is to aggregate data to support business operations.

- Limiting Data Warehouse access through secure services reduces risks for corporate information.

Implementing a Data Warehouse enables organisations to discover new applications of information not easily achieved with disparate databases. Organisations can also use these architectures to separate their maintenance databases from their delivery databases. This separation is very useful when delivering services for Online

GIS. The data warehouse architectures discussed here also make it easier to move some database components into cloud computing, where data marts can be placed into the cloud while leaving the maintenance systems on the premises.

When choosing which model to implement, consider how the models best support business needs. For some organisations, the performance of internal data processing is a priority so that internal resources are used effectively, while an online business may consider the performance of services for external clients as its priority. Decisions about the best Data Warehouse architecture depends on the business needs.

10.4   Database Processing Models

There are two common data processing models used for data management in relational database systems. These are online transaction processing (OLTP) and online analytical processing (OLAP). These processing models provide the common data manipulation capability found in Data Warehouse systems and are described in more detail below.

10.4.1   Online Transactional Processing

OLTP involves the processing of a large number of short transactions sourced from either GIS client systems, where changes are made through client-side editing, or from databases themselves. Transaction types for OLTP are *insert*, *update* or *delete* SQL statements.

Insert An insert is a process to create a new record in a database table.

```
insert [record] into [tableName];
commit;
```

Update The update process is the editing of one or more components of an existing database record.

```
update [field = value], [field = value] in [tableName]
        where objectID = [IdNumber];
commit;
```

Delete The delete process is essentially one of deleting an existing database record.

```
delete [record] from [tableName]
        where objectID = [IdNumber];
commit;
```

These three data processing methods cover the fundamental data manipulation tasks of OLTP. It is worth noting that the Insert command above is creating a new record, so it does not use an existig IdNumber as shown in the Update and Delete processes. An IdNumber, or object sequence is automatically generated by the database and applied to the new record.

When implementing OLTP, it is also essential that the transactions are applied to the Data Warehouse in the same order that they were processed in the client applications and operational databases. Maintaining this sequence is essential so that the information is kept in a consistent state. If transactions are applied out of sequence the data will be corrupted. It is also essential that the OLTP does not try to modify or delete objects that do not exist in the database, as this will result in failed transactions and the processing will need to be stopped. This ordering of transactions is termed *synchronous* or *sequential*, one after the other, in the same order the changes were created.

There are two important database capabilities that can be implemented for spatial information management when using an OLTP processing model.

The first is using OLTP for the *temporal* management of spatial objects. In this scenario, no data is ever removed from the database. Records are simply end dated using an update process which effectively retires the record and leaves it in the database. Temporal queries can then be used to show *active* records at any given point in time by supplying a *where* clause in the query that defines the date to use. By this we mean, for example, to show all land parcels that existed at this point in time. This has significant benefit for Online GIS where the change of objects over time can be rendered by doing simple temporal queries. This functionality exists today in a number of GIS systems where both GIS data and imagery are manipulated through temporal sliders to show changes over time.

The second important database capability, which can be implemented using OLTP, is the management of *persistent IDs*. Because a GIS only contains an approximation of real world objects, there are situations were spatial objects may change their location, despite being the same world feature. This may happen when the accuracy of an object is upgraded or they physically move in the real world. Objects in a GIS may also change their shape (e.g., a road may be upgraded in the GIS by adding additional vertices to its shape). In these situations, where different GIS objects represent the same real world feature, persisting (or keeping) the object identifier against the new GIS features is one effective way to retain metadata of an object as it changes over time.

10.4.2 Online Analytical Processing

OLAP involves the consolidation of many enterprise datasets into a common, often single database, for the purpose of *multi-dimensional* analysis. The process selectively extracts data patterns from these consolidated datasets and provides insights into these data from different points of view. It provides the capability for complex calculations, sophisticated data modeling and trend analysis over sequential time periods supporting end user analytical processes and activities. OLAP also allows users to generate reports from data warehouses to gain insight into the *business* meaning of data.

The data outputs from this consolidation process are also termed *Big Data*, and there are a number of vendor products now available to execute this consolidation process using traditional databases to automate report generation that can be fed into business analytics software.

For these reasons, OLAP involves the processing of more complex data queries with more computationally intensive data manipulation. However, it helps to analyse data specifically for business purposes and generates *business intelligence* from operational data.

10.4.3   Differences between OLTP and OLAP

OLTP and OLAP models are significantly different in how they process and manipulate data. In addition, the source of data for each model is different, with OLTP sourcing its data from operational systems, while OLAP consolidates data from many systems into one.

OLTP is characterised by larger volumes of transactions compared to OLAP, and the processing in OLTP usually involves fast running simple queries, while OLAP processing is more likely to involve more complex SQL with many table joins and data aggregation processes in a single query. OLAP queries may take many hours to run and are usually run at periodic time intervals to align with the timing of reports.

The purpose of OLTP is based on operational tasks needed for data maintenance, while OLAP provides business decision support. OLTP schemas are usually highly normalised while OLAP uses de-normalised schemas, supporting the storage of new data types such JSON. This de-normalisation of traditional database schemas is commonly used in data mining (§10.10) to reveal previously unavailable information.

The storage and backup of data in OLTP and OLAP is also different. OLTP are smaller in size to OLAP, as OLAP contains additional historical metadata and larger aggregated datasets. With OLTP, backup and restore is critical as these systems contain the original source data, while backup for OLAP is less critical as data can be reloaded from the related OLTP source systems.

In summary, OLTP and OLAP models play key roles in database processing and storage of spatial data, and the creation and storage of metadata for Online GIS. These include:

OLTP   can be used effectively to maintain data warehouses and data marts which directly support online map and feature services of Online GIS. The update transactions are applied in a synchronous manner through insert, update and delete methods; providing live updates to online systems.

OLAP   can be used to build complex aggregations for Online GIS to answer complex spatial queries efficiently, such as, comparing the relationships between spatial objects over time; using spatial operators to assess juxtaposition of spatial objects; perform complex routing and terrain analysis operations; just to name a few.

10.5 Considerations When Building Data Warehouses

So you want to build a Data Warehouse? Where do you start and what resources will you require? This section summarises some of the major issues, considerations and resources required to develop a data warehouse.

10.5.1 Practical Issues

The resources required in developing a Data Warehouse include the following:

Money In the development of any type of software project, funding is the biggest issue. Being an enterprise wide information service, the development of Data Warehouses can be extremely expensive. The key questions to ask here include How much will the construction of the Data Warehouse cost? What is the budget for the project? Who will pick up the difference between the budgeted amount and the actual amount? As with other IT projects, there need to be contingencies factored in to account for unforeseen costs (Gardner 1996; Inmon 1996).

Time Time is a critical consideration when developing any software system (Inmon 1996).

Users Any information system is designed for users. If the information system does not deliver the required information to end users it is practically useless. The needs of the end users are the most critical factors when developing the warehouse. The warehouse in essence is for the end users not the IT staff (Hammer et al. 1995).

People Apart from the end users, there is a wide array of additional people who need to have input into the construction of a Data Warehouse. They include support staff, database administrators, programmers, business analysts, Data Warehouse architect, help desk staff, training staff, system administrators, data administrators, data engineers, users, decision-makers, middle management and top management (Hammer et al. 1995; Inmon 1996).

Software and tools Data warehousing systems consist of three main components: an interface to the data, tools for searching and tools for retrieval of data (Green 1994). Currently there are literally hundreds of commercial and public domain software packages, datasets, and data analysis tools. The key consideration when selecting software is that the underlying database software is compatible with the selected analysis tools and other database packages. In addition, any software package should be extendable to meet the analysis and data requirements in the future (Worbel et al. 1997).

Reliability and robustness Just like any software project, the construction of a Data Warehouse goes through the systems development life cycle (SDLC). The phases that we will discuss here, however, are limited to analysis, design and infrastructure.

10.5.2 Analysis

The analysis phase is the most important phase in the construction of a Data Warehouse. Failure to correctly analyse the requirements of the system will result in failure. The key issues that need to be addressed are What data is required by the users? What data is/are the current system(s) collecting? What data is missing? What/where is the source of the data? What are the intended applications of the collected data? What legacy data is stored on legacy media (Gardner 1996)? Is this legacy data useful? Finally, one design attribute that should be built in from the start is extensibility. The Data Warehouse should be extendable in a number of ways.

First the warehouse should be able to cope with new data marts being added. Secondly, it should be possible to add new tables with little effort. Finally, data should be stored in a format that allows new analysis tools to be used to extract information. This leads to the point that the construction of the Data Warehouse should conform to some form of standard.

10.5.3 Design

The design phase transforms the findings from the analysis phase into a specification that can be implemented. During this phase designers need to decide on a warehouse architecture, database product, database models, the level of normalisation and the type of interfaces. Also the designers need to determine what users should have access to what data, what security measures are to be in place, and in what mode (online/batch) the warehouse is to operate.

There are a number of Data Warehouse products now available in the marketplace. Oracle offers a Data Warehouse plaform through their Exadata Database products. These products include a number of tools to assist in the design and deployment. IBM provides database appliances under a Neteeza[4] product name. These appliances provide Big Data and analytics capabilities against Data Warehouse implementations and include strong vector caching capabilities for spatial data. Microsoft also provides a modern Data Warehouse product, their SQL Server product.[5] This solution offers seamlessly scalable, within-memory processing of both relational data and non-relational data using their open source Hadoop[6] framework. Finally, a Data Warehouse can be implemented using any relational database, such as PostgreSQL and Informix from IBM.

These vendor products are an essential part of selecting a technology stack for implementation and the use of off the shelf products offers obvious advantages to organisations developing their own.

---

[4] http://www-01.ibm.com/software/au/data/netezza/
[5] \http://www.microsoft.com/datawarehousing
[6] Hadoop is an open source framework for the distributed storage and processing of Big Data.

10.5.4   Infrastructure

Being on an enterprise-wide scale, the chances are that an organisation will not have the required hardware and software infrastructure to support a Data Warehouse. Hardware requirements include large hard disks, backup and recovery devices; redundant hardware if the data is mission critical; and fast processors and lots of RAM to support many concurrent users and processing of the stored data. Many commercial vendors are producing software products to support data warehousing operations, and cloud technologies are now offering viable alternatives to on-premise data centres.

The emergence of new standards and protocols may make it easier to create distributed Data Warehouses in the future. For instance, the Data Space Transfer Protocol (DSTP) aims to make it simpler for different databases and systems to share data across the Internet. The idea is that people would convert datasets into a common format, using the Predictive Model Markup Language (PMML).[7] Just as the Hypertext Markup Language (HTML) allows people to place documents online in a format that can be universally read and displayed, so the aim of PMML is to achieve the same for datasets. One of the problems encountered in data mining and warehousing is that different people often use different formats to store data. We will discuss PMML further in the Data Mining section of this chapter (§10.10.1).

10.6   Data Warehouse Organisation and Operation

How do we organise information on a large scale? In many cases, the sheer scale of a Data Warehouse demands that the workload be distributed amongst many different organisations. As with any computer-based information system, the maintenance of the system is the most expensive activity in terms of both time and cost. This is especially true when the data changes often or when it needs to be updated regularly. In other cases, the warehouse needs to integrate different kinds of data (e.g., weather and plant distributions) that are compiled by separate specialist agencies.

10.6.1   Handling Legacy Data

In the early 1970s, IBM mainframes dominated business systems using software tools such as COBOL, CICS, IMS and DB2. The 1980s saw mainframes replaced with minicomputers such as the AS/400 and Digital Equipment Corporation VAX/VMS systems.

In the late 1980s and 1990s desktop computing technology become mature and was rapidly adopted by business. This era also saw an increased popularity in computer networking. Global information networks such as the World Wide Web were still growing in popularity at that time.

---

[7] http://www.dmg.org

The problem that organisations had encountered in moving from mainframe to minicomputer to desktop computing was the need for data conversion. Many legacy systems still contain valuable information. This need raises several issues:

- **De-normalise data tables:** The normalisation of database tables is common practice. The idea of data normalisation is to improve performance and conserve data storage by reducing duplication of common data, such as a street name that is used many times in the GIS (store once and use many times). So when integrating or aggregating heavily normalised data in data warehousing systems, it may be necessary to de-normalise, to ensure the correct conversion of data into the warehouse.

- **Rules for data integration:** Determine rules for matching data from different sources to allow easy data integration. In some instances simple key matching is not enough.

- **Data cleaning:** When converting data from legacy systems, it is almost always necessary to clean the data. This may include activities such as converting data to a uniform format, for example, ensuring that all characters belong to the same character set or carriage returns are consistent. The cleaning of the data may also include the addition, removal or reordering of attributes within the legacy dataset. Other activities involved in cleaning the data may include the detection of errors, identify inconsistencies, updating values, patching missing data and finally the removal of unneeded attributes.

Data conversion is an extremely important step in bringing a data warehouse online. Legacy systems are a valuable source of historic data; however, it may be quite time consuming to convert the data in these old systems. The benefit of including the historical data may outweigh costs involved in converting the legacy data.

10.6.2   Processing Objects

One of the most vexing problems with legacy data is to convert it into a form in which it can be used. This conversion can be done permanently by creating new files in the format required. However, permanent conversion may be either impractical or undesirable. For instance, the data may be used in many different ways that would require only parts of the data, each in different forms. So it may be more convenient to extract relevant portions each time they are needed.

Legacy datasets are often stored in idiosyncratic formats and require specific software to extract data from them. One example is data that was created in (say) a commercial format associated with a database program that is no longer in common use. Another case, common for older datasets, is where scientists have stored data in a manner of their own devising, but have provided a program to extract certain elements from the files. If this software is still available, then the most convenient approach might be to keep using the software with the dataset. However, if a Data Warehouse contains numerous legacy datasets, then this approach can become confusing and prone to errors.

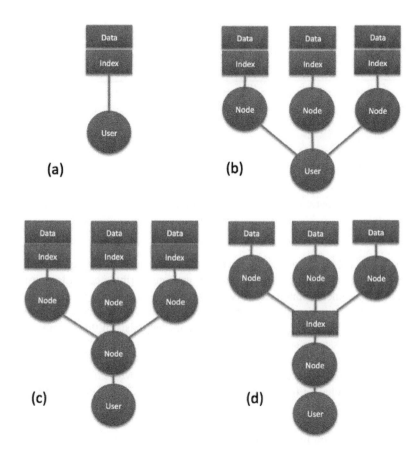

FIGURE 10.6: Some models for the configuration of online information systems. (a) A traditional centralised database, in which data and indexes reside at a single site. (b) An organised system in which the user must access different sites manually. (c) A common interface to a set of integrated, but separate databases that are loosely coupled. (d) An integrated model in which different datasets share a common, centralised index. See text for further explanation.

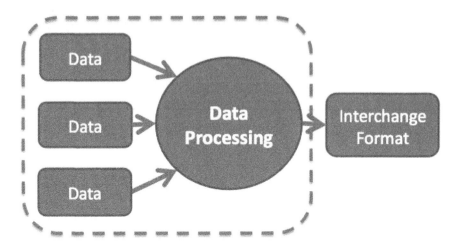

FIGURE 10.7: Operation of a warehouse object (enclosed by dashed line) to extract legacy data.

To impose order on legacy datasets, they can be treated as objects. That is, each dataset, the necessary extraction software, and any scripts needed to automate the extraction are encapsulated as a discrete object (Figure 10.7). This means that the details of the extraction process are hidden from the rest of the system. As far as users of the Data Warehouse are concerned, all that exists in the system are objects that can be accessed to supply certain information.

Note that this approach is also an effective method for organising data conversion. Suppose that a processing object (call it SET1) provides methods for outputting data in an interchange format, and that it also has methods for inputting data from the same interchange format. Then that interchange format provides an intermediate stage for converting from SET1 into any other format (SET2 say) that also has the same conversion methods available. This is what makes interchange formats so powerful.

A good example is image formats. There are at least 100 different formats in existence, many now rarely used. To provide direct converters between each pair of formats would require 9900 filter programs. However, by passing them through a single universal interchange format, the number of required programs drops to 200, one input and output filter for each format. In practice, several interchange formats are needed to take into account differences in the nature of the data, especially raster versus vector graphics.

10.6.3   Data Migration and Replication

There are two main data considerations when implementing a Data Warehouse. The first is the initial data loading of the Data Warehouse from source systems. This data migration step is often carried out using an extract, transform and load (ETL) process and source systems, may be relational databases or other files systems or applications. The ETL process involves extraction of data from one or more systems followed by the transformation of data into suitable format to be integrated into the Data Warehouse. When processing geospatial data, these transformations may include reprojecting data into a common coordinate system, resampling data to a common scale and reformatting data from disparate database models into a common schema.

There are two main approaches to data migration. The first migration method is the big bang approach. With this approach, all data is moved to the Data Warehouse in one simultaneous operation. The major benefit of this approach is its speed of data conversion. The second data migration method is the iterative approach, in which data is moved to the Data Warehouse one system at a time. So warehouses across the organisations are incrementally updated. This approach has less risk of data corruption. In the iterative approach, if something goes wrong, only a minimal number of data warehouses are damaged; however, it also requires that the datasets be held static during the migration processes, which may take many weeks or months for spatial datasets.

The second consideration is defining a process for keeping the data in the Data Warehouse maintained. This often involves transforming and loading data into a number of Data Warehouse tiers, and often across distributed Data Warehouse architectures. This replication process may use transactional data defined as add, modify and delete transactions using ascii (often XML) based records. It may also be run at different time intervals from monthly to live data streaming from source systems to the Data Warehouse.

Recent technology advancement in database replication has reduced the issues affecting data synchronisation, making the management of multi-tiered database environments simpler to administer and manage. These binary level replications work at the operating system level where the differences between source and target databases can be derived using difference files, and the differences applied to the target databases. These replications are very fast to perform and are provided as standard out of the box functionality with many database appliances.

An alternative to the replication of data across databases, as discussed above (§10.6.1 and §10.6.2), is to migrate all data from other disparate databases into an Enterprise Data Warehouse and have maintenance systems work against this new enterprise repository. One significant drawback with this approach, however, is that delivery systems exposed to the Web are using the same infrastructure as the maintenance systems, which can often result in computational conflicts as large maintenance jobs are being processed or peak periods are being experienced with Web facing delivery systems.

10.6.4   Indexing of Data Resources

Another consideration to the issues of Data Warehouse architecture and maintenance models is the question of how the data resources should be organised. There are three popular types of Data Warehouses for geospatial data.

The first of these is a *centralised* model. In this model, the entire database resides on a single server. This traditional model aggregates data from many sources into a central repository that consists of one or more schemas. The indexing of this data is maintained in this single database environment that offers complex spatial operators and spatial search and query capabilities are that greatly improved by a use of R-Tree and Quad-Tree indexing.

A second type is a *distributed* Data Warehouse where the databases are distributed across a number of servers. These servers may reside on the some domain, or across a number of disparate domains, locations and organisations. In this distributed database, information and indexes are shared across computers on each of the participating sites. In general, indexed data online can be organised in several ways (Figure 10.6), including

- **Distributed data, single centralised index:** The data consists of many items, which are stored at different sites but accessed via a database of pointers maintained at a single site.

- **Distributed data, multiple queries:** Many component databases are queried simultaneously across the network from a single interface. Many search engines adopt this approach.

- **Distributed data, separate indices at each site:** The database consists of several component databases, each maintained at different sites. A common interface (normally a Web document) provides pointers to the components, which are queried separately.

A third category of data warehousing is based on a *federated* model. In this model, the disparate databases are mashed up by clients using Web services to access the data. In this case the data is always rendered from the Data Warehouse of the custodians, providing access to the authoritative data. Client systems optimise database queries using a geographic extent within the query, constraining the query to a region or location, while running the query against a number of nodes in the federation and handling the responses asynchronously.

10.7   Data Warehouse Examples

There are many Data Warehouses now in existance, many of these containing spatial data. Two such examples are the NASA Distributed Active Archive Center (DAAC) (§10.7.1) and Walmart (§10.7.3).

10.7.1   Distributed Archive Center

NASA's Distribution Active Archive Center (DAAC)[8] was established in 1993 and consists of a number of data archiving centres. These include

- Alaska Satellite Facility SAR Data Center (ASF SDC);

- Crustal Dynamics Data Information System (CDDIS);

- Global Hydrology Resource Center (GHRC);

- Goddard Earth Sciences Data and Information Services Center (GES DISC);

- Land Processes (LP) DAAC;

- Level 1 Atmosphere Archive and Distribution System (MODAPS LAADS);

- NASA Langley Research Center Atmospheric Science Data Center (LaRC ASDC);

- National Snow and Ice Data Center (NSIDC) DAAC;

- Oak Ridge National Laboratory (ORNL) DAAC;

- Ocean Biology Processing Group (OBPG);

- Physical Oceanography (PO) DAAC;

- Socioeconomic Data and Applications Data Center (SEDAC);

Through these data centres, DAAC hosts a wide range of earth observation data from a number of disciplines. Data are included from NASA's Earth Observing System (EOS) satellites and other airborne digital sensors, and field observations and is an integral part of NASA's Earth Science Enterprise.

Historically, scientists have had difficulty conducting interdisciplinary research because locating useful data required contacting many different data custodians. EOS designed the DAAC to overcome the problems associated with data procurement needed for conducting interdisciplinary earth science research.

DAAC offered over 950 different datasets and products as early as 2002. It was unclear how many datasets DAAC data centres now manage; however, many of their datasets are publically available for free. Raster data types hosted at DAAC included

- Satellite imagery

- Digital aerial images

- Airborne sun photometer

- Advanced very high resolution radiometer

- Thermal infrared multispectral scans

---

[8] http://earthdata.nasa.gov/daacs.html

- Digital Elevation Models (DEMs)

- Field sun photometer

The DAAC allows the easy integration of data from different datasets by ensuring all data/image files are in common file formats, datasets are adequately documented, and contain sufficient metadata to accurately describe them. The Data Warehouses contain datasets collected from a range of studies over an extended period of time, held across a distributed system of many interconnected nodes. Examples of data include OTTER (Oregon Transect Ecosystem Research) and FIFE (First ISLSCP (International Satellite Land Surface Climatology Project) Field Experiment). In addition, there are periodic datasets collected by regular surveys.

The DAAC makes a large percentage of the datasets available at no cost. These datasets can be downloaded from the relevant data centre via FTP. Some datasets are available on cd-rom and can be obtained for a small fee.

An interesting point that DAAC highlights is that the management of such a data repository by one organisation is almost impossible. However, by dividing the work between various organisations and by making the organisations responsible for the management and administration of the data, such a project has become viable.

10.7.2 Earth Resource Observation and Science Data Center

Earth Resources Observation and Science (EROS) Data Center was opened in the early 1970s by the US Geological Survery (USGS),[9] a division of the Department of Interior. This Data Center hosts remotely sensed data with an archive that spans from 1937 aerial photographs to the original Earth orbits in the 1960s to current hourly updates. EROS is one of the largest civilian remote sensing data archives and its data is used by organisations and individuals from all over the world to monitor changes to the land surface over time, which effect environment, resources, health and safety.

EROS plays an important role in the processing, archiving and distribution of data recieved from EOS satellites. As such, EROS provides a Google Maps interface and many forms of searching – by address, area (arc or rectangle), predefined shape using Shape or KML input files and by time period. It also provides systems development and research capabilities and provides educational resources, information and tools to promote the use of remote sensing data.

Management of data in the Data Center requires the standardisation of policies and activities. The ESDIS Standards Office at EROS coordinates standards for metadata, including FGDC, ISO 19115, WMS and data formats, such as OGC KML. A complete list of these standards is available online,[10] and many of these are described in more detail in Chapter 9 and Chapter 6.

---

[9] http://eros.usgs.gov
[10] https://earthdata.nasa.gov/data/standards-and-references

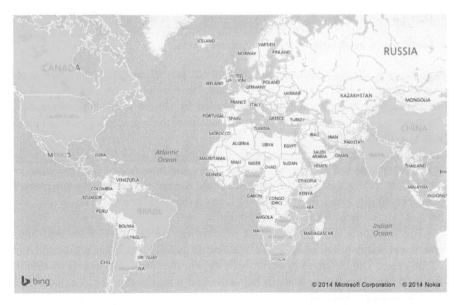

Where in the world is Walmart?

Zoom and pan within the interactive map to select a country and learn more about each location, including U.S. state details.

From our humble beginnings as a small discount retailer in Rogers, Ark., Walmart has opened thousands of stores in the United States and expanded internationally. Through this model of expansion, which brings the right store formats to the communities that need them, we are creating opportunities and bringing value to customers and communities around the globe.

Locations at a glance	
Total retail units on November 30, 2014	11,202
Walmart U.S.	4,364
Sam's Club	645
Walmart International	6,193

Today, Walmart operates over **11,000 retail units** under 71 banners in **27 countries**. We employ **2.2 million associates** around the world — 1.3 million in the U.S. alone.

FIGURE 10.8: Walmart's website containing an Online GIS component.

10.7.3 Walmart

Walmart currently operates a chain of over 11,000 retail stores across 27 countries (Figure 10.8). The organisation sells a wide range of household goods, electrical appliances, toys and other goods, and the Walmart chain also includes service stations, sports stores and discount retailers, distributed across the USA.

In the late 1980s and early 1990s, Walmart was having difficulty managing inventory, coordinating operations between the general office, distributors and retailers. In addition, Walmart's then information systems had a number of limitations. These included

- Limited data accessibility;

- Three month detail data online (limited);

- Detail history on tape for several years (with poor accessibility);

- Hundreds of generalised short term paper reports.

The overall environment was described as *"Data Rich but Information Poor"* (Hubber 1997). While Walmart has large volumes of data, the transformation of data into information was hindered by (1) the distributed nature of the databases, (2) limited time view of the operations of the organisation, and (3) limited access to the data.

The grand challenge to Walmart was to implement a system where all the data was located in one central repository. The data could be accessed from any platform; all the data were accessible; the database could support thousands of users globally; data accessibility needed to be on a 24 hours a day, 7 days a week basis; and users needed to have timely access to the data.

The solution to the problem was to construct a centralised Data Warehouse, which had standardised access. The Walmart Data Warehouse has grown to be one of the largest Data Warehouses (Hubber 1997) in the world. At the end of 1998 the Walmart Data Warehouse was 10TB in size, and it was expanding at a rate of 200–300 Mb per day. The largest table in the warehouse contained 20 billion rows and was supporting 650 concurrent users (Wiener 1997).

A number of benefits flowed from the Data Warehouse. These included

- Better markdown management;

- Improved next session planning;

- Improved stock control;

- Increased leverage during vendor negotiation;

- Improved long-term forecasting and trend analysis.

The use of a Data Warehouse allowed Walmart to gain a strategic advantage over its competitors and suppliers.

10.8   Common Publication Standards for Online GIS Warehouses

At the time of the first edition of this text, several standards were being developed that helped to simplify the process of creating an online Data Warehouse, especially for geographic information. Some of these standards are now well established and, although they were described in earlier chapters, it is worth touching on them again briefly here in the context of data warehousing. Some standards are specific to geographic information. Others are generic, but they enable the online publication and retrieval of many kinds of geographic information. These standards are derived from activities of the World Wide Web Consortium; others are specific to Data Warehouses, and others to the Open GIS Consortium (see Chapter 9).

10.8.1 Web Map Server Interface Standard

The specification for the OpenGIS Web Map Server Interface (WMSI) (OGC 2000), defined a Web Map Server (WMS) to be a site that can do three things:

1. Produce a map (as a picture, as a series of graphical elements, or as a packaged set of geographic feature data);
2. Answer basic queries about the content of the map; and
3. Tell other programs what maps it can produce and which of those can be queried further.

The importance of this early integration capability was that geographic data resources from different servers could be combined without the need for direct coordination between the two organisations involved. For instance, it makes it possible to extract a satellite image from one site and overlay geographic data (taken from a second site) of the region covered by the satellite image. This overlay data may be rendered as either vector overlays or as raster data created on the server.

The WMSI model divides data handling into four distinct stages:

1. Filtering data from a data source using SQL based query constraints
2. Generating the display elements from features in data according to the required style
3. Rendering the image
4. Displaying the image in a given format

One advantage of this hierarchy is that it allows some flexibility in the way data is delivered. For instance, if a client is capable of rendering maps that are encoded using SVG (§4.6.5), then the map could be delivered in that format, whereas a thin client, with no rendering capability at all, would need a map that had been converted into (say) a GIF or JPEG image.

10.8.2 Web Feature Server Interface Standard

The Web Feature Service (WFS) standard provides a Web service based Interface that can be used to publish and request geographic features over the internet. The standard includes support for points, lines and polygon features as well as data attribution and metadata.

WFS is based on the OGC's GML XML encoding discussed in §6.4. The importance of GML at the time was that it provided a standard for using XML to mark up GIS objects. OGC approved the OpenGIS GML Simple Features Profile[11] in 2006

---

[11] http://www.ogcnetwork.net/gml-sf

to make the WFS standard easier to implement. The WFS standard is currently at Implementation Standard version 1.1.0 and its schema[12] can be downloaded along with other OGC standards.

10.8.3 Web Mapping Tile Service

A more recent standard from the OGC has been the development of a Web Mapping Tile Service (WMTS). The purpose of this standard is to provide a standards based method of serving out digital maps through HTTP requests using predefined image tiles of around 256 x 256 pixels.

These digital map tiles are held in a pre-processed pyramid file format, where each level of the pyramid represents a higher or lower resolution map that is matched to the screen resolution of the client application to minimise the number of pixels being returned, while still providing the highest resolution image practical.

The pyramid has a predetermined number of layers (or scales) and other qualities such as tile dimensions in pixels, image format of each tile and data projection defined. As with other standards such as WMS and WFS, these tile services deploy map image tiles through Web services using Key Value Pair (KVP), SOAP(§3.1.1) and REST (§3.1.1) encoding, and the service supports commong Web service function calls, such as GetCapabilities, GetFeatureInfo and GetTile. These functions operate in a standard request response model. The WMTS standard is currently at Implementation Standard version 1.0.0 and its schema can be downloaded.[13]

10.9 Intelligent Systems

So far in this book, we have described the use of metadata to organise online sources of information for easy discovery. Although we have stressed many of the issues involved, an important concern remains. That is, how we can use that information most effectively. Knowledge-based and intelligent systems provide a natural way of working with both metadata and data content that is now available in Online GIS. Data warehouses amass data. But to realise their full potential, artifical intelligence is essential. We now look, in the most cursory way, at some of the options.

10.9.1 Knowledge-Based Systems

The essence of knowledge is the ability to do things, and doing is knowing. If information is data that has been distilled, so that essential patterns and relationships are made clear, the knowledge goes one step further. It includes details of how to use information.

---

[12] http://schemas.opengis.net/wfs/
[13] http://schemas.opengis.net/wmts/

Knowledge is usually expressed in the form of rules. Rules take the general form

$$P \implies Q$$

or

$$if\, P\, then\, Q$$

where P and Q are logical statements. The following simple examples illustrate some of the kinds of rules that might be used in an intelligent ecotourism information system.

```
If X is a park and is near a major city then
                X has ecotourism potential.

If population of X > 100,000 and X has an airport then
                X is a major city.

If the distance between X and Y is less than 200
                kilometres then X is near Y.
```

The following set of rules shows the sort of query that could be run in trying to locate websites that are worth searching for geophysical information.

```
If site X is relevant to geology then search site X.

If site X is useful and site X is near site Y, then site
                X is relevant to Y.

If site X is in list of reference sites then site
                X is useful.

If site X key words contain the word Y then site
                X is a "Y site".
```

Many systems incorporate rules, either explicitly or implicitly. Most electronic mail programs allow users to filter incoming mail. For instance, they may place copies of messages in different folders based on words in the subject line, or on the sender's address. Databases and spreadsheets can include rules (like those shown above) that convert data based on values of particular attributes.

10.9.2 Expert Systems

Expert systems are computer programs whose main function is to incorporate and use knowledge about a particular domain. For instance, a geographic expert system about ecotourism might include rules to identify sites that have potential for developing visitor programs. A geophysical expert system would contain rules about sites worthy of exploring for particular kinds of minerals. An expert system of websites would contain rules about how to locate sites that might provide particular kinds of data.

Expert systems often contain hundreds, or even thousands, of such rules. The usual search mechanism is a procedure known as backward chaining. The system starts from the statement that it needs to satisfy and works backwards. For any potential solution X it tries to form a chain of clauses

$$P(X) \implies Q(X), Q(X) \implies R(X)$$

in which the precondition $Q(X)$ in one rule is the conclusion in the previous rule. This process continues until the program finds a precondition that it can check directly.

Expert systems are usually produced using a development shell, together with an appropriate logical language. Common languages include LISP and Prolog; however, most shells include their own scripting language. A popular shell and language for developing intelligent systems online is CLIPS (Giarrantano and Riley 1989), which originated in NASA. This freeware system includes modules for incorporating expert systems into Web services.

The following short sample of CLIPS code asks the user to define a search radius (rule1). It then runs a test to determine whether a point located a certain distance from the search centre lies within a circle of that radius (rule2).

```
(defrule rule1
   =>
   (printout t "What search radius do you want? " crlf)
   (assert (radius = (read))))

(defrule rule2
   (radius ?r)
   (distance ?d)
   test (< ?r ?d)
   =>
   (assert (accept_point yes)))
```

10.9.3   Adaptive Agents

In building an expert system, the expert tries to incorporate all the knowledge that is necessary to solve a particular kind of problem before the system is used. This may involve a long development procedure. However, there are many kinds of problems for which complete knowledge is unobtainable. In such cases, acquiring new rules and data needs to be an on-going activity of the system. For example, in searching for information on the Web, new sites are always appearing and new data is constantly being added. So systems must be able to constantly add details of such items to its information base.

Adaptive agents are software programs that are intended to cope with this kind of problem. The term agent has many meanings that are relevant here. One sense covers programs that act on behalf of a user. For example, most Web search engines use software agents that automatically trawl through websites, recording and indexing the contents. Another, related definition concerns programs that automatically carry out some task or function. The term adaptive refers to software that changes its behaviour in response to its experience, that is, the problems on which it works. In general, the software agents used by search engines are not adaptive. Although they accumulate virtual mountains of data, the way they function is essentially unchanged.

One way for an agent to adapt is by adding to its store of knowledge. That is, it adds new rules to its behavioural repertoire. This idea is perhaps best illustrated via an example. Suppose that to carry out a particular kind of Web search, the user feeds the agent with a set of rules that provide explicit instructions. Under these conditions, not only could the agent carry out the query, but also it could add the instructions to its knowledge base. This way, future users would be able to carry out the same query without needing to instruct the agent how to go about it. Of course, to ensure that future users know that such a query is available, it needs to have a name by which it can be invoked.

However, agents can go further in the learning process. Let's suppose that we set the agent to work to provide a report about (say) kauri trees in New Zealand. To carry out the query we give the agent a set of appropriate rules, and we supply a name, NZKAURI say, for the query. Then this query becomes a new routine that future users can recall at any time simply by quoting its name. However, it is limited by its highly specific restriction to forest trees and to New Zealand. We can generalise the query by replacing the specific terms by generic variables. But how do we generalise terms like `forest trees` and `New Zealand`? Let's suppose that we express the query using XML. Then the start of the query might look like this:

```
<query name="NZKAUR">
    <tree>kauri</tree>
    <country>New Zealand</country>
        ... rest of definition ...
</query>
```

The way to generalise the query is now obvious. We simply remove the restrictions and turn the whole thing into a function NZKAURI (topic, country). This function now has the potential to address questions on a much broader basis. In principle, it could answer questions about (say) kauri trees in other countries, other trees in New Zealand, and potentially about any kind of tree in any country.

Of course, the success of this kind of generalisation depends on exactly how the query is implemented. For instance, if it uses a number of isolated resources that refer only to New Zealand kauri trees, then it will fail completely for any other tree, or for any other country. The generalisation is most likely to be successful if the entire system confines itself to a fairly narrow domain and if it works with services that have a widespread coverage.

Generalised pattern matching, as described above, is just one of many ways in which an online agent can acquire information. Another possibility is provided by research into natural language processing. A system might take natural language queries entered by the user and map them into a specific search dialect.

10.9.4 Support Vector Machines

Statistical learning (Ripley 1996) underlies much of today's pattern recognition and machine learning. It is the application of statistical methods for the purpose of classification (Chen and Wang 2005).

The goal of statistical learning is to find suitable functions for mapping inputs to outputs, and to automatically assign the responses where these functions can be found to exist (Steinwart and Christmann 2008).

Support Vector Machines (SVM) are one such example of statistical learning, where a set of inputs is categorised into a number of distinct classes. So, for example, the input might be a set of pixels from an image, with the classification task being to assign each set to categories of vegetation, housing, water and infrastructure. In the simplest case, T class boundaries are represented as hyperplanes. These hyperplanes separate the input classes into sets, and are learned using input/output pairs to develop decision functions that classify inputs into one of the designated classes. More complicated boundary shapes are also possible. But the goal of choosing the class boundaries is to minimise *structural risk,* which is a bound on the generalisation error, as opposed to *empirical risk,* used elsewhere, which minimises error on the training data. Similar to feed-forward neural networks (Ripley 1996), they are a *supervised learning technique,* in which a set of known classification examples is provided for training, and the categories are given in advance.

An important requirement for supervised classifiers is their capactiy to generalise. In other words, if they are presented with inputs they have never seen before, how good their classification will be. SVMs have good generalisation capabilities (Abe 2010), making them well suited to image processing tasks. It is quite feasible to create a number of classifiers based on object properties (colour, texture, shape), which can be used for feature detection of geospatial phenoma (Hope 2013), which in turn can be directly linked back to pre-existing geospatial data objects. The automation of this processing could provide GIS systems with advanced processing capabilities and increase the value of Online GIS for organisations, communities and individuals, as described in (§13.4.3).

10.9.5 Ant Model of Distributed Intelligence

One of the strongest insights to flow from research in the new discipline of artificial life is that order can emerge in a system without any central planning or intelligence. In many living systems, order emerges instead through the interactions of many agents interacting with their environment and with each other. For instance, in many insect colonies, such as ants and bumblebees, the individual insects have no overall concept of what their colony should look like. Instead they behave according to very simple rules. For instance, if you are an ant and you see a scrap of waste lying around, then you pick it up. If you are carrying waste and find some waste lying around, then you drop the scrap that you are carrying. This simple action, repeated thousands of times, is all it takes to sort the contents of an ant colony into different areas for food, for eggs, waste, and so on.

Many useful applications flow from the above simple observation. For instance, Brooks et al. (1990) applied the idea to robotics and created very effective mechanisms for controlling (say) robot walking, without any central control at all. Likewise the ant sort algorithm (Dorigo 2006) is a method by which computer systems can create order spontaneously, and without the need to apply specific sorting algorithms. In the ant sort, the ants either change or move items around, or if that is not possible, they can make it easier for other ants to find the same items again by leaving a trail of virtual pheromones.

So how do these ideas apply to online GIS? First, it is important to appreciate that the ant sort is an ideal method to apply on the Internet. The Web contains virtual oceans of data items, with more being added all the time. Suppose that virtual ants are looking for online items on a particular topic, say geographic data. Then they can "mark their trail" with virtual pheromones. These can take several forms. For example, one approach is to record a history of sites that you've visited with a score next to each one. Other ants looking at this list can see which sites have proved most relevant and useful.

There are difficulties with the ant sort model. One is that it applies best to a non-renewable resource. When real-life ants follow a pheromone trail, they take food away from the source that the trail leads to. When the source is exhausted, they stop laying down pheromones, so the trail goes cold. Not so an online resource. So there is a risk of concentrating huge amounts of activity on a single website. One potential solution to this problem is to have the ant copy the information to a cache, and delete it if it is not accessed within a particular period.

10.10 Data Mining

Data mining, as the name suggests, is the process of exploring for patterns and relationships that are buried within data. Now that you have all the data that you need in a Data Warehouse, what do you do with it? Many organisations' Data Warehouses are created and maintained for the purpose of delivering data to both internal processes and external clients and services. However, another common process performed on Data Warehouses is data mining. Data mining or knowledge discovery in databases was first suggested in 1978 at a conference on the analysis of large complex data sets (Fayyad et al. 1996a,b).

Although the field of data mining has been around for almost four decades, the emergence of data warehousing technologies has only occurred in the last 10 to 15 years. This has been largely due to the availability of the required data storage. Data mining is an interdisciplinary area of research. It draws on technologies from database technology (data warehousing), machine learning, pattern recognition, statistics, visualisation, and high-performance computing (Figure 10.9). The central goal is to identify information nuggets. These are patterns and relationships within data that are potentially revealing and useful with a variety of techniques:

- **Machine Learning:** The area of machine learning focuses on developing computer systems that can adapt and learn (induce knowledge) from examples or data (Dietterich 1996). The systems aim at moving the traditional view of computer programs from *program = algorithm + data* to the more elaborate *program = algorithm + data + domain knowledge*. The domain knowledge is acquired from prior experiences solving similar problems (Michalski et al. 1998).

- **Statistics:** Many of the machine learning techniques, and other algorithms, centre on detecting the frequency of certain patterns and relationships. Non-parametric statistics can be used to perform hypothesis and exploratory data analysis. A number of earlier references suggest that data mining is a practical application of statistics (Fayyad et al. 1996a; Eklund et al. 1998; Ester et al. 1999; Chawla et al. 2001).

- **Pattern recognition:** The pattern detection can take on a number of forms. Pattern recognition can be the determining of what things commonly appear with each other. Similarly, it can be the detection of what attributes are deterministic of other attributes.

- **High performance computing:** The central idea in data mining is to sift through large volumes of data to detect patterns. The technology has become available only in the last couple of decades and has now grown to powerful online systems that provide optimised database services and business analytics. Likewise, data mining requires fast access and lots of internal storage (RAM) to efficiently process the data in a Data Warehouse. With the increase in computer technology in the last decade, even standard desktops are now powerful enough to perform data mining and data warehousing functions (Fayyad et al. 1996a,b) as client processes.

Techniques used in data mining to extract information cover the entire gamut of machine learning and include artificial neural networks, genetic algorithms, radial basis functions, curve fitting, decision trees, rule induction and nearest neighbours algorithms (Fayyad et al. 1996a,b). Each of these techniques can be used to extract different forms of information (Kennedy et al. 1998).

Predictive Model Markup Language (PMML) fits into the general framework within this book of using XML and associated XML schemas to create general, platform independent formats and processing methods.

10.10.1   Predictive Model Markup Language

The Data Mining Group (DMG)[14] describes itself as *The Data Mining Group (DMG) is an independent, vendor led consortium that develops data mining standards, such as the Predictive Model Markup Language (PMML).* The group's online service was

---

[14] http://www.dmg.org/

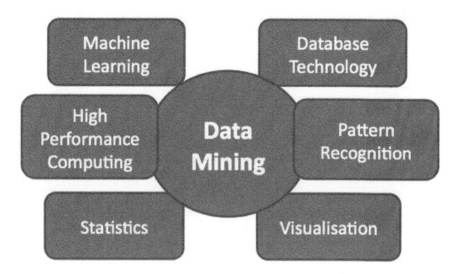

FIGURE 10.9: Disciplines associated with data mining.

hosted by the National Center for Data Mining at the University of Illinois at Chicago around 2000. Perhaps the group's chief contribution has been the development of the Predictive Model Markup Language (PMML). The group, which released PMML 1.1 back August 2000, describes PMML as follows (DMG 2000):

> *The Predictive Model Markup Language (PMML) is an XML-based language which provides a quick and easy way for companies to define predictive models and share models between compliant vendors' applications. . . . PMML provides applications with a vendor-independent method of defining models so that proprietary issues and incompatibilities are no longer a barrier to the exchange of models between applications. It allows users to develop models within one vendor's application, and use other vendors' applications to visualise, analyse, evaluate or otherwise use the models. Previously, this was virtually impossible, but with PMML, the exchange of models between compliant applications now will be seamless.*

PMML provides a capability to mark up many types of models, including statistics, normalisation, tree classification, polynomial regression, general regression, association rules, neural network, and centre-based and distribution-based clustering (DMG 2000). Under the root element, `pmml`, it includes methods for specifying data dictionaries and mining schemas. Here, for instance, is a fragment of PMML to define a simple data dictionary. In this case the data dictionary defines three kinds of variables:

- A categorical variable called `landcover`, which defines a land classification and can take three possible values, Forest, Farmland or Grassland;

- An ordinal variable called `roadtype`, which assigns ranks to different kinds of roads; and

- A continuous variable called `elevation`, which takes numbers as its values.

```
<data-dictionary>
  <categorical name="landcover">
   <category value="Forest" />
   <category value="Farmland" />
   <category value="Grassland" />
   <category value="." missing="true" />
  </categorical>
  <ordinal name="roadtype">
   <order value="Freeway" rank="4" />
   <order value="Highway" rank="3" />
   <order value="A" rank="2" />
   <order value="B" rank="1" />
   <order value=" Track " rank="0" />
   <order value="." rank="N/A" missing="true" />
  </ordinal>
  <continuous name="elevation">
   <compound-predicate bool-op="or">
     <predicate name="elevation" op="le" value="1" />
     <predicate name="elevation" op="ge" value="10" />
   </compound-predicate>
  </continuous>
</data-dictionary>
```

The main function of PMML is to define models. As an example, let's take the case of defining a decision tree model using PMML. A decision tree consists of a hierarchy of linked nodes, with a test condition at each node. For instance, a condition such as "elevation GT 1000" will return TRUE if the local elevation is greater than 1000 metres, and FALSE if not. These two outcomes (TRUE or FALSE) thus specify two branches in a tree. We can then attach other test conditions further down each branch. At the end of each branch is a so-called *leaf node*. In the decision tree values are specified at each leaf node.

Decision trees are common in classification problems, such as interpreting data in a satellite image. In the following simple example of PMML code (based on details of the language given in DMG (2000), we define a simple decision tree that is based on the values of two variables: *var1* and *var2*.

```
<?xml version="1.0" ?>
<pmml version="1.0">
  <data-dictionary>
   <continuous name="var1" />
   <continuous name="var2" />
  </data-dictionary>
  <tree-model model-id="classify01">
   <node><true/>
```

```
  <node>
    <predicate attribute="var1" op="le" value="0.5" />
    <node score="1">
      <predicate attribute="var2" op="le" value="0.5" />
    </node>
    <node score="2">
      <predicate attribute="var2" op="gt" value="0.5" />
    </node>
  </node>
  <node>
    <predicate attribute="var1" op="gt" value="0.5" />
    <node score="3">
      <predicate attribute="var2" op="le" value="0.5" />
    </node>
    <node score="4">
      <predicate attribute="var2" op="gt" value="0.5" />
    </node>
  </node>
  </node>
 </tree-model>
</pmml>
```

10.11   The Future of Geographic Data Warehouses

This chapter has focused on the design and development of Geographic Data Warehouses using relational databases. While there are also a number of alternate storage types now used for Geographic Data Warehouses, such as file based caches, multistructured data such as JSON (§8.3) and pre-rendered georeferenced map tiles, the trends driving the continued spread of data warehousing remain the same as they were a decade ago:

- The first trend has been the growth of computer storage capacity and processing speed, which makes the Data Warehouse a practical reality.

- Secondly, the increasing supply of data is now a routine, automated by-product of many commercial and professional activities.

- Finally, the growth of the Internet makes it both feasible and necessary to organise data collection for dissemination on a large scale.

Data Warehouses (including distributed Data Warehouses) are available over the Internet now and associated technologies such as data mining already constitute a significant paradigm for the online collation, interpretation and dissemination of information. In some areas of research, such as molecular biology and astronomy, the growth of large public domain repositories has revolutionised the way scientists go about their work. In other areas, especially environmental management, the development of Data Warehouses has been crucial to the future effectiveness of planning and management.

So, what might an information warehouse look like to a user in the future? Data Warehouses are becoming an integral part of the virtual Web. This transformation, also referred to as the Knowledge or Semantic Web (Berners-Lee 2014), means that the present emphasis on sites and home pages will disappear. Instead the user will simply look for information about a topic, powered by a virtual Web, and be guided to that information by an intelligent system that actually teaches you as you go. This view would also apply to (say) a world environmental information warehouse.

Suppose, for example, that a student wanted to know about conservation of plants and animals in the local area. Starting from some general heading (say plants), the system might guide the student through relevant topics (e.g., biodiversity, conservation, geographic information) at each stage providing background information and links to other information. A geographic query would involve selecting an area on a map catalogue and selecting what to be shown from a range of choices offered.

Suppose alternatively that a public servant wanted to see a report about (say) natural resources in southwest Tasmania. After selecting the exact area and topic, she might use a report generator to select the kinds of items she wanted to include. The choices might cover a standard list of resources, time period, geographic area, type of material (e.g., policy papers, scientific studies, educational material) and types of items (e.g., maps, tables, graphs, text, etc.). The system would then build a preliminary report with the option of going back and exploring any aspect in more depth.

In addition to the Knowledge or Semantic Web, Data Warehouses will also be extended to include improved data analytics, with automated tooling and standard dashboards to provide on-demand reporting and to increase value through greater access to business intelligence. They will offer improved data analytics capabilities, driven by advances in Big Data technologies that cache traditional databases and Data Warehouses into optimised data repositories for query. Data Warehouses will find their way from the current corporate data centres into elastic infrastructures that can grow dynamically based on demand. This change is highlighting the ever increasing need to have strong data security policies and practices, and accurate metadata to describe the lineage of Data Warehouse information.

Cloud computing will also continue to change the way that Data Warehouses evolve in the future. A large part of this change will be the automation of Data Warehouses, an automation we are already seeing in large cloud offerings, driven in many ways by Apache Hadoop,[15] an open source framework for distributed storage across clustered storage servers. Two of the high profile companies using Hadooop for their distributed Data Warehouses are Facebook and Yahoo, and Hadoop can be implemented on private, public and across hybrid cloud nodes.

Data Warehousing in the future will support more and more forms of online geoprocessing, including feature detection for maintenance of geospatial datasets from imagery, routing and drainage flow generation using digtial terrain models, LiDAR

---

[15] https://www.apache.org

and point density clouds, motion detection and changes to geospatial datasets over time.

There still remain some challenges for organisations managing large and dynamic geospatial datasets, when looking to deploy to Data Warehouses with nodes off premises. These challenges include the data distribution required to keep their datasets current, the need to support long transactions over the Internet and the potential growth in data volumes over time due to business needs storing full history of geospatial data changes over time. Metadata will play a key role in meeting all of these challenges.

10.12 Summary and Outlook

- This chapter extends the Data Warehouse content of the first edition of this book, re-enforcing a number of models for storing and processing geographic data in a Data Warehouse (§10.2), (§10.3) and (§10.4).

- Considerations when building Data Warehouses (§10.5) are discussed, covering practical issues, analysis and design needs, and infrastructure considerations.

- Migration of data into any newly formed Data Warehouse requires careful planning, as does the operation of the resulting database, and this section gave the reader knowledge of models (§10.6), supported by a small number of Data Warehouse examples (§10.7).

- Finally, three sections, §10.8, §10.10 and §10.11, looked at some standards used and the future of Data Warehouses. There have been some major developments by vendors at the time of writing which require further reading; however, the content of this chapter gives the reader a solid grounding in Data Warehouse technologies and implementations.

11

GIS Anytime, Anywhere

There has been significant growth in mobile applications since the first edition of this book, almost to the point that if you are in business and you don't have a mobile capability in your organisation, then the end may be near. This importance of mobile to an online organisation is reflective of the number of smart phones and tablet devices now in circulation. It is also seen in habits of users who demand information anywhere, anytime, including where and when and many mobile apps are using Online GIS to give directions on how to get there and how long it may take. However, there is more to creating a successful mobile app. This chapter highlights some of the information that organisations should be considering if they are planning to develop a useable mobile GIS capability in the marketplace.

11.1   Mobile GIS

Spatial information has long been an intricate part of our lives. From finding a vegetarian restaurant within walking distance of a hotel in a strange city to a garage which can repair a vintage Jaguar, mapping queries have so many applications when the data is available anywhere, anytime. Mobile devices are now providing this type of on-the-go access to location information.

Early versions of cut-down GIS systems first become available on palmtop computers, such as Tadpole[1] and FieldWorker.[2] While Tadpole Technologies has been acquired since their early GIS software, both companies operated on the model of providing GIS functionality on small datasets with upload/download options. Meanwhile mobile phones were starting to invade the Internet. These early wireless devices had distinct limitations: they had small, low resolution screens and very low bandwidths. Thus an alternative was to think, not in terms of a micro-GIS, but in terms of online spatial queries and rendering of location data. A new approach was needed in which spatial queries were answered with information packed down by artificial intelligence on the server. This online access to the data using a simple browser on the device lead to the formalisation of many standards to enable open access to these data and WWW services went from strength to strength, to the point that today, most of the online services people access are based on WWW service models. Online GIS in no exception.

At the same time, large mapping organisations were coming under increasing pressure to sell their data in more ways, and online publication was obviously very promising. However, given the volume and diversity of data, an effective commercial model was needed which considers issues of tracking royalties through the flow of ownership, data watermarking, discounts for customer categories such as government and issues of resale of data by third parties. This commercial drive, however, has been replaced by initiatives such as (Eggers 2007) where governments are working towards providing more open access to their data. These data have no intrinsic value and their value is only realised when it is combined with other data to improve business decision making for others.

By 2010, mobile devices, smart phones and tablets had invaded the Internet and were playing an increasing role in its use. Earlier challenges to online mapping have been overcome through improvements in how applications are developed for mobile and in the advances in data formats, free public networks, growth of the Internet, as well as data caching across the Internet and on the devices themselves.

---

[1] http://www.tadpoletechnologies.com
[2] http://www.fieldworker.com/

There are more mobile devices in the world than there are people.[3] This proliferation of devices has made a significant change to the way we live our lives. It has also changed the way companies do business. The global nature of the WWW and the number of ways that people access it means that there are significant business benefits to having mobile as part of your business portfolio. One way people are accessing data through mobile devices is by using mobile applications, or, simply, *apps*. Mobile apps are software programs that may be installed on smartphones, tablets, laptops, etc. The most popular apps provide social networking, games and weather services (ACMA 2013); however, there are a number of other apps that provide map rendering, query and searching services.

In addition to standalone apps, there are also WWW pages which have been specially constructed to allow use on mobile devices, i.e. devices with small screens, less computational power, and lower bandwidth. Such sites may use mobileOK (§8.4.2) to signify that they are so enabled.

The benefits of mobile GIS to business are wide ranging:

- Companies are using mobile GIS for fleet management, enabling the tracking and optimisation of fleets by access routing and scheduling information online, anytime, anywhere. At the time of writing in late 2014, new taxi services, such as Uber,[4] are making extensive use of mobile GIS and challenging established services.

- Mobile enables the dynamic access to information and a greater sharing of information in the field.

- It enables the use of end to end process workflows for organisations, irrespective of location.

- In addition to these benefits, there are other business uses for dynamic access to business analytics and improved business decision making in the field. When spatial context is added to these data, the benefits become much greater.

Adding to the importance of Online GIS, online businesses are dependent on location information. On one hand, the Internet exposes an organisation to a much larger market that is often global in nature. Goods and services can be delivered across the WWW, or directly to a location using address information. Privacy is still a crucial issue. The more a data server knows about a client, the more it can tailor its response. However, the client is thereby giving up personal or organisational information, perhaps unwittingly. New privacy laws, such as those which came into force in NSW Australia on July 1, 2000, ensure the client has much greater access and control. A client needs assurance that his personal information will not be spread around the Internet if she entrusts it to some agent, for which mechanisms do not yet adequately exist. It is essential that the client has access and control of the data stored and there are mechanisms to prevent its unauthorised transfer.

---

[3] http://gsmaintelligence.com
[4] https://www.uber.com

This chapter provides an overview of technologies used for developing mobile Online GIS applications and the issues that need to be considered to deliver successful mobile apps. It does this by outlining the decisions supporting a mobile strategy and explaining the technical direction that may be taken in order to develop a first mobile app. It outlines the current situation for the development of such apps and provides guidance on technology, infrastructure and regulation to be considered to future proof mobile application development. The technologies discussed are then validated through a small number of sample mobile GIS applications.

11.2   Mobile GIS and Business

There is no shortage of evidence that *mobile* is dramatically changing the styles of engagement between customers (users), corporations and governments, and that this brings with it both significant opportunities and challenges. This shift has been enabled by the decreasing cost and increasing power of mobile devices, and the increasing speed of data traffic (fixed and wireless).

In particular, mobile apps are becoming more innovative, available on a range of devices, and are quickly becoming mainstream media and communications activities. Apps are closing the gap between mobile and desktop computing capabilities by allowing users to have many functions similar to those they would use on a desktop computer available on a mobile device.

The trend of moving towards mobile devices is also changing the way organisations work. The first of these trends is Bring Your Own Device (BYOD) and more recently Corporate Owned, Personally Enabled (COPE). It is now common for workers to have more than one mobile device (mobile phone + tablet), either their own device (BYOD) or a company issued one (COPE), and they are demanding to connenct these to corporate applications, networks and data anywhere, anytime. A summary of the Gartner comparison of these technologies is available from MSPMentor.net.[5] There is no doubt that mobile technology is changing the way people work and the impact on organisations that offer mobile technologies to staff and customers is making their workforce more dynamic and responsive.

One quotation which summarises just how much organisations are changing due to mobile computing came from Richard Leyland, founder of WorkSnug. He is credited with the statement that *Work is no longer a place we go, but a thing we do,*[6] and the ability to work from anywhere, anytime empowers this paradygm shift.

---

[5] http://mspmentor.net/mobile-device-management/cope-receives-big-blow-gartner-byod-gains-momentum

[6] http://www.iapa.com/index.cfm/travel/blog.article/blog/techguide/art/Augmented-reality?C=1A

11.2.1 Role of Mobile GIS in Strategic Planning

Government is a business, just like any organisation. It provides products and services to its constituents as organisations do and is becoming increasingly aware of the importance of mobile computing to provide these products and services. For example, in Australia, the NSW Premier endorsed *NSW 2021*, as his government's 10 year plan to guide government policy (NSW Government 2014). More specifically, Goal 31[7] of this plan has the objective to

> *Improve government transparency by increasing access to government information.*

The plan sets long-term goals for government using measureable targets, and outlines immediate actions that will help achieve these goals. One of the enablers will be the development of mobile apps to provide up to date information, where and when customers need it, to improve the way people use government services and help them make more informed decisions. In addition, the government policy clearly outlines a number of actions to realise these benefits by increasing the number of mobile phone applications that allow people to access government data.

As well as how governments are approaching mobile strategies, any organisation working on developing a mobile application should consider a number of roles and outcomes of mobile development. These include the purpose of mobile to their organisation and major tasks and outcomes for the app development work being proposed. Some examples may include

Purpose To provide mobile access to a diverse range of users; to grow market share; to gain a competitive edge.

Major Tasks and Activities Develop a mobile strategy and identify the solutions that are to be made available through mobile facilities, such as crowd sourcing of spatial data; provide mobile facilities through the development or purchase of mobile solutions based on HTML that can be deployed to a number of different devices.

Outcomes and Deliverables For example, provide mobile access to information online anywhere and at anytime in order to extend the reach of an organisation's products and services; provide services and spatial data to mobile users in the field; provide the ability to increase revenue through mobile facilities and the extension of online product purchasing and access facilities.

11.2.2 Mobile Markets

Mobile technology allows governments to make public facing services more easily accessible by citizens, create interactive services accessible on-the-go, and promote

---

[7] https://www.nsw.gov.au/sites/default/files/initiatives/nsw_2021_goals.pdf

government accountability and transparency. Mobile also allows for internal labour efficiencies, providing email, software, cloud, big data and mobile commerce services, thereby allowing innovation in marketing, human resources (HR) and IT operations.

As with Goal 31, and something that will be similar to many organisations globally, the NSW Government in Australia goes on to explicitly support the growth of mobile technology as part of its long term ICT Strategic Plan.[8] In that stategy, the use of mobile technology is to enable the *delivery of high quality services that facilitate the interaction with government data irrespective of the time of day or location of the user*. However, it should also be remembered that any government adopter of mobile technology must also consider privacy, security, legacy systems and adherence to government policy, discussed by Azoff and Singh (2013) from Ovum Research.

11.3   Regulations Governing Mobile Implementations

A number of regulatory controls should be considered when developing mobile apps. The content of this section is sourced from publications by the Australian Communications and Media Authority (ACMA), in particular, Mobile Apps: Emerging issues in media and communications, Occasional paper 1, May 2013 (ACMA 2013).[9] While these regulations relate to the Australian broadcasting and telecommunications, many other countries will have similar controls.

The apps market did not exist when the current regulatory frameworks for telecommunications and broadcasting were introduced in Australia. Similarly, the apps environment has introduced business models and business practices that did not exist when measures covering personal information and consumer protection were developed. This challenges the effectiveness of current models for implementing and ensuring compliance with regulatory safeguards. There are also business impacts in terms of compliance costs, where incremental measures have been adopted to address different aspects of the apps market operation.

11.3.1   A Complex Mix of Industry-Specific and Economy-Wide Regulation

To date, there has been a piecemeal approach to accommodating technological and market developments in the regulatory framework, preserving technology and sector-specific approaches. The apps market is currently subject to a range of whole-of-economy and sector-specific regulatory measures, including

- Australian Consumer Law, administered by the Australian Competition and Con-

---

[8] http://finance.nsw.gov.au/ict/
[9] http://www.acma.gov.au

sumer Commission (ACCC)[10] and state and territory fair-trading agencies, which provides for refunds for faulty goods, consumer guarantees to ensure that items are fit for any particular purpose which is represented by the supplier, and protections where suppliers engage in activities that are likely to mislead or deceive consumers.

- The e-Payments Code and related measures administered by the Australian Securities and Investments Commission (ASIC), which covers the online electronic payments associated with app purchase (ASIC 2014). The code includes remedies for credit card charges associated with goods and services that were not received, such as the situation where a charge was imposed by an application store but no application was received by the consumer.

- The Privacy Act 1988, which provides protection for citizens in the way entities covered by the Act handle personal information through the lifecycle of collection, use (including data matching and analytics), disclosure, storage and destruction. App developers whose business model relies on using personal information to sell advertising are likely to be covered by the Privacy Act (Office of the Austraion Information Commissioner 2013). Recent amendments to the Act provide enhanced protections for personal data that is collected from Australian citizens and stored outside Australia.

- The Online Content Scheme, established under Schedule 5 and Schedule 7 to the Broadcasting Services Act 1992[11], provides mechanisms for dealing with apps and other online content, with reference to the National Classification Code. The arrangements for classification of content were recently reviewed by the Australian Law Reform Commission,[12] which has recommended reforms aimed at achieving greater consistency across offline and online platforms.

The Telecommunications Act 1997[13] establishes a regulatory framework for specific segments of the market. This framework regulates providers of communications *infrastructure* (for example, carriers) in a particular way[14] as well as regulating providers of communications *services* in a particular way (for example, carriage services providers).[15] The Act also regulates *content service* providers in a particular way[16] to protect the interests of telecommunications users.

Previously discrete service sectors, such as communications and financial payments, are also merging in the apps business model. This can create uncertainty about which entity in the supply chain may have particular regulatory obligations and, just

---

[10] https://www.accc.gov.au

[11] http://www.alrc.gov.au/publications/2-current-classification-scheme/broadcasting-services-act

[12] http://www.alrc.gov.au

[13] http://www.alrc.gov.au/publications/71.+Telecommunications+Act/telecommunications-act-1997-cth

[14] Defined in Section 7 of the Telecommunications Act 1997 as a holder of a carrier licence.

[15] Defined in Section 87 of the Telecommunications Act 1997.

[16] Defined in Section 97 of the Telecommunications Act 1997; see also Section 15, which provides for the definition of content service to be amended by ministerial determination.

as importantly, the coverage of safeguards, such as those dealing with personal information and privacy.

Both consumers and business benefit from regulatory certainty and clarity. While the apps market is subject to a range of regulatory measures, how they are applied may not always be clear to consumers and industry participants. For industry there are additional costs in dealing with multiple agencies and different regulatory frameworks. For a consumer, knowing how to take action, to whom to complain and how to resolve their apps concerns is complex, further complicated by an environment where there are multiple agencies involved and potential gaps in the safeguards that address their issues of concern.

With the ongoing development of separate responses to emerging apps practices, there is the risk of an overall loss of regulatory coherence with consequences for industry participants in terms of increased compliance costs. For consumers, increased complexity can make it more difficult to manage their apps experience. As apps become a mainstream activity, there would be benefits to apps issues being addressed within a single coherent regulatory framework, which could cross international boundaries.

11.3.2 Public Interest Outcomes Remain Relevant in the Apps Environment

Apps illustrate some of the disconnect between traditional regulatory concepts and contemporary communications. However, there are points of ongoing public interest that can inform discussion about the need, or not, for intervention in the apps market. ACMA consideration of the apps environment has identified the following particular matters of interest relevant to the delivery layers of an information economy:

- Market standards and redress mechanisms, including those which apply to apps payment mechanisms

- Digital information management, particularly as it relates to users' personal data

- Content safeguards, which protect children and reflect community values

- Ensuring access to emergency services to protect individuals and communities

11.4 Technology and Implementation of Mobile Apps

In addition to the regulatory considerations for mobile development, there are several technology and implementation options for organisations to consider when moving into mobile app development. There are a number of mobile operating systems on the market. These include Apple mobile (IOS) for iPhones and iPads, Android developed by Google, BlackBerry OS, WebOS, Symbian, Windows Mobile Professional (touch screen), Windows Mobile Standard (non-touch screen) and Bada. Mobile development for any organisation should not be limited to developing applications for specific mobile operating systems. Organisations should be developing applications that may be installed on any smartphone, tablet, laptop or other device capable of supporting apps.

In addition to decisions on what operating systems will be supported, the organisation should also invest in skills that can evaluate the mobile application functionality. This includes consideration of the response design requirements for easy reading and navigation with a minimum of resizing, panning and scrolling across a wide range of devices (from mobile phones to desktop computer monitors) and development guidelines to use when designing their apps. The W3C Mobile Web Application Best Practices list of 32 best practices may be a useful resource for guideline development (W3C 2014a).

The W3C has also published a range of standards, tools and resources under the WWW of Devices (W3C 2014a). It is also focusing on technologies to enable WWW access using any device. This includes mobile phones, other mobile devices and consumer electronics, printers, interactive television, and even automobiles.

11.4.1 Build Technologies

As discussed, there are multiple technologies that can be used to build a mobile app, with ongoing industry debate between using native and HTML5 applications and code. There are three principal ways to develop a mobile solution, listed below:

Native Referring to a specific platform (iOS, Android), that is, the development is done in an environment native to either of those platforms

HTML5 A core Internet technology and language that can be leveraged across multiple platforms (§5.3)

Hybrid Usually refers to applications that are developed largely in HTML for the user interface, but rely on native codes to access device-specific features (Traeg 2013)

Many organisations have elected to use a hybrid solution for app development, as a balance between native and WWW development. This will allow the app to be written in HTML and then wrapped for deployment to any device. A hybrid method allows these organisations to remain flexible in development as technology rapidly evolves (Azoff and Singh 2013) so that they are not locked to a specific vendor. It will also allow them to utilise the native device functions (such as diagnostics, camera, device orientation and GPS) by creating specific plug-ins.

11.4.1.1 Adobe PhoneGap Framework

When developing mobile apps to support a mobile business strategy, organisations should look for vendor neutral solutions that can deploy to many platforms easily. It is also inefficient for organisations to have programmers with language skills for each of the platforms they want to deploy to; thus a well-regarded approach for hybrid app development is to use a product such as **PhoneGap** (Web 2014). Understanding these three statements will be crucial to the success of any mobile deployment, so it is important that we delve a little more into what these statements mean.

Traditionally, building WWW apps meant that developers had to develop seperate applications for each device or operating system (in the case of mobile, this would involve iPhone, Android, Windows Mobile, etc.). Early development efforts often meant that the resulting apps might have different behaviours for different platforms, or worse, might only function on the platform on which they were first developed. This form of vendor lock-in does not work well for the deployment of mobile business apps. To get around this issue, it often meant that the development process required different frameworks and different languages to support a cross-platform deployment. This is not ideal.

PhoneGap is an open source framework from Adobe that allows developers to quickly build cross-platform mobile apps. In addition to this framework, Adobe also offers an open-source version known as Apache Cordova (Gartner 2014). PhoneGap has the significant advantage in that it uses standards based WWW technologies to build WWW applications that can be deployed to mobile devices from many vendors. Develop once, deploy to many devices. This is critical when organisations want to develop an application that will run on all the popular mobile device types in the market.

There are a number of nice features built into the PhoneGap framework. Phone-Gap implements an HTML5 Canvas object to enable free style and touch-based drawing. The framework can also access mobile device features using JavaScript and it takes care of device configuration for the developer. The PhoneGap framework also comes with a number of plug-ins, providing geospatial capability such as geo-location and file upload. With PhoneGap, apps are like HTML5 WWW applications running in a full screen browser and PhoneGap provides full support for CSS3.

In addition to the available frameworks, there are a number of Mobile Application Development Platforms (MADPs) in the market. Some of these MADPs are tied to specific vendor platforms; however, at the time of writing, many of these development platforms were using the frameworks from Adobe PhoneGap and Apache Cordova, making them the de facto standard being used by companies to wrap their plaform code in, for deployment and exectution. Risks associated with use of MADPs will be discussed in more detail in §11.6.4.

11.4.2 Mobile App Functionality and User Experience

Consumers in today's society need their information now, and they want it presented simply in a way they understand. For those at the helm of the site development game, keeping up to date with technology and functionality is the only way to stay relevant to online communities. Across the variety of WWW browsers and operating systems, mobile types and tablet devices, understanding how to create and code responsive design is key to enhancing a user's experience (UX). Having a responsive site that will resize in accordance with the device it is being used on can be the difference between keeping a consumer long enough to view your content, or losing them instantly.

Developing apps for mobile is not the same as developing applications for desktops. On a mobile device, more consideration needs to be given to the ease of use, bandwidth requirements and simplicity of the interface. Interfaces should be uncluttered, fast to load and require less data storage due to the limited memory on mobile devices. The UI/UX should also be adaptable across many devices and scalable.

Workshops to firm up business needs, training or buying in resources to provide mobile graphics design expertise and good development practices will all contribute to a user friendly app that can be used on many mobile devices. Leveraging existing functionality, services and APIs should be encouraged, rather than creating applications from scratch. For example, there are a number of freely available map APIs which can be re-used to create lightweight mobile applications in conjunction with the ESRI JavaScript API, using HTML5, CSS3, and JavaScript.

User experience encompasses all aspects of the end users interaction with the WWW site, its services and products, by giving users a simple environment to navigate, allowing them to view and achieve their objective easily. A crucial part of creating an exceptional user experience is having a site that can be viewed on multiple devices without creating barriers. The number of consumers that are choosing to browse websites on their mobile and tablet devices is growing exponentially, changing the way sites are being designed and how they function.

There are several considerations for delivering mobile UX. Applications should

- Interact with maps and views via touch events as opposed to mouse clicks;

- Use a paradigm of combining a set of mobile views to transition from one screen function to the next;

- Use responsive design principles to display and function effectively on multiple platforms and devices;

- Use responsive design techniques to handle multiple form factors.

UX design must also consider how it can be coded for implementation. There is little point in creating an intricate design that can't be coded or that requires a lot of code to implement as this would make the app hard to download and appear cluttered. With the release of the latest iPhone 6 in 2014, Apple notably increased screen size on both the iPhone 6 and 6 plus models. The 6 plus model has a screen resolution of 1920 x 1080 pixels, allowing better implementation of split-screen modes for both apps and the home screen. Samsung also responding to market demands by increasing screen size and resolution of its models, with the Galaxy Note 4 having a high resolution 2560 x 1440 screen to provide high contrast display, making WWW browsing a sensory experience. This increase in size in phone devices is a strong indicator that screen real estate is becoming more important to users as they use their devices for many purposes, and data streamed through Online GIS is one of these.

11.4.3   Mobile App Lifecycle Considerations

At this point, it may be useful to consider the lifecycle of a mobile app. Development should always comply with the existing Systems Development LifeCycle (SDLC) being used by the organisation, and would generally follow these five steps:

Setup The setup phase requires two main steps. Firstly, to establish development environments for each mobile platform being supported. The two common platforms are:

- Android RAD, Android Virtual Devices, Android devices.
- iOS Xcode, iOS Simulator, iOS devices.

Secondly, establish developer accounts for each of the platforms. These are simple to do and for Android and iOS would involve creating accounts with

- iOS developer using a company email such as fred@mmy-co.com.
- Google Play publisher.[17]

Development While this is the fun part for programmers, it should only start once the concept for the app is really clear. The extent to which the full functional requirements and detailed designs are completed, in advance, varies greatly with software engineering methodolgy. Agile methods[18] take the iterative interaction between users and developers to a high level.

Outcomes from this step should ensure

- The mobile app implements good UI/UX design, is small, fast to load, responsive and easy to use
- That all functional requirements are implemented as described in the documentation
- All resources to use (icons, splash screens, images, etc.) have been created by graphics design specialists
- Object Oriented (OO) development principles should also be used wherever possible (§1.2), commensurate with speed and efficiency

If the development phase follows good software engineering principles, this step should only take up to around 20 percent of the pre-deployment lifecycle time. There are some schools of thought, however, that a lot of mobile development for apps is often trial and error, more than structured software, so these times may not apply.

---

[17] https://play.google.com/apps/publish/signup
[18] http://www2.computer.org/portal/\www/computingnow/archive/june2008

Testing Testing is a critical stage of any development lifecycle for delivering a robust app. There are a number of testing stages that need to be used, including unit tests, regression testing, integration and performance testing, and user acceptance testing. A number of online resources can be found that provide useful guidelines to testing mobile applications:

- App quality testing:
  - Android Core App Quality Guidelines and Tablet App Quality Checklist[19] [20]
  - iOS App Quality Alliance (AQuA) Testing Criteria for Apple iOS applications [21]
- Acceptable under platform policies and guidelines
  - Android link [22]
  - iOS link [23]
- Performance testing and optimisation (memory, battery, storage usage)
- Alpha testing on local devices

Deployment The deployment process for mobile apps involves the submission of the completed app to the mobile stores offered by vendors. This is a different model to that which most organisations would use, where their apps are deployed by internal IT staff to desktops and WWW servers. This deployment process should only happen after the successful completion of rigorous testing. It requires

- Prepare app store promotional material (graphics, screenshots, videos)
- Build app, including versioning of the release, and upload to app store
- Perform the beta release testing. Information on Android and iOS beta testing can be found at
  - Android [24]
  - iOS [25]
- Complete app product information
- Final checks and app submission

---

[19] http://developer.android.com/distribute/googleplay/quality/core.html

[20] http://developer.android.com/distribute/googleplay/quality/tablet.html

[21] http://www.appqualityalliance.org/

[22] http://developer.android.com/distribute/googleplay/policies/index.html

[23] http://developer.apple.com/appstore/guidelines.html

[24] https://support.google.com/googleplay/android-developer/answer/3131213?hl=en

[25] https://developer.apple.com/library/ios/documentation/IDEs/Conceptual/\newline\ldotsAppDistributionGuide/TestingYouriOSApp/TestingYouriOSApp.html

    – Android[26] [27]

    – iOS[28]

Post Deployment This stage is about the management of the mobile app after it has been published. Because the apps are hosted external to the organisation, there are consoles and WWW sites that can be used to manage the application in much the same way as managing systems in the cloud. For Android and iOS, these tools can be found at the Android Google Play Developer Console[29] and iOS iTunes Connect.[30]

These platform vendors also provide a number of tools that provide user support, services to monitor usage, reviews and crash reports as well as app statistics and quality improvements of the platform. Finally, information about the application can still be updated after it has been deployed through WWW sites and consoles provided by the platform hosting companies.

11.5   Mobile Security

There are always security risks associated with any online mobile app. These include unsafe data storage or data transmission practices, unauthorised logins or payments, or unauthorised system modification.

    A recent situation in Hong Kong involving student protesters demonstrating for democratic rights highlights why security on mobile platforms is essential. The protests began in September 2014 to demand the right for Hong Kong citizens to chose their own head of state, known as the Chief Executive. These protests became known as the Umbrella Revolution when social media captured scenes of protesters using umbrellas to protect themselves from tear gas.

    What happened during the demonstrations was that an online mobile app was developed and used by protestors to help organise locations and times for rallies, etc. Or at least that was how it appeared. What actually transpired was that the app was infected with malware that, when activated, opened a dialog box stating "Application Updates, please click to install". Once users accepted the install they had given the app permission to install code to access SMS messages, address books and call logs,

---

[26] http://developer.android.com/tools/publishing/publishingoverview.html

[27] http://developer.android.com/tools/publishing/preparing.html

[28] https://developer.apple.com/library/ios/documentation/IDEs/Conceptual/\newline\dotsAppDistributionGuide/SubmittingYourApp/SubmittingYourApp.html

[29] http://developer.android.com/distribute/googleplay/publish/console.html

[30] https://developer.apple.com/library/ios/documentation/IDEs/Conceptual/\newline\ldotsAppDistributionGuide/UsingiTunesConnect/UsingiTunesConnect.html

photos and, most significantly, time and location of the phone at all times and sending these data to other sites.

The app worked on both Android and iOS devices, capturing a lot of the mobile device market. Some online WWW sites reported that the app used a form of Remote Access Trojans (RATs) [31] to collect information from the device about the user activities, contact lists, text messages, photos, call logs and movements. This is not the only instance where attackers have used malware to access personal information, and it does show the invasive nature of mobile and how much information can be stolen from unsuspecting users.

11.5.1 PhoneGap Security Considerations

When using PhoneGap, the major security considerations are very similar to the WWW applications that an organisation has experience with developing and deploying.

Data Encryption is concerned with securing sensitive data that may be stored on the device by the app. As it may utilise a number of WWW services to source its content, and authentication would be handled by a back end service, there may be no need to store sensitive data on the device. However, data may be automatically cached by proxy servers or the browser, which is the case for all WWW applications, not just mobile implementations. Where there is a specific requirement for the application to store sensitive data, options to secure the data include

- *Hardware based encryption,* the encryption of all data stored on permanent media, such as tapes and drives. Some of the encrytion techniques used for hardware encryption are

  - iOS Keychain encrypts all data stored. Access would be via the Phone-Gap plug-in that allows access to the iOS Keychain[32]
  - iOS Entitlements confers specific capabilities or security permissions on an app[33]
  - Android Account Manager provides similar functionality to iOS Keychain[34]

- *Software based encryption,* the encrytion of all data as it passes through a computer network. Also referred to as in-flight data, the encryption techniques used include

---

[31] http://www.voanews.com/content/china-declares-cyber-war-on-occupy-central/2474288.html

[32] http://phonegap.com/2010/11/06/ios-keychain-plugin-for-phonegap/

[33] https://developer.apple.com/library/mac/documentation/Miscellaneous/Reference/EntitlementKeyReference/Chapters/AboutEntitlements.html

[34] http://developer.android.com/reference/android/accounts/AccountManager.html

- JSAES – a compact JavaScript implementation of the AES block cipher, supporting 128, 192, 256 bit encryption[35]
- Crypto-js – a collection of standard and secure cryptographic algorithms implemented in JavaScript[36]

App Permissions App permissions refer to the hardware functions that an app is allowed to use. Both Android and iOS have robust permissions systems. However, it is essential that the app is only given the bare minimum permissions that it requires. This ensures any attack only has access to the application features, not the full functionality of the device.

Plug-ins PhoneGap (§11.4.1.1) makes use of plug-in interfaces to enable JavaScript components to communicate with the device native functions. The core PhoneGap API features are implemented as plug-ins, with much more functionality available in various plug-in repositories. In order to make use of any plug-ins developed by a third party, it is imperative that they are vetted to ensure there will be no nefarious behaviour.

Reverse Engineering This is the risk that a user may reverse engineer the Phone-Gap application to gain access to the JavaScript source code to reuse it. A more extreme example would be to add malicious JavaScript code to this reverse engineered code, which in turn is then re-packaged and resubmitted to the app store (a process known as app phishing). The resulting app could then be executing malicious code without the developer being aware. One way to reduce this risk of reverse engineering, which could occur with any hybrid or native application, as it is a relatively simple task to decompile Java and Objective-C, could be through the use of code obfuscation to confuse the JavaScript source code. This means that any reverse engineering of the PhoneGap wrapper will only provide built source code, which itself would then need to be reverse engineered in order to be of any use to the attacker.

Code Obfuscation Code obfuscation is a process that takes program source and generates a functionally equivalent source file, but with smaller variable and function names. The resulting file is much harder to read as meaningful variables such as "minimumLineLength" may be replaced with "xm" or "t". This obfuscation will also improve the performance of the apps by minimising its footprint.

Server Side Execution In another approach to further alleviate the risk of reverse engineering, sensitive code should be executed on a WWW server, rather than on the client device, and only the results returned to the client device.

---

[35] http://point-at-infinity.org/jsaes/
[36] https://code.google.com/p/crypto-js/

11.5.2 OWASP Security Considerations

The Open WWW Application Security Project[37] identifies the most important attack vectors that should be addressed in a WWW application.[38] These risks and how they relate to an organisation's mobility project include

Injection Flaws such as SQL, OS, and LDAP injection occur when untrusted data is sent to an interpreter as part of a command or query. The attacker's hostile data can trick the interpreter into executing unintended commands or accessing unauthorized data. Protections against injection must be implemented on the server side, as it is a server side risk. As such, this risk is not specific to mobile applications, applying to all an organisation's WWW-based applications and the WWW services that they utilise.

Cross Site Scripting (XSS) XSS flaws occur whenever an application takes untrusted data and sends it to a WWW browser without proper validation and escaping. XSS allows attackers to execute scripts in the victim's browser which can hijack user sessions, deface WWW sites, or redirect the user to malicious sites. As with any WWW-based application, a PhoneGap app is subject to XSS. This is negated by sanitising HTML content to remove any script tags that could initiate an XSS prior to rendering data in the application. The risk of injection can be negated by sanitising user submitted content with JSON.parse[39] (§8.3) and some custom code based on JsHtmlSanitizer.[40] PhoneGap also has additional XSS prevention in the form of white listed (a security method listing trusted websites, allowing them to be accessed – the opposite of black listed sites) domains from which the client can request resources.

Broken Authentication and Session Management These application functions are related to authentication and session management, and are often not implemented correctly, allowing attackers to compromise passwords, keys, session tokens, or exploit other implementation flaws to assume other users' identities. Organisations should look closely at products that are available from the market for authentication, authorisation, and session management for WWW applications. In the instance that a mobile product was to require access to secure services, these third party products can be leveraged to provide access to secure REST (Restful) services.[41,42] There is also potential for the use of OAuth[43] in conjunction with market products. OAuth is an authorisation framework that allows a resource owner to grant permission for access to their resources without the sharing of

---

[37] https://www.owasp.org/index.php/AboutOWASP
[38] https://www.owasp.org/index.php/Top102010-Main
[39] https://developer.mozilla.org/en-US/docs/web/JavaScript/Reference/GlobalObjects/JSON/parse
[40] https://code.google.com/p/google-caja/wiki/JsHtmlSanitizer
[41] http://www-01.ibm.com/support/docview.wss?uid=swg27038292
[42] http://blogs.perficient.com/ibm/2012/04/24/securing-rest-services-with-datapower/
[43] http://oauth.net/2/

credentials and to provide limited access to resources hosted by WWW-based services accessed over HTTP[44].

Insecure Cryptographic Storage Many WWW applications do not properly protect sensitive data, such as credit cards, SSNs, and authentication credentials, with appropriate encryption. Attackers may steal or modify such weakly protected data to commit identity theft, credit card fraud, or other crimes. This kind of sensitive data should not be stored on client devices; instead it should be stored and secured (encrypted and not stored in plain text) at the back end server infrastructure. The use of authentication mechanisms such as OAuth and third party products would facilitate the use of this sensitive data for the application.

Insufficient Transport Layer Protection Applications frequently fail to authenticate, encrypt, and protect the confidentiality and integrity of sensitive network traffic. When they do, they sometimes support weak algorithms, use expired or invalid certificates or do not use them correctly. This risk is not specific to a mobile application it affects all WWW-based applications using an organisation's WWW services. Therefore, it is essential that an organisation's Information and Communication Technology group (ICT, which handles all computer and network matters) ensures valid certificates are correctly installed for all SSL-based services and resources.

11.6 Future Proofing Mobile Development

Investing in new technologies is expensive for any organisation and making the most of these investments is critical to ensure a strong return on investment. It's also important to protect an organisation's reputation with new products, so longevity of applications should be high on the requirements. Finally, given the massive growth in mobile application in Online GIS, a successful mobile GIS deployment can have significant benefits for organisations, including maintainers of large spatial information datasets.

There are a number of principles that should be considered by organisations when planning development of mobile apps. These are discussed next, along with frameworks, functionality and development risks that apply to mobile GIS development.

11.6.1 General Principles

The following general principles for application development should be considered by organisations when embarking on any software development project. These prin-

---

[44] http://www.ibm.com/developerworks/websphere/library/techarticles/1208rasmussen/1208rasmussen.html

ciples apply even more to mobile GIS development than many traditional developments because of the significant resources, vendor based and freeware, that are available to development teams. In all cases, such projects should avoid re-inventing the wheel, as there are many libraries existing that can be used to reduce development times and improve product design and robustness. Thus general principles for mobile app development are

- Leveraging off-the-shelf products wherever possible, as this saves time and money.

- Focusing on customising existing code wherever possible, rather than starting custom code development from scratch using customised code which has already been widely used, as this code should have fewer bugs, although the HeartBleed Bug is a cautionary tale.[45]

- Considering the impact across the lifecycle, e.g., use/develop how to deploy and support in production templates

- Developing a mobile app guideline within the organisation that

  - Outlines the pros and cons for using native, HTML5 and hybrid applications and code;
  - Guides the decision on internal versus external development, including graphic design principles;
  - Outlines how app development can be managed across the lifecycle where the development work is outsourced. This should include who owns the IP and what templates to use in the development;
  - Clarifies the support and communication planning required so that an app goes smoothly into production.

11.6.2 Business Principles

There are some fundamental business principles that, if adhered to, will result in targeted mobile apps that will serve an organisation well into the future. These principles may make sense to development teams more experienced with mobile technologies, however, they may not be so obvious to business teams that are used to dealing with existing products. They are

- Target Clients: Is the target user group clearly identified? It is a direct client of the organisation?

- Target Services: What information or service is the app providing?

- Location Based Optimisation (Chapter 12): Mobile apps enable you to take advantage of a customer's current context and situation. Is this information or service more valuable when provided contextually?

---

[45] http://heartbleed.com/

- Reward Clients: What will motivate people to use the app? Does the app use your business information to solve their business needs?

- Integrate Services: Is the app an extension of other delivery platforms? Does it provide additional customer value or extend existing services?

11.6.3 Frameworks and Functionality

We have already discussed the importance of selecting a cross-platform tool such as PhoneGap (§11.4.1.1) for developing mobile apps which can run on more than one platform. In addtion, there are a number of UI/UX frameworks for mobile development. These include jQuery Mobile,[46] PhoneJS,[47] EnyoJS,[48] Kendo UI Mobile,[49] Sencha Touch Chocolate Chip UI,[50] to name a few. These frameworks provide developers with a number of functions to enable navigation and transition beween different views of the app. The frameworks commonly provide built-in widgets for developing navigation bars, slider popup, header and footer components; and support for customised widgets and plug-ins, giving developers great flexibility for creating mobile apps tailored for an organisation's individual requirements. This capability allows developers to either use existing functions or to develop their own where they need different behaviours in the app.

Selecting the most appropriate mobile UI framwork for an organisation depends on organisational needs. That is, how unique are the apps they want to build and how much control do they need to retain over the developed apps? The less custom coding that is done, the less testing is required, as the third party frameworks should be more robust than one-off code development; however, using these frameworks does create a dependency for the organisation on external framework providers.

Some questions to consider when chosing frameworks and functionality are

- Which approach is best for the organisation - native, HTML or hybrid? What skills do we have now, and what will we need in the future?

- Which back-end systems will be accessed? Are any of these high risk?

- Does the desired functionality (API) already exist? What modification does it require?

- What are the security needs of the mobile app?

Further reading for UI/UX frameworks suited to mobile development can be found in a post by Jim Cowart at the DE (Developer Economics) blog.[51]

---

[46] https://www.jquerymobile.com
[47] http://propertycross.com/phonejs/
[48] https://www.enyojs.com
[49] http://www.telerik.com/kendo-ui
[50] http://www.sencha.com/products/touch/
[51] http://www.developereconomics.com/look-4-mobile-ui-frameworks/

11.6.4 Development Risks and Considerations

The following list outlines a rationale to better inform an organisation when considering mobile development risks. A key to success is to start out with vendor neutral technologies where it can and once it has developed some mobile capability it can investigate the use of a more formal mobile application development platform (MADP) to promote mobile apps up to a more robust enterprise standard. Some key points that should be considered:

- There is no real vendor leader in MADP at this time.

- It is not known how much mobile usage will grow, or what role companies and organisations will have in the field of mobile.

- MADPs are rapidly evolving and one of the key recommendations from the Gartner research is to avoid vendor lock in at this stage[52].

- By continuing to use PhoneGap, organisations will continue to invest in HTML5 technologies which are likely to be applicable to any future developments.

- There does not appear to be an easy migration path from one MADP to another, so once an organisation has chosen any particular technology, it could be difficult to change.

- There is an easy migration path from PhoneGap to many of the available MADPs, as a large number are based on PhoneGap.

- This approach positions organisations to adopt a suitable MADP at the time when it becomes necessary. This is achieved without locking into a MADP prematurely when an organisation may have very little experience in this space.

- Millions of apps have successfully been deployed to app stores without being developed using a MADP. MADPs become particularly relevant when developing and maintaining multiple apps.

By using a lower level library to begin with, developers will get a more thorough understanding of the technologies involved, as opposed to clicking a button in a MADP to perform a function without understanding what it does or how it works. Having this better understanding of the underlying technology and functions will help organisations avoid vendor lock should they need to switch MADPs in the future.

---

[52] http://www.gartner.com/technology/reprints.do?id=1-20SX3ZK&ct=140903&st=sb

11.7   Summary and Outlook

Mobile computing is an exciting growth area of Online GIS with organisations needing to think about more than just the interface when planning for mobile capabilities. The importance of location to mobile computing is discussed in §11.1.

- Many businesses should be going mobile if they want to stay competitive. §11.2 discusses the role of mobile in strategic planning and marketing.

- However, because of the ubiquitous nature of mobile computing and the geolocation capabilities of mobile devices, there are a number of regulations in many countries that need to be considered by technology implementers. These are discussed in §11.3. These considerations are also important for mobile security (§11.5), as devices can contain sensitive information, including where the device is, or where the device has been.

- Technology often provides options when it comes to implementation, and mobile computing is no different (§11.4). Given the rate of change in technologies, §11.6 has given the reader some things to consider when future proofing their mobile solutions.

12

Location-Based Services

CONTENTS

Contributed by Christoph Karon

The ubiquitous nature of Location-Based Services (LBS) combined with their ability to provide location aware business intelligence and new business opportunities is making LBS an exciting growth area of Online GIS. This chapter introduces the concepts of LBS, including application domains, which will benefit from context-based location services in the future.

12.1 What Are Location-Based Services?

Location-Based Services (LBS) are defined by localisation and usage of current position information. While the simple usage of location-related information is the main aspect of such services, its influence varies in different applications. Some use it as the main feature, such as navigation or transportation applications, while others just use it as an additional feature, such as social or weather services. LBS come in several different forms, such as in special software, mobile applications, or Web services.

Beside their usual name and usage context, there are distinct aspects of LBS. Location awareness is just one example. In so-called Location-Aware Services, the location information of the user is simply utilized to adapt the service. For that reason

location-aware services are a special case of location-based services (Kaasinen 2003)

On the other hand, a broader definition of LBS is obtainable when looking at context-aware services, because

optimal adaptation often depends on the contents usage semantics and the user's context (Mohomed et al. 2006).

Context parameters cover different attributes such as

- network bandwidth (how much information can be transferred?)

- screen size (how much information can be displayed?)

- device type (which features should be offered?)

- user experience (how many features should be offered?)

- time (what are the opening times of the locations?)

- location (which locations are nearby?)

Location information is only one possible context dimension and may be one of the most important, but combining it with other dimensions increases the relevance to the user. If you search for restaurants, just looking at the relevant locations next to you is not enough, if you do not check the opening times.

12.1.1 Historic Development

In the 1990s, different precursors of LBS were developed which used location information for simple applications such as call-forwarding services, active maps, or mobile shopping assistants (Chen and Kotz 2000). In an exemplary call-forwarding service, telephone calls were forwarded to the nearest phone of a user, and in an exemplary active map, positions of people in a building were highlighted in order to locate them quickly. Basic methods for the localisation and the general infrastructure were relevant topics in this period. Until 2000, only a few services were published in the LBS context, but the situation changed massively at the beginning of the new millennium.

Different aspects increased the growth of LBS and their infrastructure during this period. One important aspect was the regulations for emergency calls. Basic requirements for required positioning infrastructure were released in different parts of the world such as Europe and the USA (Junglas and Watson 2008; Liutkauskas et al. 2004). The USA even stated accuracy requirements and policies for positioning techniques in order to improve emergency responses (Junglas and Watson 2008). Commercial LBS benefited from these infrastructure regulations as well, where customers pay for such services compared to the free positioning services for emergency situations.

12.1.2 Topics in LBS Research

In the academic world, infrastructure related topics were most relevant between 2000 and 2005 (Karon 2015). During this period, researchers focused on issues such as general architecture, network management, handover techniques, and middleware applications. These issues focused on integration in the current technical landscape and methods to develop new technical frameworks. Essential questions covered aspects such as the configuration of the right hardware, dealing with low transmission rates, and systematic ways to transfer data.

Afterwards, data processing topics became important until 2008 (Karon 2015). Aspects such as data mining, location prediction, trajectory management, and spatial queries were mainly developed during this time in the LBS context. In this period, the main issue was the handling of position information and ways to get additional benefits out of it. Methods to reduce inaccuracy of positioning, deriving behaviour patterns from trajectories, and predicting future behaviour were central questions.

In the following periods, technical aspects became less important while user related issues were raised (Karon 2015). Topics such as privacy protection, identity management, user preferences, recommendations, and participation became important. This shift to a non-technical focus took the needs of the user to the center of attention as well as aspects that occur during LBS usage. Dealing with potential security and privacy risks, focusing on the interests of users, and allowing proactive interactions between users were main drivers at the later periods.

12.1.3 Importance of Smartphones

Usage of LBS at the current time is strongly related to smartphone distribution, because of the tracking technologies implemented therein and the availability of apps. However, smartphone distribution differs strongly between different countries. While the USA (44%), Australia (52%), and Singapore (62%) had rather high shares in 2011, other countries such as Germany (29%), Poland (24%), and Hungary (16%) had lower shares (Google 2013). Smartphone users are often LBS users as well. Two-thirds of German smartphone owners used LBS in 2012 (Goldhammer et al. 2013), while this share was as high as three-quarters in the USA (Zickuhr 2014). The continuous increase of smartphones provides additional chances for LBS as well.

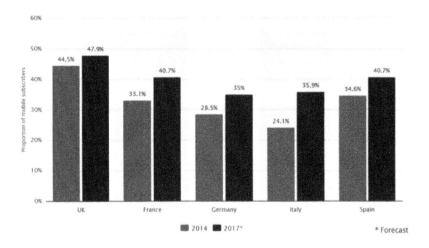

FIGURE 12.1: Percentage of mobile subscribers using LBS in 2014 and 2017.

12.1.4 Current LBS Usage

LBS usage and the motivation to use LBS strongly differ. The TNS Mobile Life study in 2012 revealed that (TNS 2012)

> almost one fifth (19%) of the worlds six billion mobile users are already using LBS, with more than three times this number (62%) aspiring to do so in the future.

The most important motivation worldwide is navigation, but social related aspects such as check-ins in platforms like Foursquare and Facebook have become more relevant recently. Latin American LBS users use friend finders most often (39%), while this share is lower in Europe (20%), India (11%), and North America (9%) (TNS 2012). General usage of LBS even differs when looking at different European countries. In 2014, only 24% of mobile subscribers in Italy and 29% in Germany used LBS, while this share was almost 45% in the United Kingdom (Statista 2012). Therefore, chances for additional LBS usage varies between different places in the world. Figure 12.1 [1] shows the percentage of users (from Statista (2012), reproduced with permission).

12.2 How Do LBS Work?

There are several different ways to derive current position based on the device used and in the required environment:

---

[1] Reprinted from Statista with permission (Statista 2012).

- The most common situation is the localisation of a mobile device in an outdoor environment. Basic non-GPS solutions cover methods such as Cell of Origin, Time of Arrival and Angle of Arrival (Dao et al. 2002). Cell of Origin uses telephone network cells in order to approximate the current location while Time of Arrival calculates time arrivals of at least three base stations for position determination. Angle of Arrival, on the other hand, computes the angle of signals received in order to derive current position relative to the base stations.

- Localisation, by using the global positioning system (GPS), is the most usual way for outdoor tracking. GPS, a space-based positioning, navigation, and timing system was developed by the USA Department of Defense (McNeff 2002). Twenty-four satellites provide signals for the localisation using a triangulation method. GPS positioning is available in outdoor environments only and has problems in proximity to high buildings (Drab and Binder 2005).

- For indoor tracking, different technical approaches exist such as using Wi-Fi signals, Bluetooth, RFID, or ultrasonic waves. If no tracking possibilities are available, device positions can be estimated as well by using other methods such as self-reported positioning, location approximation and IP address derivation.

12.2.1 Use Case

In order to demonstrate basic functionalities and aspects of LBS, an example location-based restaurant service is shown. Three important aspects are relevant in this scenario: How does the service get the position information; how does the service get relevant information about restaurants nearby; and how are user preferences selected?

Some tracking technology has to be used to receive relevant position information, as mentioned before. Let's imagine a common GPS tracking in an outdoor situation. Based on a triangulation of satellite signals, the position of the user can be calculated and used. If correct positioning is not possible, because the user is in the middle of a city with several buildings next to him, a fallback solution is required. In this scenario, the user has the option for self-reported positioning where the current city and street name can be typed in manually or selected by placing a point in a map. The disadvantage of self-reported positioning is the requisite knowledge about the current place and location.

The position information is then transferred to a database server, which determines the restaurants close by. Usually the user has selected a maximum distance range to his current position or this range has been automatically set by the service. Based on this information, a result set is sent back to the user with a list of restaurants containing additional information for each of those entries such as current distance to the user, opening times, ratings, kitchen types, etc. If the query of the database server brought an empty result set or if the user has changed his location, while the process has to be repeated with possible changes.

Finally, the user is able to filter restaurants based on his preferences by using the

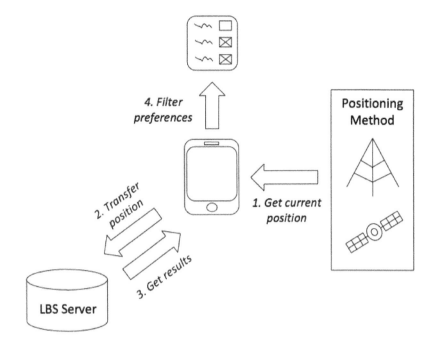

FIGURE 12.2: LBS usage flow diagram.

additional information in the result set. A possible filtering may be the selection of all vegetarian restaurants which are currently open, or which will open in half an hour sorted by their ratings in descending order. Filter options could be transferred to the database server at an earlier stage. It is even possible to store all the required restaurant information directly in the service instead of asking a server for any information. The advantages of this solution are that there is no mobile traffic and privacy problems are negated. The disadvantages are the high demand for storage space on the device and possible out-of-date information. Figure 12.2 shows this usage flow.

12.3   What Are Application Domains for LBS?

Application areas are manifold in the LBS context and specific groups of LBS are aggregated into sub-categories. Various approaches for the classification of sub-categories exist such as the categorisation by use case (Fritsch and Muntermann 2005; Jensen et al. 2001) or by functionality (Virrantaus et al. 2001). The following classification approach is based on the differentiation by use case and includes the following sub-categories: location-based advertising, location-based games and education, location-based health services, location-based information and navigation services, and location-based social networks.

12.3.1 Location-Based Advertising

This sub-group includes use cases which are mainly based on commercial and advertising activities. Information requests can be executed in two ways: by a pull action or by a push action. In a pull request scenario the user is able to ask for information on his own, while in a push request scenario the service is automatically triggered if the user enters a specific area. In an exemplary location-based coupon service, the user has the choice to search for coupons of shops next to him using the pull method or the service can send appropriate coupons to the user if he walks by these shops when using the push method. Pull-based services are less intrusive and reach higher perceived benefits and intentions to sign up than push-based services, which also cause higher privacy concerns (Unni and Harmon 2007).

The first commercial services used SMS as the favorite messaging technology. Later on MMS became important where researchers found that (Oh and Xu 2003)

"multimedia location-aware advertising messages lead to more favorable attitudes and increase the intention to reuse the mobile advertising service."

In the smartphone area, possible message technologies were extended by apps.

12.3.2 Location-Based Games and Education

Games and education related use cases are covered in this sub-group. Location-Based games developed from regular video games and are mainly produced for fun and entertainment. Three gaming types can be distinguished: games that are based to meet players, games where geo-location of players is important, and games where real world surroundings take part in the game (Nicklas et al. 2001). *Can You See Me Now* (Benford et al. 2003) is an example of a game where the real world is included. This mixed reality game requires online players and professional runners where the runners carry special equipment to localise the online players in a virtual city. The position of the online players in the virtual map is reflected as the same position in the real world and the runners have the goal to catch them by reaching the same coordinates.

Location-Based education services have important influences from the electronic and mobile learning area. They have the target to teach their users while using LBS functionalities. An example use case is the historical knowledge trip of pupils in Amsterdam (Huizenga et al. 2009). Two groups were built up in this scenario: a city team that walked through the city and a headquarters team that assisted and guided the city team from a computer base station. Both teams had to work together in order to solve the required tasks. The results were that more knowledge transferred for those who participated in this game compared to those who received regular project-based instruction in school. In several use cases, location-based education services were found to provide good knowledge transfer and high entertainment as well.

12.3.3   Location-Based Health Services

This sub-group summarizes use cases for people in emergency situations and for disabled, ill, or elderly people. The general task of location-based emergency services is to localise people in need in order to send them help as fast as possible. Improving the general infrastructure of such services was the issue of task forces such as the Co-ordination Group on Access to Location Information by Emergency (CGALIES) in the EU that dealt with topics such as the requirements for location technologies and common interfaces for public safety answering points (Salmon 2003).

High accuracy and low costs in infrastructure are important topics in this sub-group. An exemplary use case is the indoor tracking system for the elderly called *Matilda's House* (Helal et al. 2003). Four ultrasonic transmitters at each corner of a building and two receivers on the shoulders of the elderly person are used for positioning, reaching accuracies up to one centimeter. Different application areas are described as remote monitoring, attention capturing, and blind guides.

12.3.4   Location-Based Information and Navigation Services

The earliest and most classical use cases in LBS are covered in this sub-group, including services such as tourist guides, direction assistants, and route guiding systems. These services have the task to navigate users in unknown environments and to provide additional information to real world locations. Tourist guides, for example, can be rated by the following aspects: context, adoption, and customisation. A verification of nine guides based on these attributes revealed that many services suffer from (Schwinger et al. 2005)

> *"considerable lacks with respect to standards, reusability, extensibility and interoperability."*

An exemplary museum guide called *MUSEpad* is demonstrated by Waite et al. (2005) which has the task to support persons with special needs. Additional information of objects next to the user are displayed in order to assist him/her. Calculating the closest object is not always possible and therefore awareness zones are discussed as alternatives to fixed zones.

12.3.5   Location-Based Social Networks

The last sub-group is a rather new discipline that combines location-sharing aspects with social networks in order to recommend interesting places to other users. Friend finder applications are an early implementation of such services. Users can share comments and position information with relevant people next to them or just coordinate meeting locations. An important aspect in this sub-group is the analysis of trajectories in order to predict location and friendship relations. Cranshaw et al. (2010) analyzed the popular social network Facebook and found that users with many different location visits have more social network ties. Type of places and regularity of routines are strong predictors for the number of friends as well.

Another study analysed the location-based social network Brightkite and derived a prediction model that is able to forecast a user's location with a median accuracy of 49% (Li and Chen 2009). In 17% of all cases, a user's current position can be forecast with even 99% accuracy. The results revealed the requirement for transparent communication about the information collected and the high sensitivity for privacy protection.

12.4 What Is the Future of LBS?

The growth of LBS depends on the development of technical infrastructure. Smartphone distribution and mobile traffic varies between countries and therefore preconditions for each of these countries are different. In many parts of the world smartphone distribution, data transfer rates, and technologies for accurate positioning improve while location information is added to several applications.

Many LBS use GPS as an outdoor tracking standard, but no indoor tracking technology has reached even an informal standard yet. Using WLAN signal strength is a promising solution, but requires additional effort in configuration in order to provide acceptable accuracy levels

An important challenge in this context is the handling of privacy needs. People do not like to be tracked all the time. Therefore, they need easy ways to verify when they are tracked, what information is stored about them, what information is used for which scenario, and how they are able to turn off the tracking. Governmental regulations are just one possible solution to increase the trust of users, but the industry has to define privacy standards as well in order to prevent misusage of personal information. In several applications, storage of profile information is not required in order to offer standard functionalities and services have to provide simple ways to turn off the unnecessary collection of information.

Finally, LBS must show their additional value in order to be successful. Intelligent solutions combine LBS features with daily activities such as calendars that proactively use location information of dates or shopping assistants that benefit from location patterns. The combination of social and location information is a promising area as well due to the current success of social networks such as Facebook. Social networks already offer options for location sharing and social features could be easily implemented in LBS as well.

Services that provide augmented reality features are another important topic for the future where additional information is added to the current world such as in the case of Google Glass. Route information is projected directly on the street instead of on the display of the device, which reduces the complexity of abstraction for the user. Also other use cases, such as augmented reality games, have a high potential for future success.

12.5   Summary and Outlook

Location-Based Services are an exciting, fast growing area of Online GIS.

- §12.1 sought a precise description of what constitutes LBS.

- §12.2 went on to consider how they work.

- The application's domains were considered in §12.3.

- Finally, §12.4 considered the future of LBS.

13

Future Directions of Online GIS

CONTENTS

Imagine a world where countries of all different economic standing can access geographical information over the Internet for free. Where these data are maintained by authoritative spatial data organisations, using a range of different sources and are accessible around the clock, hosted in powerful data centres that users don't need to establish and maintain. Where the data is accessible through standards-based Web service calls that can be easily integrated into existing business functions and workflows. Many of the technologies discussed in this book have now matured to the point where they provide global solutions to governments, private sector and citizens. These solutions are now commonplace for querying and rendering through a wide range of services and mediums and both individuals and organisations can now combine (mash-up) these data and apply them to solve a wide range of problems. This chapter discusses a number of current trends and the technology directions which will underpin the next wave of Online GIS.

13.1 Realisation of a Global GIS

In the first edition of this book, we looked at elements of technology that could be used for placing geographic information services on the World Wide Web (WWW). In 2000 these elements were in their relative infancy. Since that time, there has been significant advancement in these technologies and the number of implementations of Online GIS. This has largely been driven by two changes. The first is increasing globalisation where more and more organisations demand that managers, planners and policymakers in every sphere of activity must put their decision making into a global context. Business is increasingly international, not only through multi-national corporations, but also through global electronic commerce (e-commerce), stock markets and currency exchange. The second is the technologies that support truly global infrastructures.

The same is true in the areas of health, society, environment and government. In health, for instance, international travel means that diseases such as Ebola are not only regional concerns, but quickly escalate to worldwide problems. Enhanced communications mean that culture and social values are becoming universal. In this context, it is highly desirable to create an online system that encompasses geographic information about any issue, anywhere.

In today's world, users want data, and they want it now. For organisations to retain these users as customers, they need to provide these types of services and capabilities. These users often operate globally. One example of their global nature is they may be after information to inform their travel plans. They will want to book accommodation for the destination online, they want live weather data, need transport information and want to know what they can do when they get there.

To take another example, suppose that a young couple living in Edmonton, Canada look to investments as a way of boosting their savings and securing their future. So, they look at prospects not only within Canada, but also around the globe. In the evening they go online and check out the stock exchanges in Australia, Tokyo and Hong Kong. In the morning they do the same for London and Bonn. If they find companies they are interested in, then they naturally want to find out more. So they might be looking at such widely spread prospects as a tour company based in Dunedin, New Zealand; a chain of micro-breweries in Portland, Oregon; or a company building intelligent robots in Edinburgh, Scotland.

In each case they would probably want to access detailed local information, not only about the company, but also about the area, local competition, and so forth. In short, they need to be able to access detailed geographic information from all over the world. And the need for on-the-spot, up-to-date information is no longer confined to large organisations. A new generation of technically savvy users now armed with smart devices that connect seamlessly with the Internet through satellite communications, Wi-Fi and bluetooth.

It has often been said that if a user cannot access the information they want within three clicks, they will go to another website. This is also true for people looking to access spatial information. We could find similar stories from many other areas of activity as well. They all point to the need for rapid access to global geographic datasets, in easy to use formats and viewers, accessed from all over the world.

13.1.1   Digital Earth

The logical endpoint to putting geographic information online is to create a comprehensive, global GIS. It has been understood for some time that if information from different sources can be seamlessly combined into a single resource, then there is nothing in principle to prevent such a system from being developed. But what should this system look like and what is needed to develop it?

Most of the data needed to power a global GIS for commerce, trade, environment and tourism already exists. Government organisations around the world are making this data available through online services that, based on many of the standards covered in previous chapters of this book, can be integrated into existing systems for rendering and query. These datasets provide coverages or queries for particular themes or data layers, using geographic layers of many different kinds, such as physical, biological, economic and political. They also provide a hierarchy of data, from global datasets covering environmental and cultural data, to regional and localised datasets in formats including spatial coverages, spreadsheets and reports.

Now it could be argued that there is no need to set out to build a global GIS from scratch. Such systems already exist. Some of the services referred to above are very impressive in the range and depth of information they supply. However, most of the current examples are responding to a particular commercial need and opportunity. There are many kinds of studies for which existing information resources are totally inadequate. Some concerns are already being covered by international cooperation between governments. For instance, the Global Biodiversity Information Facility (§7.6.2) is a model framed in the context of international agreements on biodiversity conservation. Again, such networks are responding to a perceived need, this time environmental, rather than commercial.

The challenge for a global GIS online is how to integrate all of these resources into a single overriding service. This is not a new idea, since there are plenty of precedents to work from. Integrated online services, many of them global, already exist. They provide striking proof that a comprehensive global GIS is a practical possibility.

Building a global GIS will require contributions by a lot of organisations, either through collaboration, or through independently working towards this common goal. There are a number of global initiatives driving these visions for a digital earth or global earth. One of the strongest organisations is the International Society of Digital Earth (ISDE).[1] ISDE is a non-political, non-governmental and not-for-profit organisation set up to create a virtual representation of the world. A second is the academic sector, which supports development of a Digital Earth through their research initiatives and project work, for example, the Cooperative Research Centre - Spa-

---

[1] http://www.digitalearth-isde.org/

tial Information (CRC-SI) in Australia.[2] Other global initiatives include the creation of platform technologies that deliver 3D globe based viewing technologies such as Google Earth. Finally, there have been a number initiatives which have been replicated around the world, such as Open Government and Open Data that are providing the geospatial data to power these technologies through open standards.

In addition to the development of technologies and standards to support Online GIS, some of the barriers to online publication have also been lessened, if not removed. The need for complex purchasing models has diminished, with the benefits of sharing data online providing significant economic return to communities, which is far beyond the intrinsic value of the data itself. Concerns about data security have also been reduced, as many datasets do not pose risks to issues such as personal security and identity, and there are copyright models now in place such as Creative Commons that can be used to protect ownership of data.

Such concerns indicate the tension that always exists in information studies between top-down and bottom-up approaches to problems. In 2000, the best way to create a global GIS was through a balanced combination of both methods. Governments (and other organisations, such as W3C) now provide basic top-down guidelines that both permit and encourage individual, bottom-up initiatives, but at the same time need to ensure that they conform to certain basic principles.

So what of the future? The problem is too big to be resolved by any one organisation, or system. The successful creation of a global GIS which is sustainable requires automated mechanisms for dynamic data aggregation and presentation. These mechanisms need to be ubiquitous in nature, constantly trawling the Internet for GIS related resources, updating existing data and aggregated services from discoverable portals, auto-building global GIS through the use of intelligent systems. We will cover several of these topics later in this chapter.

13.2 Future Benefits of Online GIS Systems

One fundamental question that should continually be asked is why is a global GIS needed? The obvious answer is that in an era of increasing globalisation, people need to be able to access and combine many different kinds of detailed information from anywhere on Earth. Also, another aspect of globalisation is that virtually every activity impinges on everything else. So the proponents of a commercial venture need to know (say) about environment, social frameworks and politics so they can be prepared for possible impacts and repercussions. Likewise, in (say) conservation, managers and planners need to be aware of the potential commercial, political and other consequences of banning development in particular regions. All of the answers to this question, however, should consider the planet's sustainability as a key driver

---

[2] http://www.crcsi.com.au/

and with the continued growth and expansion of the Internet, a global GIS will play an important role in managing such issues in the future.

There are many benefits of using Online GIS. It can give non-spatial organisations access to key foundation datasets that can be used to provide spatial context to their own datasets. The data can be collected and reused in the field to coordinate campaigns to fight infectious diseases. More importantly, Online GIS can provide easy to use systems for non-GIS specialists, such as medical workers, veterinarians and community workers. Two areas to benefit from Online GIS, discussed below, are environmental management (§13.2.1) and disease control (§13.2.3).

13.2.1 Problem of Global Environmental Management

One of the great challenges facing mankind at the turn of the millennium is how to manage the world's environment and its natural resources. The problem is immense. The planet's surface area exceeds 509,000,000 square kilometres. Simply monitoring such vast tracts is a huge task. The total number of species is estimated to be around 8.7 million (Sweetlove 2011; Mora et al. 2011). At the current pace it would take another 480 years (May 2011) of taxonomic research simply to document them all. Modern technology can help with these tasks, but at the same time generates huge volumes of data that must somehow be stored, collated and interpreted.

The problem is also acute. As human population grows, the pressure on resources grows with it. We have now reached a point where virtually no place on earth is untouched by human activity, and where it can be questioned whether the existing resources can sustain such a large mass of people indefinitely. Slowly we are learning to use resources more carefully.

Given the size and urgency of the problem, piecemeal solutions simply will not do. We have to plan and act systematically. Governments, industry and conservation all need sound, comprehensive information from which to plan. The problem is so huge that nothing less than the coordinated efforts of every agency in every country will be adequate. Our ultimate aim should be nothing less than a global information warehouse documenting the world's resources. Until recently such a goal was unattainable. Collating all available information in one place was simply not possible. However, improvements in communications, and especially the rise of the Internet as a global communications medium, now make it feasible to build such a system as a distributed network of information sources. Online GIS will be a key enabler of this form of data publication and integration of global geospatial datasets, where environmental data can be combined and shared, providing answers to complex spatial and temporal queries.

13.2.2 Role of Online GIS in Emergency Response

There have several global efforts in recent years to provide Online GIS resoures for emergency response and humanitarian aid. These efforts have included, most significantly, communities of people using online mapping systems to coordinate information gathering and distribution, all with a spatial context. People in the field can

FIGURE 13.1: The OpenStreetMap website collecting live emergency response information, withthis image captured from their website on the 30th of April, just 4 days after the earthquake struck.

be reporting the location of incidents using smartphones and images, sharing these invaluable forms of information needed for planning. The spatial context that can be provided through these forms of logging can greatly assist organisations in informing them where to concerntrate their efforts to optimise any resources available through Online GIS Portals.

The most recent Online GIS efforts supporting humanitarian aid are those in response to the 7.9 magnitude earthquake which struck Kathmandu Valley in Nepal in April 2015. There have been reports of significant loss of life and damage to homes and infrastructure. There is an urgent need for emergency shelters as many people are sleeping out in the open, and a critical need for clean drinking water, warm clothing, blankets and food. Because the area is also an international destination for trekking and mountain climbing, and there were already a large number of foreign aid workers and volunteers in the region, it has a truely global impact.

There are resources also available online[3] from OpenStreetMap. There are several themes of data being collated at this website (Figure 13.1). The green map markers shown on the site are verified information, while the red markers show data collected from anonymous users which should be independently verified.

The map tiles in Figure 13.1 have been provided by the Humanitarian OpenStreetMap Team (HOT).[4] This team is coordinating priority areas for mapping, to secure satellite imagery and other mapping data, publication and prioritisation of tasks being worked on and other key communications providing up to date information in a spatial context. Information on ativities can be followed on Twitter (§13.4.5) using @HotSOM.

---

[3] https://www.openstreetmap.org/relation/184633 (c) OpenStreetMap contributors.
[4] http://hot.openstreetmap.org/updates/

The United Nations also has a project called the Humanitarian Data Exchange (HDX),[5] which provides an online data catalogue of mapping data which can be accessed using a number of data formats. This website makes significant use of metadata standards to publish its content. To follow HDX live, use @humdata.

There are also a number of media sites using Online GIS to report the earthquake. One example, BBC news[6], has published images gathered using Online GIS to show the devastation and extent of the earthquake. Publication of these forms of information is important in raising the awareness of such incidents.

As with many forms of communication using the Web, access to information can be live, detailed and accessed through many devices. This type of geospatial information, in which Online GIS will play an increasing role in the future, can only improve the quality and success of emergency response and humanitarian aid efforts through involvement of organisations and individuals and their input and sharing of data.

13.2.3 Role of Online GIS for Disease Control

There are also significant benefits that can be gained from the use of Online GIS systems for disease control. The integration of information from this distributed network can help to provide immediate access to live data from labs, researchers, incident teams and hospitals, which will improve decision making for disease outbreaks, providing the most current data to organisations anywhere they can get an Internet connection. We will look briefly at two examples of where disease control using Online GIS had a significant positive impact.

13.2.3.1 Eradication of Equine Influenza in Australia

An outbreak of Equine Influenza (EI), otherwise known as horse flu, was discovered in Australia in 2007. A considerable effort to eradicate the disease was undertaken by authorities, veterinarians and owners.

On 24 August 2007, a veterinarian reported to NSW Department of Primary Industries (NSW DPI) that he had observed sick horses at Centennial Park in Sydney. The report followed an outbreak of Equine Influenza (EI) in Japan, the import of breeding stallions from Japan into quarantining and reports that some of these stallions at the Eastern Creek Quarantine Station were showing signs of EI.

NSW DPI and Rural Lands Protection Board (RLPB) veterinary staff began an immediate investigation and later the same day, laboratory testing at the NSW DPI veterinary laboratory confirmed that the horses at Centennial Park were infected with EI. The outbreak that eventuated was the most serious emergency animal disease Australia has experienced in recent history. At its peak, 47,000 horses were infected in NSW on 5943 properties, and horse owners and industry workers were facing dark times with major impacts on their livelihood and lifestyle.

The campaign led by NSW Department of Primary Industries (DPI) to eradi-

---

[5] https://data.hdx.rwlabs.org
[6] http://www.bbc.com/news/world-asia-32479909

cate the disease was the largest of its type ever undertaken in Australia, using the latest laboratory, vaccine, surveillance, mapping and communication technologies. The disease was eradicated within six months well ahead of predictions and by July 2008 horse industry operations had returned to normal.[7]

Because of Australia's geographic isolation and a long history of quarantining imports, the discovery of EI was a significant event. Infection spread quickly across a wide region, from Sydney to the North West, due to contact between horses and through the airborne spread of the virus, which could be up to 10 km. It became critical to identify the location of all positive test results within this range. At its peak, EI had spread to 18 Restricted Areas which were quarantined, with 488 infected horses on 41 known properties across NSW.

It was this mapping and geo-locating of information where Online GIS contributed towards the eradication of EI in Australia in 2008. Data was collected by NSW DPI and geo-referenced, and then provided as nightly updates for publication to an online portal (Figure 13.2) for everyone to access. The portal supported standard location based queries and showed EI related information, contained metadata about test outcomes and enabled property based searches, including imagery, address and mapping information.

There were many other contributing factors to the successful campaign, including careful management of horse movements, the efforts of NSW DPI staff and vets, and the cooperation of horse owners however, Online GIS played an important role through the monitoring and recording of EI spread and its ultimate eradication.

13.2.3.2 Control of Human Infectious Disease

Online GIS can also provide significant benefits in the management and control of infectious diseases in humans. One global epidemic which has significant media attention, is the outbreak of Ebola in West Africa in 2014. This recent outbreak was the largest on record and due to the movement of health workers leaving affected areas, cases have also been recorded in Spain, the US and the United Kingdom.

One organisation providing infectious disease information online is the Center for Disease Control (CDC)[8] in the US. At this point in time, CDC provides map resources in a limited number of formats, such as PDF. These mapping resources include a World Health Organisation (WHO) Ebola Response Roadmap showing total cases mapped onto West African countries, and regions where new cases have been reported within the last 21 days. The site also makes reference to vaccination trials being started with around 6000 health and frontline workers in Sierra Leone. CDC Director Tom Frieden stated,

We hope this vaccine will be proven effective but in the meantime we must continue doing everything necessary to stop this epidemic – find every case, isolate and treat, safely and respectfully bury the dead, and find every single contact.[9]

---

[7] http://www.dpi.nsw.gov.au/agriculture/livestock/horses/health/
general/influenza/summary-of-the-200708-ei-outbreak
[8] http://www.cdc.gov
[9] http://www.cdc.gov/media/releases/2015/p0414-ebola-vaccine.html

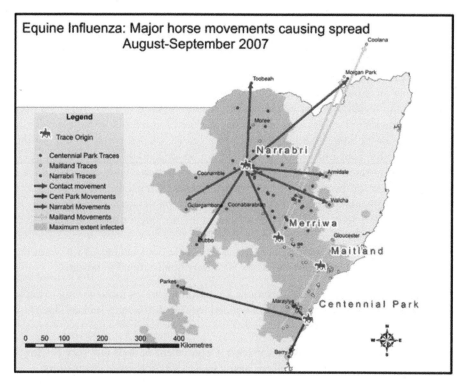

FIGURE 13.2: A GIS image of the tracking of affected areas and horse movements during August–September, 2007.

There appear to be significant parallels in the types of information being tracked and recorded for the management of Ebola cases to those during the EI outbreak in Australia. This information (location and temporal based records) is ideally suited to management in an Online GIS system, where records could be manually entered based on simple collection methods such as digital image records that are geo-located automatically. This would give a spatial reference at a point in time, and could be uploaded to the Online GIS by simply dragging and dropping onto a map canvas in a standard browser. Tracking every single contact is a more complex problem. Online tools, available on many devices, may provide powerful infectious disease control in the future.

13.3   Online GIS Organisational Trends

At the start of the new millennium, the creation of Online GIS systems was little more than an aspiration to most organisations. Leaders of organisations wanted to leverage the benefits of Online GIS to improve business processes. Business needs included tracking fleets, asset management and increasing profit through new online business. On the other hand, GIS specialists feared what it meant to expose their data externally, concerned that users might not understand the fundamentals of GIS and would question the currency, accuracy and precision of the data. Would it be *fit-for-purpose* for the many different ways users would try to use it?

Fortunately, many of the concerns regarding the publication of an organisation's data on the Web, have subsided. Organisations are no longer asking questions about *Should we use Online GIS?* but they are asking themselves how does Online GIS fit into their business operating model and what role does Online GIS have in their customer and internal business systems. The top organisational trends are

1. **Willingness to Share:** Organisations are becoming much more willing to deploy datasets online through Web services, aided by global strategies such as Open Government and Open Data.

2. **Mobile and Location Based Services:** Deployment of apps and services for the Mobile market is now essential for businesses to stay relevant to their users.

3. **Online GIS Mash-Ups:** Standardisation of map projections, symbology and the standardisation of Web services has made it viable to mash-up data from disparate datasets at the client through a series of asynchronous requests to backend services.

4. **GIS in the Cloud:** The cloud offers fast implementation of scalable (elastic) GIS services that provide access to robust Online GIS data and services.

5. **Richer GIS Content:** Publication of richer GIS datasets, higher resolution imagery and 3D data content.

6. **Crowd Participation:** With the impact of mobile devices, crowd sourcing is changing the way users interact with Online GIS websites, services and apps. It is extending the reach of Online GIS to a much larger number of users.

7. **APIs and Shared Libraries:** APIs (§4.7) provide a powerful way for developers to access existing Online GIS libraries and functions which can be integrated on clients or servers. Many of these APIs are now freely available.

8. **Big Data:** The trend towards Big Data provides rich data sources for Online GIS tools, including business analytics and intelligence.

9. **Online GIS Processing:** Some vendors are developing Online GIS geoprocessing capabilities which can provide geospatial data and image processing.

It is expected that these trends will continue in the near future, as the Internet continues to improve and standards mature, and as organisations develop innovative uses for Online GIS. There are many organisations and projects, as discussed in Chapter 9, that are providing rich online experiences for organisations and users of Online GIS.

13.4 Online GIS Technology Trends

This section provides a brief introduction to some of the technology trends that are influencing the future of Online GIS.

13.4.1 Cloud Computing

What is Cloud computing? When an organisation outsources its hardware and/or software to another company offering services accessed over the Internet, the organisation is using cloud based services.

Cloud service providers are often global companies, so users of Cloud computing won't necessarily know or care where their hardware and/or software is running. They just need to be able to access it online, anywhere, anytime. Because of the global nature of Cloud computing, the cloud is another important trend for Online GIS systems and the growth in cloud uptake appears to be proportional to the growth of the Internet. This trend is similar to the on-premise use of mainframe computers in the 1970s, before the shift towards low cost desktop environments, and many organisations are finding that buying computer services as cloud offerings is much more efficient that procuring, building and maintaining their own hardware and/or software. There are several benefits to using Cloud computing to hosting IT services:

Minimal Capital Expenditure Cloud computing is often pay-as-you-go, reducing the need for a large capital investment to be able to access data centre resources.

Low Cost Set-up and Maintenance Set-up and maintenance costs are based on monthly usage and the cloud infrastructure can grow and shrink to cater for peaks and troughs.

High Availability Cloud service providers have developed very robust environments that can automatically replicate and shut down servers using worker agents as required.

Fully Managed Services The cloud service providers manage the hardware and/or software used in their services, including patching and software updates. This removes a lot of the normal data centre administration needed for internally hosted data centres.

Fast Set-up and Deployment The set-up and deployment of a system to the cloud can generally be done in hours and the resulting deployment can grow and shrink dynamically as required.

Secure Access Cloud security can be managed using standard internet security protocols, through the use of service level agreements (SLAs) and through data encryption. There must still be a level of trust between the service provider and consumer, however, and there are other longer term security implementations to consider.

There are three main types of clouds. These are

1. Private Cloud: belongs to a organisation and is commonly running on the organisation's intranet. It is likely that the site is accessible from outside the intranet, but it is a capital asset owned by the organisation, for use by the organisation.

2. Public Cloud: is located somewhere on the Internet. It is made available by the service provider to other organisations. The capital investment is made by the "cloud service provider" and used under license by others through the Internet.

3. Hybrid Cloud: is made up of more than one type of cloud, public and private or combinations of both.

Enterprise needs for hardware and software can be procured through online services. Storing device details automatically in clouds gives access to such data anywhere there is Internet access. There are three common types of cloud services available from online service providers:

1. Infrastructure as a Service (IaaS): the buying *access* to raw computing hardware that you access over the Internet.

2. Software as a Service (SaaS): using complete software solutions that are hosted on someone else's network.

3. Platform as a Service (PaaS): used for developing Web-based applications that run over the Internet on software and hardware provided by another organisation.

There is one major technology strength that comes with Cloud computing. This is the *elasticity* of cloud implementations, in particular, with public clouds: the way the cloud can automatically grow, or stretch, as the demands on an application grow. Users of cloud technology can set tolerances that enable the cloud to automatically adjust servers to meet client demands. These types of peaks and troughs are common when major events happen, such as stock market releases, sporting events and natural disasters, in fact, anything that generates significant interest in the community. Online GIS is one of these, as many people are not using the Internet to follow what's happening and *where*.

Gartner[10] has stated that enterprises should be designing agility, flexibility and

---

[10] http://www.gartner.com/binaries/content/assets/events/keywords/
data-center/lsce11/gartner_dc_trip_report_2014.pdf

scalability into both their technology choices and business strategies over the coming year. Organisations need a clear technology roadmap that includes hybrid data centres. This also applied to Online GIS companies. Online GIS are data rich applications, especially when you include imagery, surface models and point density clouds, so minimising storage costs while reducing capital expenditure can be important to their financial viability. Thus Cloud computing is an attractive option. Many organisations appear to be choosing one cloud provider for their computational needs and another for their storage needs to minimise IT spending.

Cloud computing is not without its weaknesses that need to be considered when organisations are planning to move services to the cloud. Some of the obvious down sides of using a cloud infrastructure are

Dependency on Cloud provider for resolving problems quickly and cost effectively.

Proprietary cloud vendor systems overhead of migration to and from service providers proprietary platforms.

No control over long term software support which could influence organisation infrastructure.

Privacy and security guarantee The implications of outsourcing your business intellectual property to a cloud. Clouds may be operating in another country under different laws and provisions and data may be vulnerable if they go out of business.

Other Internet access at all times is essential.

Closely related to Cloud computing is the term Utility Computing (UC) (Foster and Kesselman 2003). UC is a cloud related term that was first used to describe infrastructure and services that can be metered, just like utility companies meter gas and electricity. In fact, UC is a form of Cloud computing Irrespective of the terminology. Cloud computing should figure in any organisational architecture design, especially as infrastructure to support Online GIS operations, and it is expected that the continued growth of cloud service provides will have a large impact on future directions of Online GIS.

13.4.2 Mashing-Up Geospatial Data and Services

One of the trends that will continue to drive future Online GIS systems will be the ability to *mash-up* spatial datasets. This mashing-up of data, as the term suggests, is the displaying of spatial data from disparate datasets, often from different organisations, into a cohesive display of information. The data can be sourced from different data warehouses and can include historical, current and future data predictions.

There were a number of successful mash-ups of geospatial data a decade ago. These involved the use of a single base map, a geocoding capability for searching and an interface for display. The data were commonly accessed through online published Web services, based on open standards as described throughout this book, with the client storing its data in cache and not local to the device. The Web services are often accessed asynchronously; however, the client application still needed to preserve order of the datasets when rendering.

The ability to make asynchronous calls to multiple Web services created a number of issues. The most prominent was what to do with the data as it was returned, and this issue placed rendering limitations on the amount of datasets and datatypes that could be displayed at any one time. The perception of transparency, as described by Westland et al. (2012), was one method for displaying data and a method devised by Hope and Miller (2009) used Cascading Style Sheets (CSS) to achieve transparency in the client viewer that could be manipulated using a JavaScript slider. This manipulation of mashed-up datasets enabled better mash-up of multiple imagery and vector geospatial datasets.

The real value of geospatial mash-ups will be realised when geo-processing capabilities are developed for the processing of mashed up data. This combinatorial geo-processing of spatially coincident data that was previously unlinked will unlock significant value for Online GIS users. To that end, vendors are starting to offer business intelligence and in-memory online analytical processing (§10.4.2) as part of their geospatial mash-up products.[11] Online Geospatial Marketplaces will also play an important role in the discovery of datasets that can be used for mashing up geospatial datasets.

13.4.3   Online Geo-Processing

One future growth area of Online GIS will be online geo-processing that combines vector and raster data processes. GIS functionality is already available online for some online geo-processing tasks, such as the generation of buffer queries, radial searches and network analysis. However, high resolution digital imagery captured from airborne digital sensors has now become widely available to organisations, and at a reduced cost. This provides new opportunities for organisations to utilise this imagery to improve their spatial maintenance processes and workflows. It will also enable users of geospatial data to find new ways of analysing and solving spatially related problems.

Object detection based on combinatorial processing and statistical image processing models is showing potential for the automated maintenance of spatial data, and is one way to increase return on technology investment (capture once – use many). Add to this, the rapid growth of the Internet, the Open Standards now available to support online publication and the global movement towards Open Data, and the demands for sharing geospatial information between organisations and individuals is set to explode. With this increased demand, the ability to run automated feature and detection processes on images and then link these geospatial datasets will have significant benefits to geospatial data users.

There are two growth areas contributing to the development of online geo-processing.

---

[11] http:www.gartner.com/it-glossary/geospatial-mashups

Research in the area of machine vision and image processing has been extensive over the last four decades. However, much of this research as been in the domain of imagery only. Utilising high resolution imagery for feature detection of environmental and man made features to maintain attribution and metadata of geospatial datasets has many applications (Hope 2013). Some of these have included the development of a number of methods:

Colour-Based Classification: Colour based classifications have been successfully using threshold based segmentations, nearest neighbour classifications and region optimisation using an edge detection algorithm. The use of these techniques to capture GIS features such as buildings and swimming pools, and to identify storm damage. The ability to locate these features has many applications and would be well suited to Online GIS.

Texture-Based Classification: Texture based classifications have been used successfully to locate features such as trees in digital imagery and have many uses for environmental management and urban planning. These classifications use several image processing techniques including mean shift filtering segementations, colour co-occurence histograms and Gaussian weighted histogram intersection.

Co-occurring Pattern Methodology: Feature detection has also been successfully demonstrated using co-occuring pattern methodologies: image segementation with bi-level thresholding and edge detection, indentifying changes with post processing to classify features of interest. This was successful in locating man made features over water under certain conditions.

Support Vector Machines and Colour-Based Segmentation: The use of colour-based Support Vector Machines (SVM) (§10.9.4) for locating man made features in digital imagery shows resilience to colour variations and has application across a broad number of Spatial Data Infrastructure roles including emergency services, urban planning and legistative compliance.

Secondly, the merging of GIS platforms and imagery platforms by vendors is providing increased opportunity to develop integrated systems to carry out these forms of geo-processing online. The ability to perform image processing tasks on mashed up raster datasets and relating the process outcomes back to vector datasets will have immediate benefits to communities and individuals. Locating buildings or trees in aerial imagery and relating these back to land ownership can better inform urban planners' authorities on changes over time, including changes such as building construction and tree removal. The location of swimming pools in digital imagery can be located within property footprints, where GIS attribution such as ownership and postal address, can provide local authorities with automated workflows to monitor water safety regulations.

While many of the existing applications of Online GIS do not use this type of processing, the discovery of new information from remote sensing data mashed up online with vector datasets, and the integration of these as Online GIS data and services, will provide valuable search and discovery capabilities in the near future. Making these types of services available online, where they can access large amounts of data for processing, will further extend the benefits to communities and economic growth. In the future, there will be significant use of geo-processing to discover previously unknown information to feed into decision making, from localised areas to entire continents. The potential application of online geo-processing is almost endless.

13.4.4   DIG35 Metadata and Online GIS

A metadata standard for improving the *semantic interoperability between image devices, services and software* was developed by a consortium of digital camera manufactures and released in 2007. This standard, known as the *DIG35 Specification*,[12] is now used in many low cost digital cameras and, combined with their inbuilt GPS, provides metadata for each image. These metadata tags include file format, date and time, focus levels, title, comments, etc., and also include location information in the form of global coordinates.

Coordinates for the image location are embedded in the header file of each image, and can be easily accessed to geolocate each image. One example of accessing and using the location information was a drag and drop functionality developed at the Land and Property Information in NSW Australia. With this application, digital images could be simply dragged from a Windows Explorer window onto a browser based Online GIS mapping canvas. The image coordinates were simply extracted dynamically and used to geolocate the image, zooming the user to the location where the image was taken. This capability was demonstrated at Digital Earth Symposium in Perth (Hope et al. 2011) and was a great example of the value of metadata. It also extended the implementation of the DIG35 metadata standard for SDI use and opened up a new world of georeferencing digital images against SDI layers.

The application demonstrated at ISDE7 (Hope et al. 2011) was implemented using the *exchangeable image file format* (EXIF) metadata, which is one of 10 popular implementations of the DIG35 specification. Others include file formats such as TIFF and PNG. EXIF metadata has been used for both image and audio files. Because the metadata forms part of the header file of these image formats, the metadata is always available as part of the image file. That is, the metadata always travels with the image pixel data.

When handling any location elements from the DIG35 specification metadata, such as those found in EXIF, there are a few issues to be aware of. Some camera manufacturers have not coded for negative latitudes in their implementation. This bug places images taken in Australia to somewhere off the coast of Japan. There is no provision for time-zone information, so date and time metadata is only localised. The standard allows manufacturers to include proprietary tags which are non-standard and can be difficult to retrieve. These data can also be encrypted by manufacturers to protect IP. There are limitations to the size of EXIF metadata in some image file formats and because of the location information that can be retrieved, plus the unique ID number of the device, there may be privacy issues to consider. DIG35 specifica-

---

[12] http://digit.nkp.cz/knihcin/digit/vav23/\digtf.pdf

tion metadata can identify where the photo was taken, at what time and whose device it was taken from, and non-technical users can easily view this location online once they have the image file.

13.4.5 Blogs and Tweets

There are many social media and networking tools now available on the Web. Internet Blogs are a great source of information about the current trends in Online GIS. Blogs range from online work diaries of industry professionals to technical sites from practitioners and vendor organisations with lots of source codes samples. A number of these may be sponsored by GIS organisations. However, a lot of blog content is from Online GIS interest groups and individuals. Some of the popular blogs at the time of writing were:

Geographika is an independent blog that contains a lot of practical experiences in developing Online GIS technologies, including Open Layers.[13]

GIS and Science is written by Matt Artz from ESRI. His blog presents a more scientific view of GIS, including Online GIS, and contains a number of links to scientific research and implementations.[14]

James Fee GIS Blogs is from another eponymous independent blogger who continues to post a lot of interesting articles relating to open source, proprietary software and his work experiences.[15]

EdParsons.com is slightly different in that posts are accessed through a key word search.[16]

Mapperz Blog contains information and links to many of the best online mapping sites available.[17]

Twitter is another great source of information for Online GIS (and many other topics) communities. Users can simply create a free online Twitter account,[18] adding a personal profile to attract followers. Topics of interest can then be followed, users can tweet interesting information or retweet other people's tweets that may be of interest. A simple search for Online GIS at the time of writing returned hundreds of active feeds on topics ranging from emergency response mapping to supporting the Nepal earthquake, to new online data services, website, jobs and links to tutorials for simple to advanced online geo-processing. Tweets can contain links to other websites and embedded images, and as they are limited to 140 characters, Twitter is creating an abbreviated language all its own.

---

[13] http://www.geographika.co.uk
[14] http://www.gisandscience.com/
[15] http://www.jamesfee.us
[16] http://www.edparsons.com
[17] http://mapperz.blogspot.com.au
[18] http://www.twitter.com

13.5   Virtual Worlds

What is better than looking at a map or photograph of where you want to holiday? The answer is obvious, being there. But if you can't be there in person, then the next best thing is to be virtually there. When an architect wants to show clients what a new development will look like, the building plans are only part of the story. Any major work includes artist's drawings of what the place would look like, and even models that clients can explore and look at from different angles.

It has now become common practice for architects to make virtual reality models of the buildings they plan to build. That way, the owners can experience what the finished product will be like before building even begins. They can also try out design variations and potential colour schemes. The same kind of technology has been used to build virtual reconstructions of ancient buildings. Good examples of this approach are systems that have been developed to help people assess environmental impacts of logging and other activities. The program Smart Forest, for instance, combines GIS with forest models and virtual reality (Orland 1997). It allows users to define various scenarios and to experience the consequences.

There also exist 3D models of many major capital cities of the world, showing building heights and facades. These models, many available online through open standards, have many uses. These include uses for town planning where it is important to project shadows from manmade structures, to modelling sun reflection off glass buildings.

Many computer games place the user in a virtual world. In games such as Sim-City and SimEarth, for instance, the user acts as manager for entire cities or even the entire planet. The games are based around a GIS that allows the user to select regions and zoom in to the level of individual buildings. In other games, players can drive through virtual landscapes following road maps (e.g., Fast and Furious Showdown and Destiny) or negotiate their way through entire fictional 3D artificial worlds (e.g., Call of Duty: Black Ops). Having combined GIS and virtual reality, it is now commonplace for gamers to participate in online sessions with other players, where they can play with or against each other from distributed *nodes*. This virtual worlds movement continues to develop online virtual reality as a means of developing virtual communities and other activities, such as business meetings.

But if Internet users can explore alternate universes online, and interact with other explorers in the process, then why not virtual reproductions of the real thing? Many areas, such as city centres, are already mapped in great detail. By that we mean that the exact form of the buildings is well known. Tourist maps of (say) downtown London sometimes include pictures of the buildings. Why not extend these globe based models to online virtual worlds so that tourists or students, for instance, could explore an area to get the feel of where things are before they even set foot there, or explore remote areas, such as the Andes or Himalayas, in greater depth than they could ever do in real life. They could even take an interactive virtual tour of Abbey Road Studios, complete with historical memorabilia and facts.[19]

13.6 Envoi (Online GIS)

As this book goes to press, Nepal has just experienced another 7.4 magnitude earthquake, not long after the devastating 7.8 earthquake we discussed in §13.2.2. GIS facilitated a fast and effective response, requiring many dedicated people, with many volunteers, fighting difficult terrain. The techniques we covered in this book emphasise the growing importance of Online GIS and the metadata which glues it together. Still emerging, though, is the Internet of Things, where spatially aware devices will be one of the big growth areas of Online GIS.

So, what would the response look like for the great Asian earthquake of 2020. We are now firmly in the domain of the Semantic Web and the Internet of Things. A pivotal element of the response now is a series of encrypted authorisations (§8.5.2), which can be released immediately to people and things to start the recovery option as soon as possible. Sensors provide early warning indicators directly to important international relief and humanitarian organisations. Encrypted authorisations are generated immediately. Countries which have signed up to crisis protocols will now accept these authorisations for rapid clearance of people, equipment and fungible resources through customs.

Cloud AI services are used to pool information from different sensors, from earthquake detectors in the ground to satellite observation, and generate metadata search terms for organisations and autonomous equipment to be mobilised.

Autonomous agents (§10.9.3) spring into action, using the principle of *mixtures of experts*, to collate reports from previous earthquakes and identify actions prioritised. In Nepal getting the mobile network back was a high priority. Telecom companies are queried for mast distribution and activity. Agents use cloud services for mobile tower distribution and determine which towers to get back in service first. The closest services for providing temporary telecom services and Internet are identified and deployed.

Drone rescue services are notified in order of accessibility to the site. Funding code authorisations are put in place and helicopters loaded with search drones head to the disaster area. The same drones are equiped with infra-red cameras to help with this rescue effort. LIDAR is used to compare surface maps before and after to target areas of greatest human risk and to increase the chance of locating survivors. Large supplies of GPS collars for search and rescue dogs are freighted to rescue teams gathering on the ground, using spatially tagged uploaded requirements data. The GPS collars receive drone directions to direct rescue teams to trapped people.

Many more ideas and opportunities will develop rapidly as the Internet of Things

---

[19] https://insideabbeyroad.withgoogle.com/en

takes off. But the whole immensely complicated recovery process will need immediate, unfettered access to spatial data of all kinds. It will need to be able to find it and to find online services to use it immediately with high levels of artificial intelligence and advanced geospatial processing. The technologies we have described in this book form the foundations for the next generation of spatial applications.

13.7　Summary and Outlook

This chapter has discussed the likely directions of Online GIS to give the reader a vision of how Online GIS, now a fundamental Web technology, will benefit society.

- §13.1 discusses how Global GIS systems are being built, now, providing a future pathway for the technologies described in this book.

- The use of Online GIS for environmental management, emergency response and disease control have unmistakeable benefits to society, and form the content of §13.2.

- To provide insight into what organisations are thinking, organisational trends are briefly discussed in §13.3, with a top 10 list provided to outline what organisations are currently planning with their Online GIS strategies and implementations.

- Online GIS technology trends are then discussed in §13.4, and include insights into how technologies are influencing Online GIS.

- §13.5 looks towards the future where virtual worlds will be commonplace on the Web, providing world simulations used in training.

- Finally, §13.6 introduces a possible scenario, where a great Asian earthquake has resulted in a significant natural disaster, and shows how the technologies described in this book will play an important role in search and rescue efforts.

Glossary

Numbers following each definition indicate the chapter and section in which the relevant discussion is to be found.

ACCC Australian Competition and Consumer Commission (§11.3.1)

ACMA Australian Communications and Media Authority (§11.3)

ANZLIC Australia-New Zealand Land Information Council (§7.2), (§9.2.1.1), (§9.6.3)

API Application Programming Interface, a set of routines, protocols and tools for building software applications (§2.3.4), (§4.7)

ArcGIS A major GIS package from the company ESRI Inc. (§9.6)

AS Application Schemas (§6.5.2)

ASDD Australian Spatial Data Directory (§9.6.3.4)

ASDD Australasian Spatial Data Directory (§7.2)

ASIC Australian Securities and Investments Commission (§11.3.1)

BIN Biodiversity Information Network (§7.2)

BIOCLIM Used extensively for species distribution modelling. Bioclim is the classic "climate-envelope-model" (§2.4.2)

BYOD Bring Your Own Device (§11.2)

C A general purpose, imperative computer programming language (§5.5), (§3.3)

CAM Computer-aided manufacturing is the use of computer software to control machine tools and related machinery in the manufacturing of workpieces (§4.6.5)

CEN Comite European de Normalisation – a standards organisation to foster the European economy in global trading, the welfare of citizens and the environment (§9.6.2)

CEOS Committee on Earth Observation Satellites (§9.4)

CERCO Comite European des Responsables de la Cartograpltie Officielle (§9.6.2)

CERN European Organization for Nuclear Research - also the birthplace of the World Wide Web (§7.3)

CGALIES Coordination Group on Access to Location Information by Emergency (§12.3.3)

CGDI Canadian Geospatial Data Infrastructure (§9.6.1.4)

CGI Common Gateway Interface (§3.3), (§3.3.1)

CICS Customer Information Control System (§10.6.1)

CIESIN Center for International Earth Science Information Network (§7.6.2)

CKAN Comprehensive Kerbal Archive Network (§9.6.1.4)

CLIPS A popular shell and language for developing intelligent systems online (§10.9.2)

COBOL Common Business-Oriented Language (§10.6.1)

COLLADA Collaborative design activity (§6.3)

Component A modular software building block comprising a number of interrelated objects (§6.5.1)

COPE Corporate Owned, Personally Enabled (§11.2)

CSDGM Content Standard for Digital Geospatial Metadata (§9.4.1)

CSV Comma Separated Value (§8.6)

CVS Concurrent Version System, a version control system designed for software projects (§9.6)

CSW Catalogue Services for the Web (§9.4.3)

DAAC Distributed Archive Archive Centre (§10.7.1)

Data mining Techniques for knowledge discovery in huge datasets. Often associated with Data Warehouses and Big Data (§10.10)

DB2 A family of database server products developed by IBM (§10.6.1)

DC Dublin Core (Dublin, Ohio) (§7.5.1.1)

DCMI Dublin Core Metadata Initiative (§8.2)

DEM Digital Elevation Model (§10.7.1)

DFSI Department of Finance, Services and Innovation, NSW Australia (§9.6.3.3)

DGN Design File, a proprietary file format owned by Hexagon (§9.6.1.6)

DIVERSITAS An international research programme aimed at integrating biodiversity science for human well-being (§7.6.2)

DPI Department of Primary Industries (§13.2.3.1)

DR Description Resources (§8.4.1)

DSM Digital Geospatial Metadata (§9.6.1.4)

DSTP Data Space Transfer Protocol (§10.5.4)

DTD Document Type Definition (§5.2.4)

DWG Domain Working Group (§6.5.3.1)

DMG Data Mining Group, an independent consortium for defining data mining standards (§10.10.1)

Exif Exchangeable Image File Format (§13.4.4)

ECMA European Computer Manufacturers Association (§8.3.1)

ECMA Script A scripting language widely used for client-side scripting on the Web (§4.1)

ECW Enhanced Compression Wavelet – a proprietary wavelet compression image format optimised for aerial and satellite imagery (§2.2.2)

EI Equine Influenza (§13.2.3.1)

EIONET European Environment Information and Observation Network (§7.6.2)

EMBL European Molecular Biology Laboratory (§7.2)

EMPs Emergency Management Professionals (§9.6.1.4)

ENV Euro-Norm Voluntaire (§9.6.2)

EOS Earth Observation Satellites (§9.4)

Esri Shape files (*.shp) A proprietory data format for geographic information system (GIS) software, developed by Esri (§9.6.1.5)

EU A politico-economic union of 28 member states that are located primarily in Europe (§12.3.3)

EURADIN European Addresses Infrastructure (§9.6.2.1)

EuroSDR European Spatial Data Research Network (§9.4)

EUROSTAT Directorate-General of the European Commission (§9.6.2)

Extensible Markup Language (XML) A system for marking up documents and data using tags that indicate structural elements. XML is a recommendation of the W3C for making up documents, a later variant of SGML (§5.2), (§5.5)

FGDC Federal Geographic Data Committee (§9.4)

FIFE First ISLSCP (International Satellite Land Surface Climatology Project) Field Experiment (§10.7.1)

FIND Australian National Spatial Data Catalogue (§9.6.3.4)

FOAF Friend of a Friend (§8.2.2)

FTP File Transfer Protocol (§2.2.1)

GBIF Global Biodiversity Information Facility (§7.2)

GEOJSON A format for encoding a variety of geographic data structures. A GeoJSON object may represent a geometry (§7.5.1)

GET A method of transferring form data to a server by embedding the data within a URL (Web address) (§3.3.1.1)

GDDD Geographical Data Description Directory (§9.6.2)

GEE Google Earth Engine (§6.3.2)

Geographic Information System A computer system for storing, displaying and interpreting geographic information (§1.1)

Geographic Markup Language A system of XML tags for marking up geographic information (§6.4)

GeoRSS Geographically Encoded Objects for RSS feeds (§9.6.1.4), (§6.4)

GIE Group d'Interet Economique (§9.6.2)

GIF Graphics Interchange Format – a bitmap image format that was introduced by CompuServe in 1987 (§3.6)

GIPSIE GIS Interoperability Project Stimulating the Industry in Europe (§9.4)

GIS See Geographic Information System (§1.1)

GME Google Maps Engine (§6.3.2)

GMES Global Monitoring for Environment and Security (§9.6.2.1)

GML See Geographic Markup Language (§6.4)

GMT A freeware package of Generic Mapping Tools (§2.4.1)

GOS Geospatial One-stop Shop (§9.6.1.1)

GPS Global Positioning System (§6.1), (§7.5.2.1), (§12.2)

GPX GPS Exchange Format (§6.2)

GSDI Global Spatial Data Infrastructure Association (§9.5)

HEAD A container element for metadata (data about data) as the start of an HTML document (§8.2.3)

HTML HyperText Markup Language, a system of tags (based on SGML) used for marking up documents for the World Wide Web (§5.1), (§2.7.3.1)

HTTP Hypertext Transfer Protocol, The client-side protocol used in transferring data between servers and clients on the World Wide Web (§3.1.1)

HTTPD HTTP Daemon (§3.1.1)

Hypermedia A system in which various forms of information including text, graphics, video and audio, are linked together by a hypertext program (§2.3.4.1)

IaaS Infrastructure as a Service (§13.4.1)

ICSM Intergovernmental Committee of Surveying and Mapping (§6.5.2)

ICT Information and Communication Technology (§11.5.2)

IETF Internet Engineering Task Force (§8.3.1)

ILDIS International Legume Database and Information Service (§7.6.2)

IMS Information Management System (§10.6.1)

IN Information Network (§7.1)

INSPIRE Infrastructure for Spatial Information in the European Community (§9.6.2.1)

Internet A global computer network, linked via the Internet Protocol (IP) and governed by the Internet Society (§1.4)

IOPI International Organisation for Plant Information (§7.2)

IRI International Resource Indicators (§8.4.1)

ISBN International Standard Book Number (§8.2.1.1)

ISOC Internet Society (§1.4)

ISO International Organization for Standardization (§9.1), (§9.4)

ISPRS International Society for Photogrammetry and Remote Sensing (§9.4)

IUFRO International Union of Forestry Research Organisations (§7.2)

JRC EU Joint Research Centre for the European Commission (§9.6.2)

JPIP JPEG2000 Interactive Protocol – a compression streamlining protocol that works with JPEG200 to produce an image using the least bandwidth required (§2.2.2)

JSON Javascript Object Notation (§8.3)

KDD Knowledge Discovery in Databases, the discovery of useable knowledge in data mining (§10.1.1)

KVP Key Value Pair is a set of two linked data items: a key, which is a unique identifier for some item of data, and the value, which is either the data that is identified or a pointer to the location of that data. (§10.8.3)

KML Keyhole Markup Language (§9.6.1.3)

KMZ Keyhole Markup Language files when compressed (§9.6.1.3)

LBS Location-Based Services (§12.1)

LDAP Lightweight Directory Access Protocol (§11.5.2)

LDSII London Datastore Mark II, a data clearinghouse (§9.6.2.3)

LPI Land and Property Information, NSW Australia (§6.5.2)

MADP Mobile Application Development Platform (§11.6.4)

Mashup Combining geospatial data from multiple sources at the client to create some new functionality (§13.4.2)

MD5 A message-digest algorithm commonly used to verify data integrity (§8.5.2)

MDB Microsoft Access Database file extension (§9.6.3.2)

MDS Message Data Services (§8.4)

MEGRIN Multipurpose European Ground Related Information Network (§9.6.2)

META Tags used in HTML to provide structured metadata about a Web page (§8.1)

MPEG Moving Picture Experts Group – a working group formed to set standards for audio and video compression and transmission (§3.1.2)

MMS Multimedia Messaging Service is a standard way to send messages that include multimedia content to and from mobile phones (§12.3.1)

mobileOK POWDER denotation to denote a Web page or site suitable for mobile devices (§8.4.1)

Namespace A collection of terms and definitions that describe the usage and sometimes meaning of XML tags (i.e., the semantics of the markup language) (§1.4)

NCSA National Centre for Supercomputer Applications (§1.4)

NFIS National Forest Infrastructure System (§10.2)

NGDC National Geospatial Data Clearinghouse (§9.5)

NGOs Non-Governmental Organization (§9.6.2.3)

NII National Information Infrastructure (§9.6.2.2)

NMAs National Mapping Agencies (§9.6.2)

NSDI National Spatial Data Infrastructure (§9.1), (§9.5)

OCLC Online Computer Library Center (§8.2)

OECD Organisation for Economic Co-operation and Development (§7.2)

OGC Open Geospatial Consortium (§9.4.3)

OGL Open Government Licence (§9.6.1.5)

OLAP Online Analytical Processing (§10.4.2)

OLTP Online Transaction Processing (§10.4.1)

OMG Object Management Group (§9.4)

OO Object Oriented (§1.2)

ORDBMS Object Relational Databases (§10.1)

OTTER Oregan Transect Ecosystem Research (§10.7.1)

P3P Platform for Privacy Preferences Project (§8.1)

Paas Platform as a Service (§13.4.1)

PDF Portable Document Format originally created by Adobe systems in 1993. PDF documents are easily readable on any operating system (§9.6.1.4)

PERL A family of high-level, general-purpose, interpreted, dynamic programming languages (§3.3)

PHP A server-side scripting language designed for Web development but also used as a general-purpose programming language (§3.3)

PICS Platform for Internet Content Selection (§8.4)

PMML Predictive Modelling Markup Language (§10.10.1), (§10.5.4)

POST Request method supported by HTTP Protocol – designed to request that a Web server accept the data enclosed in the request message's body for storage (§3.3.1.2)

POWDER Protocol for Web Description Resources (§8.4), (§8.4.1), (§8.4.2)

Pyramid file format A hierarchical data format for images and maps (§10.8.3)

RAM Random-access memory (§10.5.4)

RAT Remote Access Trojans (§11.5)

RDF Resource Description Framework (§8.5)

REST Representational State Transfer, a software architecture style consisting of guidelines and best practices for creating scalable Web services (§3.1.1)

RFID Radio-frequency identification the wireless use of electromagnetic fields to transfer data (§12.2)

RDBMS Relational Databases (§10.1)

SaaS Software as a Service (§13.4.1)

SABE Society for the Advancement of Behavioral Economics (§9.6.2)

SDI Spatial Data Infrastructure (§2.3.2), (§9.5)

SDLC System Development Life Cycle (§11.4.3)

SGML Standard Generalized Markup Language, a general approach to marking up the structure and format of text according to a document type definition (DTD) (§5.1)

SHP An Esri Shapefile file format which stores geospatial vector data for GIS (§9.6.1.5)

SLA Service Level Agreement (§13.4.1)

SLEEP A scripting language executed on the Java platform (§3.6)

SMS Short Message Service – a text messaging service component of phone, Web, or mobile communication systems (§12.3.1)

SOAP Simple Object Access protocol, a specification for exchanging structured information in the implementation of Web services in computer networks (§3.1.1)

SPARQL A recursive acronym for SPARQL Protocol and RDF Query Language (§8.6)

ssh Secure SHell is a protocol for securely accessing one computer from another (§2.2.1)

SSL Secure Sockets Layer (§11.5.2)

Stability The capacity of online sites and services to keep functioning and to remain at the same Web address (§7.5.4)

Standardisation The process of making information and services conform to standards (§1.4), (secRefinfonet-standardise)

Stateless Any process in which no state (i.e. a record of history) is recorded (§2.7.3.1)

SVG Scalable Vector Graphics (§4.6.5)

SWG Standards Working Group (§6.3)

TIFF Tagged Image File Format (§6.5.2)

TIGER Topologically Integrated Geographic Encoding and Referencing database (§2.3.1)

TSV Tab Separated Value (§8.6)

UML Universal Modelling Language (§1.3.3)

UNCED United Nations Conference on Environment and Development (§9.6)

UNEP United Nations Environment Programme (§7.6.2)

UNIX A trademarked family of multitasking, multiuser computer operating systems (§5.1)

URI Uniform Resource Indicators (§5.4)

URL Uniform Resource Locator (§2.2.1)

USDA ITIS United States Department of Agriculture Integrated Taxonomic Information System (§7.6.2)

UX User experience (§11.4.2)

VRML Virtual Reality Modeling Language (§6.5.2)

VT Virtual Tourist, an early information network with a geographic basis (§7.3), (§1.4.2), (§2.3.4.1)

W3C World Wide Web Consortium (§1.4)

WCMC World Conservation Monitoring Centre (§7.6.2)

Web Client A computer program that retrieves information from World Wide Web servers (§3.1)

Web Server A server that delivers information in response to HTTP requests across the Internet (§3.1)

WFS Web Feature Service – provides an interface allowing requests for geographical features across the Web using platform-independent calls (§2.3.2), (§10.8.2)

WGS84 World Geodetic System 1984 (§6.2)

WGS84 EGM96 World Geodetic System Earth Gravitational Model 1996 (§6.3.1)

WMS Web Map Server – a standard protocol for serving georeferenced map images over the Internet that are generated by a map server using data from a GIS database (§2.3.2), (§10.8.1)

WMSI Web Map Server Interface, an OpenGIS standard (§10.8.1)

WMTS Web Mapping Tile Service (§10.8.3)

WLAN Wireless Local Area Network (§12.4)

XLS A proprietary binary file format called Excel Binary File Format used by Microsoft Excel up until 2007 as its primary format (§9.6.1.5)

XLSX Microsoft Excel file created using the Open XML standard. It is stored as a compressed Zip archive (§9.6.3.2)

XML See Extensible Markup Language (§8.1)

XSS Cross Site Scripting (§11.5.2)

Bibliography

Abe, S. (2010). *Support Vector Machines for Pattern Recognition.* Springer-Verlag London Limited, Second Edition.

ACIL-Tasman (2008). The value of spatial information: The impact of modern spatial information technologies on the australian economy. Prepared for the CRC for Spatial Information and ANZLIC - the Spatial Information Council, Melbourne.

ACMA (2013). Mobile apps: Emerging issues in media and communications. Occasional paper number 1. http://www.acma.gov.au/theACMA/Library/researchacma/Emerging-regulatory-issues/emerging-issues-and-future-thinking.

Alschuler, L. (1995). *ABCD...SGML.* International Thomson Publishing, Boston.

Anon (2014). Size of the world wide web. http://www.worldwidewebsize.com/.

ASIC (Accessed Oct 2014). E-payments code. http://www.asic.gov.au/asic/asic.nsf/byheadline/ePayments-Code:download.

Azoff, M. and C. Singh (2013). 2014 trends to watch: Software development and lifecycle management. Ovum Research, London.

Benford, S., R. Anastasi, M. Flintham, A. Drozd, A. Crabtree, C. Greenhalgh, N. Tandavanitj, M. Adams, and J. Row-Farr (2003). Coping with uncertainty in a location-based game. *IEEE Pervasive Computing 2*(3), 34–47.

Berners-Lee, T. (accessed Nov 2014). Linked data. http://www.w3.org/DesignIssues/LinkedData.html.

BIN21 (Accessed 2013). Biodiversity information network. http://anbg.gov.au/bin21/bin21.html.

Bossomaier, T. and D. Green (2000). *Complex Systems.* Cambridge University Press, Cambridge.

Brickley, D. and L. Miller (accessed Nov 2014). FOAF vocabulary specification 0.99. http://xmlns.com/foaf/spec/20140114.html.

Brooks, R. A., P. Maes, M. J. Mataric, and G. More (1990). Towards a new frontier of applications. In *Proc. IROS, IEEE International Workshop Intelligent Robots and Systems.*

Bryan, M. (1988). *SGML: An Authors Guide to the Standard Generalized Markup Language.* Addison-Wesley, Reading, MA.

Burdet, H. (1992). What is IOPI? *Taxon 41*, 390–392.

Busby, J., J. McMahon, M. Hutchinson, H. Nix, and K. Ord (Accessed Dec 2014). Species distribution models – BIOCLIM. `http://ecobas.org/www-server/rem/mdb/bioclim.html`.

Cathro, W. (1997, August). Metadata: An overview. Standards Australia Seminar. `http://www.nla.gov.au/nla/staffpaper/cathro3.html`.

Chawla, S., S. Shekhar, W. Wu, and U. Ozesmi (2001). Modeling spatial dependencies for mining geospatial data: An introduction. In H. Miller and J. Han (Eds.), *Geographic Data Mining and Knowledge Discovery.* Taylor and Francis, London.

Chen, C. and P. Wang (2005). *Handbook of Pattern Recognition and Computer Vision.* World Scientific Publications, Singapore.

Chen, G. and D. Kotz (2000). A survey of context-aware mobile computing research. Technical Report TR2000-381, Dartmouth Computer Science Technical Report.

Clark, J. (1999). XML namespaces. FIND NEW LINK ON W3C. `http://www.jclark.com/xmVxmlns.htm`.

Cranshaw, J., E. Toch, J. Hong, A. Kittur, and N. Sadeh (2010). Bridging the gap between physical location and online social networks. In *Proceedings of the 12th ACM International Conference on Ubiquitous Computing*, pp. 119–128.

Dao, D., C. Rizos, and J. Wang (2002). Location-based services: Technical and business issues. *GPS Solutions 6*, 169–178.

Dietterich, T. (1996, December). Machine learning. *ACM Comput. Surv. 28*(4es).

DMG (2000). Pmml 1.0 - predictive model markup language. Data Mining Group (DMG). `http://www.dmg.org/html/pmml_v1_1.html`.

Dorigo, M. (2006). Ant colony optimisation and swarm intelligence. In *Proc. of the 5th Int. Workshop ANTS, Brussels.*

Drab, S. A. and G. Binder (2005). Spacerace: A location-based game for mobile phones using assisted GPS. In *Second International Workshop on Gaming Applications in Pervasive Computing Environmnets at Pervasive 2005.*

Eggers, W. D. (2007). *Government 2.0: Using Technology to Improve Education, Cut Red Tape, Reduce Gridlock, and Enhance Democracy.* Rowman & Littlefield, Lanham, MD.

Eklund, P. W., S. D. Kirkby, and A. Salim (1998). Data mining and soil salinity analysis. *International Journal of Geographical Information Science 12*, 247–268.

Ensign, C. (1997). *SGML: The Billion Dollar Secret*. Prentice Hall, New Jersey.

Erdos, P. and A. Renyi (1960). On the evolution of random graphs. *Math. Inst. Hungarian Acad, 5*, 17–61. In Hungarian.

ERIN (1995). Species mapper. `http://www.environment.gov.au/`.

Ester, M., H.-P. Kriegel, and J. Sander (1999). Knowledge discovery in spatial databases (invited paper). In *23rd German Conference on Artificial Intelligence (KI '99). Bonn, Germany*.

Evans, D. (2011). The internet of things. how the next evolution of the internet is changing everything. Technical report, CISCO.

Fayyad, U., G. Piatetsky-Shapiro, and P. Smyth (1996a). From data mining to knowledge discovery: An overview. In *Advances in Knowledge Discovery and Data Mining*, pp. 1–34. AAAI Press, Menlo Park.

Fayyad, U., G. Piatetsky-Shapiro, and P. Smyth (1996b). The KDD process for extracting useful knowledge from volumes of data. *The Communications of the ACM 39*(11), 27–31.

Foster, I. and C. Kesselman (2003). *The Grid: Blueprint for a New Computing Infrastructure*. Morgan Kaufmann, Burlington, MA.

Fritsch, L. and J. Muntermann (2005). Aktuelle Hinderungsgründe für den kommerziellen Erfolg von Location Based Service-Angeboten. In *Konferenz Mobile Commerce Technologie & Anwendungen (MCTA)*, pp. 143–156.

Gamma, E., R. Helm, R. Johnson, and J. Vlissides (1995). *Design Patterns: Elements of Reusable Object-Oriented Software*. Addison-Wesley, Reading, MA.

Gardner, C. (1996). *IBM Data Mining Technology*. IBM Corporation, Stamford, CT.

Garfinkel, S. (1995). *PGP: Pretty Good Privacy*. O'Reilly & Associates, Sebastopol, CA.

Gartner (2014). Magic quadrant for mobile application development platforms. `http://www.gartner.com/technology/reprints.do?id=1-20SX3ZK&ct=140903&st=sb`.

Giarrantano, J. and G. Riley (1989). *Expert Systems: Principles and Programming*. PWS-KENT Publishing, Boston, MA.

Goldfarb, C. and P. Prescod (1998). *The XML Handbook*. Prentice Hall, N.J.

Goldhammer, K., C. Link, J. Tietz, and M. Hochhaus (2013). Location-based services 2013. In *Presented at the Local Web Conference 2013, Nnberg*.

Google (Accessed March 2013). Global Business Map. Think Insights. Available online `http://www.google.com/think/research-studies/global-business-map.html`.

Green, D. (1993). Emergent behaviour in biological systems. In D. Green and T. Bossomaier (Eds.), *Complex Systems – from Biology to Computation*, pp. 24–35. lOS Press, Amsterdam.

Green, D. (1994). Databasing diversity – a distributed public-domain approach. *Taxon 43*, 51–62.

Green, D. (1995). From honey pots to a web of SIN – building the world-wide information system. In P. In Tsang, J. Weckert, J. Harris, and S. Tse (Eds.), *Proc. of AUUG'95 and Asia-Pacific World Wide Web '95 Conference*, pp. 11–18.

Green, D. (1996). A general model for on-line publishing. In T. Bossomaier and L. Chubb (Eds.), *Proc. AUUG'96 and Asia-Pacific World Wide Web '96 Conference*, pp. I52– I58. Australian Unix Users Group, Sydney.

Green, D. (2000). Coping with complexity – the role of distributed information in environmental and resource management. In H. Salminen, J. Saarikko, and E. Virtanen (Eds.), *Proc. Resource Technology '98 Nordic.*, Finish Forestry Research Institute, Rovaniemi, Finland.

Green, D., P. Bristow, J. Ash, L. Benton, P. Milliken, and D. Newth (1998). Network publishing languages. In *Proc. 7th International World Wide Web Conference*.

Green, D. and J. Croft (1994). Proposal for implementing a biodiversity information network. In D. Canhos, V. Canhos, and B. Kirsop (Eds.), *Linking Mechanisms for Biodiversity Information*, Proc. Workshop for the Biodiversity Information Network, pp. 5–17. Fundacao Tropical de Pesquisas e Tecnologia "Andre Tosello," Campinas, Sao Paulo, Brazil.

Hammer, J., H. Garcia-Molia, W. Labio, J. Widom, and Y. Zhuge (1995). The Stanford data warehousing project. *Data Engineering Bulleting. Special Issue on Materialized Views and Data Warehousing 18*(2), 41–48.

Hardy, G. (1998). The OECD's megascience forum biodiversity informatics group. FIND NEW LINK `http://www.oecd.org//ehslicgb/BIODIV8.HTM`.

Helal, S., B. Winkler, C. Lee, Y. Kaddoura, L. Ran, C. Giraldo, S. Kuchibhotla, and W. Mann (2003). Enabling location-aware pervasive computing applications for the elderly. In *Presented at the Proceedings of the First IEEE Conference on Pervasive Computing and Communications*, pp. 531–545.

Hope, B. A. (2013). *Spatial Data Infrastructures: Extending Digital Image Processing Techniques and Managment Models for Spatial Data Infrastructure Improvement.* PhD Thesis, Charles Sturt University, Bathurst, Australia.

Hope, B. A., P. Harris, and D. Miller (2011). Opening up Enterprise GIS: A web 2.0 spatial architecture. In *Proc. 7th International Symposium on Digital Earth (ISDE7), Perth, Australia.*

Hope, B. A. and D. Miller (2009). A multi-layered web deployment model for spatial information. In *Proc. 6th Int. Symp. on Digital Earth, Beijing China*.

Hubber, H. (1997). A success story: Wal-mart stores. In *Proc. of the First State of Florida Data Warehousing Conference*. UNPUBLISHED.

Huizenga, J., W. Admiraal, S. Akkerman, and G. Dam (2009). Mobile game based learning in secondary education: Engagement, motivation and learning in a mobile city game. *journal of Computer Assisted Learning 25*(4), 332–344.

IETF (accessed Nov 2014a). Internationalised Resource Identifiers. Internet Engineering Task Force RFC3987. http://tools.ietf.org/html/rfc3987.

IETF (accessed Nov 2014b). Uniform Resource Identifiers. Internet Engineering Task Force RFC3986. http://tools.ietf.org/html/rfc3986.

Inmon, W. (1995). What is a data warehouse? *Prism 1*(1).

Inmon, W. (1996). The data warehouse and data mining. *Communications of the ACM 39*(11), 49–50.

ISO (Accessed Dec 2014). International standards organisation. http://www.iso.org/iso/home/about.htm.

ISOC (2000). The internet society (ISOC). HomePage:http://www.isoc.org/.

IUBS (1998). Species 2000. howpublished= International Union of Biological Sciences. http://www.sp2000.org/.

IUFRO (1998). Global forest Information service. International Union of Forestry Research Organisations. http://www.gfis.net. http://iufro.boku.ac.at/.

Jensen, C. S., A. Friis-Christensen, T. B. Pedersen, D. Pfoser, S. Saltenis, and N. Tryfona (2001). Location-based services – a database perspective. In *Proc. 8th Scandinavian Research Conference on Geographical Information Science*, pp. 59–68.

JSON (2014). The JSON schema specification. http://json-schema.org/documentation.html.

JSON (Accessed 2014). JavaScript Object Notation. http://tools.ietf.org/html/rfc7159.

Junglas, I. A. and R. T. Watson (2008). Location-based services. *Communications of the ACM 51*(3), 65–69.

Kaasinen, E. (2003). User needs for location-aware mobile services. *Personal and Ubiquitous Computing 7*(1), 70–79.

Karon, C. (2015). *Adoption and Acceptance of Innovation in Location-Based Services*. Ph. D. thesis, Charles Sturt University.

Kay, M. (Accessed Dec 2014). Saxonic products. `http://www.saxonica.com/products/products.xml`.

Kennedy, R., Y. Lee, B. van Ray, C. Reed, and R. Lippman (1998). *Solving Data Mining Problems Through Pattern Recognition*. The Data Mining Institute and Prentice Hall.

Knuth, D. (1984). *The TeX Book*. Addison-Wesley, MA.

Krol, E. (1992). *The Whole Internet User Guide and Catalog*. O'Reilly & Associates, Sebastopol, CA.

Lamport, L. (1986). Latex: User's guide & reference manual.

Lapp, J. Robie, J. and D. Schach (accessed Nov 2014). Xml query language (xql). `http://www.w3.org/TandS/QL/QL98/pp/xql.html#Query%20Operators`.

Larman, C. (1998). *Applying UML and Patterns*. Prentice Hall, New York.

Lassila, O. and R. Swick (1999). Resource description framework (RDF) model and syntax specification. W3C. `http://www.w3.org/TR/REC-rdf-syntax`.

Li, N. and G. Chen (2009). Analysis of a location-based social network. In *Proceedings of International Conference on Computational Science and Engineering*, pp. 263–270.

Liutkauskas, V., D. Matulis, and R. Pleŏstys (2004). Location-based services. *Elektronika ir elektrotechnika 3*(52), 35–40.

Malhotra, A. and M. Maloney (1999). XML schema requirements. W3C. `http://www.w3.org/TR/NOTE-xml-schema-req`.

Manning, J. and N. Brown (2003a). Positional frameworks for SDI. In *Developing Spatial Data Infrastructure – From Concept to Reality*, pp. 111–118. Taylor and Francis, New York.

Manning, J. and N. Brown (2003b). SDI Diffusion - a regional case study with relevance to other levels. In *Developing Spatial Data Infrastructure – From Concept to Reality*, pp. 80–94. Taylor and Francis, New York.

May, R. M. (2011, 08). Why worry about how many species and their loss? *PLoS Biol 9*(8), e1001130.

McNeff, J. G. (2002). The global positioning system. *IEEE Transactions on Microwave Theory and Techniques 50*(3), 645–652.

Michalski, R., I. Bratko, and M. Kubat (1998). *Machine Learning and Data Mining Methods and Applications*. John Wiley, New York.

Mohomed, I., J. C. Cai, S. Chavoshi, and E. de Lara (2006). Context-aware interactive content adaptation. In *MobiSys 06, Proceedings of the 4th International Conference on Mobile Systems, Applications and Services*, pp. 42–55. ACM.

Mora, C., D. P. Tittensor, S. Adl, A. G. B. Simpson, and B. Worm (2011, 08). How many species are there on earth and in the ocean? *PLoS Biol 9*(8), e1001127.

Mozilla (Accessed Oct 2014). Open badges. http://openbadges.org.

Nebert, D. and A. Whiteside (2005). Ogc 04-021r3: Catalog services specification, version 2.0.1. Open Geospatial Consortium Inc., USA.

Nebert, D., A. Whiteside, and P. Vretanos (2007). Ogc 07-006r1: Catalog service specification, version 2.0.2. Open Geospatial Consortium Inc., USA, p. 218, http://portal.opengeospatial.org/files/index.php?artifact_id=20555S.

NGDC (2000). Webmapper interface. National Geophysical Data Center. http://www.ngdc.noaa.gov/paleo/.

Nicklas, D., C. Pfisterer, and B. Mitschang (2001). Towards location-based games. In *Presented at the Proceedings of the International Conference on Applications and Development of Computer Games in the 21st Century*, pp. 61–67.

NSW Government (Accessed Oct 2014). NSW 2021 http://www.2021.nsw.gov.au/sites/default/files/NSW2021.

Office of the Austraion Information Commissioner (2013). Mobile privacy: A better practice guide for mobile app developers. http://www.oaic.gov.au/news/consultations/mobileprivacyappguide/Mobile.

OGC (2000). OpenGIS abstract specification. Open GIS Consortium. http://www.opengis.org/.

Oh, L.-B. and H. Xu (2003). Effects of multimedia on mobile consumer behavior: An empirical study of location-aware advertising. In *Presented at the Proceedings of Twentyfourth International Conference on Information Systems*, pp. 679–691.

OMG (1997). The object management group (OMG). http://www.omg.org/.

Orland, B. (1997). Forest visual modeling for planners and managers. In *Proc. ASPRS/ACSMI RT'97, Seattle*, Volume 4, pp. 193–203. American Society for Photogrammetry and Remote Sensing, Washington.

Plewe, B. (1997). *GIS Online*. Onward Press, Albany NY.

Putz, S. (1994). Interactive information services using world-wide web hypertext. *Computer Networks and ISDN Systems 27*(2), 273–280.

Richardson, L. and S. Ruby (2007). *RESTful Web Services: Web services for the real world*. O'Reilly Media.

Ripley, B. (1996). *Pattern Recognition and Neural Networks*. Cambridge University Press.

Salmon, P. H. (2003). Locating calls to the emergency services. *BT Technology Journal 21*(1), 28–33.

Schwinger, W., C. Grü, B. Pröll, W. Retschitzegger, and A. Schauerhuber (2005). Context-awareness in mobile tourism guides: A comprehensive survey. *Statista* (Accessed Dec 2014), 1–20. http://www.statista.com/statistics/294314/.

Sprigman, J. (2015). Oracle v. Google: A high-stakes legal fight for the software industry. *Comm. ACM 58*(5), 27–29.

Stallings, W. (1995). *Network and Internetwork Security*. Prentice Hall, Boston, MA.

Statista (2012). Location-based service usage in European countries 2014–2017. Available online http://www.statista.com/statistics/294314/share-of-mobile-subscribers-using-location-based-services.

Steinke, A., D. Green, and D. Peters (1996). Online environmental and geographic information systems. In H. Saarenma and A. Kempf (Eds.), *Internet Applications and Electronic Information Resources in Forestry and Environmental Sciences*, EFI Proc. No. 10, Joensuu (Finland), pp. 89–98. European Forestry Institute.

Steinwart, I. and A. Christmann (2008). *Support Vector Machines*. Springer-Verlag, New York, NJ.

Sweetlove, L. (2011). Number of species on Earth tagged at 8.7 million. *Nature*. http://www.nature.com/news/2011/110823/full/news.2011.498.html.

TNS (2012). Two thirds of worlds mobile users signal they want to be found. http://www.tnsglobal.com/press-release/two-thirds-world\%E2\%80\%99s-mobile-users-signal-they-want-be-found.

Traeg, P. (2013). Best of both worlds: Mixing HTML5 and native code. http://mobile.smashingmagazine.com/2013/10/17/best-of-both-worlds-mixing-html5-native-code/.

(UNEP), U. N. E. P. (1995). Background documents on the clearing-house mechanism (CHM). In *Proc. Convention on Biological Diversity*, Jakarta, Indonesia. FIND NEW LINK http://www.biodiv.org/chm/info/official.html.

Unni, R. and R. Harmon (2007). Perceived effectiveness of push vs pull mobile location-based advertising. *Journal of Interactive Advertising 7*(3), 28–40.

US Census Bureau (1994). The TIGER mapping system (retired). `http://tiger.census.gov/,http://tigerweb.geo.census.gov.`

Virrantaus, K., J. Markkula, A. Garmash, V. Terziyan, J. Veijalainen, A. Katanosov, and H. Tirri (2001). Developing GIS-supported location-based services. In *Proceedings of WGIS2001 First International Workshop on Web Geographical Information Systems*, pp. 423–432.

W3C (Accessed Nov 2014a). RDF 1.1 N-triples. `http://www.w3.org/TR/n-triples/.`

W3C (Accessed Nov. 2014b). SPARQL 1.1 overview. `http://www.w3.org/TR/sparql11-overview/.`

W3C (Accessed Nov 2014b). XML path language (XPath). `http://www.w3.org/TR/xpath/.`

W3C (Accessed Oct 2014a). Mobile web applications best practices. `http://www.w3.org/TR/2010/REC-mwabp-20101214/:bp-interaction-uri-schemes.`

W3C (Accessed Oct 2014b). PICS superceded by POWDER. `http://www.w3.org/2009/08/pics_superseded.html.`

W3C (Accessed Oct 2014c). POWDER protocol for web description resources. `http://www.w3.org/2007/powder/.`

W3C (Accessed Oct 2014a). Web of devices. `http://www.w3.org/standards/webofdevices/.`

Waite, T., J. Kirkley, R. Pendleton, and L. Turner (2005). MUSEpad: supporting information accessibility through mobile location-based technology. *TechTrends: Linking Research & Practice to Improve Learning 49*(3), 76–82.

Web (Accessed Oct 2014). Phonegap. `http://www.phonegap.com.`

Weibel, S., J. Kunze, and C. Lagoze (1998). Dublin Core metadata for simple resource discovery. Dublin Core Workshop Series. (Online) `http://purl.oclc.org/metdata/dublin_core_elements/draft-kunze-dc-02.txt.`

Wessel, P. and W. Smith (1995). New version of the generic mapping tools released. Eos Trans. American Geophysical Union. `http://www.agu.org/eos_elec/95154e.html.`

Wessel, P. and W. H. F. Smith (1991). Free software helps map and display data. *Eos Trans. American Geophysical Union 72*(41), 441–446.

Westland, S., C. Ripamonti, and V. Cheung (2012). *Computational colour science using MATLAB* (2nd ed.). John Wiley & Sons, Hoboken, NJ.

Whalen, D. (1999). The cookie FAQ. Cookie Central. `http://www.cookiecentral.com/faq/`.

Wiener, J. (1997). Data warehousing: What is it? and related Stanford DB research. Stanford Database Research Laboratory.

Williamson, I., A. Rajabifard, and M. Feeney (2003). *Developing Spatial Data Infrastructure - From concept to reality*. Taylor and Francis, New York.

Williamson, I. P. and D. M. Grant (1999). United Nations-FIG, Bathurst declaration on land administration for sustainable development: Development impact. In *Proc. Workshop on Cadastral Infrastructures for Sustainable Development, Bathurst, NSW*.

Worbel, S., D. Wettschereck, E. Sommer, and W. Emde (1997). Extensibility in data mining systems. In E. Simoudis and J. Han (Eds.), *Proc. of the 2nd International Conference On Knowledge Discovery and Data Mining*. AAAI.

Zickuhr, K. (Accessed Nov. 2014). Three-quarters of smartphone owners use location-based services. Technical report, Pew Research Centre's Internet & American Life Project.

Index

Printed and bound by CPI Group (UK) Ltd, Croydon, CR0 4YY

01/11/2024

01782617-0014